W9-CRW-572

INTERMEDIATE ALGEBRA

Derek I. Bloomfield, Ph.D.

Orange County Community College
Middletown, New York

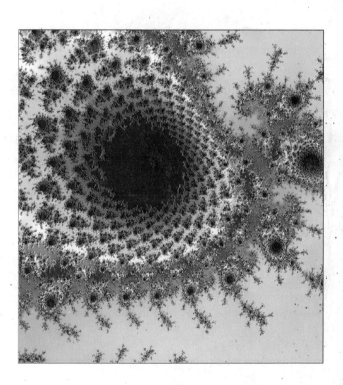

West Publishing Company

Minneapolis / St. Paul ▌ New York
Los Angeles ▌ San Francisco

DEDICATED TO SWEET MARCELLA

Copyeditor: Susan Gerstein

Cover and Text Design: Susan Guillory, TECHarts
Boulder, Colorado

Proofreader: Andres Delgado

Art: Christine Bentley, Visual Graphic Systems

Chapter Openers and Cover Image Art: Kristopher K. Hill,
FinalCopy

Compositor: Clarinda Company

Index: E. Virginia Hobbs

Production, Prepress, Printing and Binding
by West Publishing Company.

Copyright © 1994

By WEST PUBLISHING COMPANY
610 Opperman Drive
P.O. Box 64526
St. Paul, MN 55164-0526

All rights reserved

Printed in the United States of America
01 00 99 98 97 96 95 94   8 7 6 5 4 3 2 1 0
Library of Congress Cataloging-in-Publication Data

Bloomfield, Derek I., 1940-
    Intermediate algebra / Derek I. Bloomfield.
      p.   cm.
    Includes index.
    ISBN 0-314-02892-7 (soft—student version)
    ISBN 0-314-02817-X (soft–annotated teacher's edition)
    I. Algebra.   I. title.
  QA 154.2.B56   1994
  512.9—dc20   93-42966
                                          CIP

West's commitment to the environment

In 1906, West Publishing Company began recycling materials left over from the production of books. This began a tradition of efficient and responsible use of resources. Today, up to 95 percent of our legal books and 70 percent of our college and school texts are printed on recycled, acid-free stock. West also recycles nearly 22 million pounds of scrap paper annually—the equivalent of 181,717 trees. Since the 1960s, West has devised ways to capture and recycle waste inks, solvents, oils, and vapors created in the printing process. We also recycle plastics of all kinds, wood, glass, corrugated cardboard, and batteries, and have eliminated the use of styrofoam book packaging. We at West are proud of the longevity and the scope of our commitment to the environment.

About the cover and chapter opener images

More than eye-catching computer graphics, these images are known as fractals. Fractals are mathematical characterizations of the complexities found in nature. Jagged coastlines, fluffy clouds, the bark of trees, and snowflakes all exhibit these intricate patterns. These fractals are from the Mandelbrot and Julia sets. They are named for Benoit B. Mandelbrot, who introduced the concept of fractals in the early 1970s, and French mathematician Gaston Julia, who studied self-similar boundaries with Pierre Fatou during the first world war. An interesting property of fractals is their iterative nature. No matter how much you enlarge the fractal, it looks uncannily similar to lesser enlargements. The simple yet complex beauty of nature and our universe is represented in the colorful rings, whorls, and spirals.

CONTENTS

PREFACE

Purpose and Style

This book is the third in a series of mathematics texts which include:

Basic Mathematics
Introductory Algebra
Intermediate Algebra

It is intended for students who have completed a first course in algebra in high school or the equivalent college introductory algebra course.

This book is complete insofar as it starts at the beginning of algebra and then proceeds through the usual intermediate algebra topics. The topics which are normally considered in an introductory algebra course are presented here at a faster pace and with more of the unusual cases pointed out. It is intended to prepare a student to move ahead to the next course in college algebra, finite mathematics, business mathematics, liberal arts mathematics or a course in brief calculus.

This book may be used in any of a variety of instructional modes:

1. a conventional lecture-type class;

2. a self-study program in which the student works at his or her own pace;

3. a mathematics laboratory in which available video tapes or computer based instructional materials are used.

The books consists of 9 chapters, and the material in each chapter is presented in the following manner: A concise explanation of the fundamental concepts for the particular topic is given; examples illustrating these fundamental concepts are worked out step-by-step; and an ample number of similar problems are given in the exercise sets enabling the student to master the concepts involved. The exercises progress from simple to more difficult in an effort to give the student confidence in the ability to solve problems. The answers to odd-numbered exercises are given at the end of the text.

The major goal of this method of presentation is a correct understanding of the topics and maximum skill in performing the mathematical operations.

The material is presented in measured amounts so that the student can complete a topic before moving on to the next concept. An achievement test is given at the end of each chapter with answers to all exercises given in the answer section.

Outstanding Features

- **Number Knowledge:** At the beginning of each chapter is a "human interest" mathematical interlude.
- **How to be Successful at Mathematics:** Hints to help students are scattered throughout the first two chapters.
 - **Warning Signs:** These help students avoid frequently committed errors.
- **Critical Thinking Exercises:** Students are required to find errors in worked-out solutions to problems

• **In-Your-Own-Words Exercises:** Students are required to think through and to write out definitions and procedures in their own words. This lets them know whether they really understand a concept or not.

• **Living-in-the-Real-World Exercises:** These represent real-world applications of mathematics in every day situations.

• **Right Triangle Trigonometry:** For many students, intermediate algebra is the last mathematics course they will ever take. For this reason, many instructors wish that their students could be exposed to at least a brief introduction to trigonometry. Appendix B contains a concise explanation of right triangle trigonometry which is meant to be covered in about one week's time.

 • **Calculator Exercises:** Appendix D is included on how to use a calculator, and calculator exercises are scattered throughout the text. These exercises stand alone and may or may not be included at the individual instructor's discretion.

• **Chapter Summary:** At the end of each chapter is a list of definitions, rules, and procedures, each of which is accompanied by a worked-out example illustrating that procedure.

• **Chapter Review Exercises:** A set of exercises, reviewing each section in the chapter, is included at the end of each chapter.

• **Achievement Test:** A test is included at the end of each chapter with answers to all exercises to show the student whether or not he or she has mastered the concepts of the chapter.

• **Cumulative Reviews:** At the end of every third chapter is a cumulative set of exercises to ensure that students remember previous material.

• **Pedagogical Use of Color:** It's not just decoration.

• **Workbook Style with Perforated Pages:** These may be handed in if the instructor so desires.

• Over 800 worked-out examples with step-by-step explanations are presented, with important steps highlighted in color.

• Over 5,000 exercises have been carefully chosen to clarify explanations and to provide necessary drill.

• Sufficient space is provided for working out the exercises right in the text. This also provides a good reference when it comes time to review.

Ancillaries

• **Westtest 3.0:** A computer-generated testing program. Problems may be selected and mixed and matched to suit the instructor's needs. There are versions for both MacIntosh and IBM PCs or compatibles running DOS or Windows 3.0.

• **Instructor's Manual with Test Bank** containing a test bank, review sheets, and chapter test.

• **Instructor's Solutions Manual** containing worked-out solutions for all even-numbered problems in the test.

• **Student's Solutions Manual** containing worked-out solutions for all odd-numbered problems in the text.

• **Annotated Instructor's Edition** containing answers to all exercises written adjacent to the problem for easy instructor reference.

• **Mathens Tutorial Software** containing multiple examples of the basic problem types in *Intermediate Algebra*. The program is easy to use, interactive with the student, keyed to each section in the text and versions for both IBM PCs and compatibles and the MacIntosh are available.

• **Video Tapes** keyed to the text section-by-section.

I would like to thank all of the reviewers for their many helpful and constructive suggestions. These include:

- Melvern K. Taylor—*Willmar Community College*
- Dennis C. Ebersole—*Northampton County Area Community College*
- Kathryn C. Wetzel—*Amarillo College*
- Sharon Edgmon—*Bakersfield College*
- Robert L. Maynard, Jr.—*Rockingham Community College*
- Norma F. James—*New Mexico State University*
- James D. Blackburn—*Tulsa Junior College*
- Carol J. Page—*St. Charles Community College*
- Rebecca W. Eller—*Blue Ridge Community College*
- Virginia Morgan—*Montgomery Community College*
- Sylvia Kennedy—*Broome Community College*
- Alice Grandgeorge—*Manchester Community College*
- Gael T. Mericle—*Mankato State University*
- Richard N. Dodge—*Jackson Community College*
- William M. Mays—*Gloucester County College*
- Brian Hayes—*Triton College*
- Arthur P. Dull—*Diablo Valley College*
- W. Arlene Jeskey—*Rose State College*
- Mary B. Cabral—*Middlesex Community College*
- Lana Taylor—*Siena Heights College*
- Jack Rotman—*Lansing Community College*
- Susan Hahn—*Kean College*
- Karen L. Schwitters—*Seminole Community College*
- James D. Blackburn—*Tulsa Jr. College*
- Gwen Huber Terwilliger—*University of Toledo*
- Barbara Sallach—*New Mexico State University*
- Lois Norris—*Northern Virginia Community College*
- Vivian Dennis—*Eastfield College*
- Debbie Millard—*Florida Community College at Jacksonville-Kent Campus*

Acknowledgments

I want to thank the people at West Publishing Company and everyone who assisted with this book. Special thanks go to my editors Ron Pullins and Denise Bayko for their friendship and for their guidance and encouragement through many steps along the way; Laura Mezner Nelson and Poh Lin Khoo, my production editors, who took special care to ensure that every detail in the finished product would look just right; Doug Abbot for his marketing efforts; and Susan Gerstein, my copy editor, who does her work as well as anyone can. Max Bloomfield prepared both the Instructor's and the Student's Solutions manuals. Andres Delgado proofread the final manuscript and prepared the Instructor's manual. My most sincere thanks to all of them.

I am grateful to the administration, my colleagues, and especially to the students at Orange County Community College for providing a stimulating and positive environment. It's a great place to work. My wife Marcella typed the manuscript and made innumerable suggestions that improved the book greatly. Marcella and my children, Jennifer, Max, Derek, Jr., and David, gave up many Sunday afternoons. I thank them for their incentive, their encouragement, their understanding, and their love.

A Note from the Author to the Student.

There is a lot of truth in the old adage, "There's no success like success." I learned a long time ago that the first problem on a math exam should be one that *everyone* can do. It leads to confidence and success. Rearranging the same problems on the same exam so that the first problem is an especially difficult one can so destroy a student that you can't even spell your own name. Your anxiety level goes up, your confidence level plummets.

So, in my writing, I have attempted to alleviate anxiety, build confidence, and ensure mathematical success in the following ways:

1. I speak personally to you, the student, letting each of you know that I care about you and about your success.

2. I explain how to solve problems in a way that you can understand.

3. I very, very carefully make sure that the exercise sets progress slowly from simple to the more difficult ones. If you start at the beginning and work through the set in sequence, you will learn a tremendous amount, sometimes subtly, without even knowing what is happening to you. And, as on my exams, *everybody* can do the first few problems.

4. Nothing in the book is incidental. There are no fillers. In that regard, it's like a Mozart symphony—every note means something. In my book, every word counts.

5. My style is convincing and encouraging. In general I have more confidence in you as a student than perhaps you have in yourself. I speak to that throughout.

This book is written for you and I'm genuinely interested in what you think about it. Please write to me and let me know what you like about it or how you think it can be improved.

Thank you.

Derek Bloomfield
R.D. #4, Box 834
Middletown, NY 10940

INTERMEDIATE ALGEBRA

CHAPTER 1

Fundamental Definitions and Concepts

INTRODUCTION

Before going to work on solving equations and involving ourselves in the solution of practical problems, we first must take a look at some of the preliminary definitions and concepts and how to operate with the **real numbers,** the working language of algebra. This will be our task in this chapter.

NUMBER KNOWLEDGE—CHAPTER 1

Some Astonishing Facts About Numbers

1. If you purchase a house for $185,000, put $35,000 down, and borrow $150,000 for 30 years at a rate of 11%, the total of your payments for the 30 years will be $514,258.20. You will actually pay back over half a million dollars, about three and one-half times the amount that you borrowed.

2. If, instead of purchasing Manhattan Island from Chief Manhasset for $24.00 in the year 1626, Peter Minuit had invested his money in Chase Manhattan Bank at a rate of 6% compounded quarterly, it would now be worth $74,535,172,950—almost 75 billion dollars. Do you think he made a wise purchase?

3. You want to ensure that your newborn little girl will have a healthy retirement at age 65. How much will you have to invest today so that she will have one million dollars at age 65? Assuming monthly compounding and depending on the interest rate, the amount of your one-time investment is amazingly small.

RATE	AMOUNT INVESTED
8%	$5612.59
12%	$425.92
16%	$32.60

As you can see, the amount you must invest is very dependent on the interest rate. If you were fortunate enough to find an interest rate of 16%, you would only have to make a one-time investment of $32.60 to have $1,000,000 in the bank at 65 years later. This sounds unbelievable, but it's true!

1.1   Sets

The concept of sets is basic to all areas of mathematics and serves to unify the various branches. A knowledge of the following definitions and operations will prove useful in any further study that you undertake in mathematics.

We use the word **set** to mean any group or collection of things. Sets are usually named by using a capital letter, such as A, B, or C, and the individual items in the set are referred to as **elements** or **members** of the set. The elements of a set are enclosed in braces and the symbol $\in$ is used to indicate that an element belongs to the set.

EXAMPLE 1

(a) The set B containing the first five counting numbers is written
$B = \{1, 2, 3, 4, 5\}$.

(b) Since 4 is an element of the set B, we can write $4 \in B$.

(c) We indicate that 6 is *not* an element of the set B by writing $6 \notin B$.

(d) The set of letters in the name of the capital of Nebraska is written
$C = \{L, I, N, C, O\}$. The last two letters, L and N, in LINCOLN are not written, since elements in sets are not repeated.

(e) The set of counting numbers greater than 7 is $H = \{8, 9, 10, 11, \ldots\}$. The three dots after the 11 indicate that the numbers keep going in the same pattern. This notation is used when it is not practical (or not possible) to list all of the elements of the set.

▮ ▮

Cardinal Number of a Set

• • • • •

The **cardinal number** of a set is the number of elements in the set. The cardinal number of the set $A = \{a, b, c, d\}$ is 4 and we write $n(A) = 4$.

EXAMPLE 2   **(a)** If $B = \{2, 4, 6\}$ then $n(B) = 3$.

(b) If $T = \{4\}$ then $n(T) = 1$.

(c) If $Q = \{0\}$ then $n(Q) = 1$.   ▌▌

▌ **Finite Set** • • Any set that has a cardinal number equal to a whole number (zero or a counting number) is called a **finite set.** Keep in mind that finite sets can be very large. For example, the set $T = \{1, 2, 3, 4, \ldots, 1{,}000{,}000{,}000\}$ and has 1 billion elements in it, it has a cardinal number equal to 1,000,000,000, which is a whole number. Therefore, T is a finite set.

▌ **Infinite Set** • • Any set that is not a finite set is an infinite set. An example of an infinite set is $E = \{2, 4, 6, 8, \ldots\}$. The cardinal number of an infinite set is not a whole number. (Why not?)

▌ **Empty Set or Null Set** • • The **empty set** is a set that contains no elements; it is denoted by the symbol $\varnothing$. It has a cardinal number equal to 0: $n(\varnothing) = 0$.

EXAMPLE 3   Label the following sets as finite or infinite.

(a) $R = \{6\}$ Finite

(b) The set of odd counting numbers   Infinite

(c) $V = \{a, e, i, o, u\}$ Finite

(d) $\varnothing$ Finite   ▌▌

A common error is to try to write the empty set as $\{\varnothing\}$. The set $\{\varnothing\}$ cannot be the empty set because it contains *one* element, $\varnothing$. It has cardinal number equal to 1. The empty set has *no* elements. It has a cardinal number equal to 0.

Subsets

• • • • •

A set B is a **subset** of a set A if every element of the set B is also an element of the set A. This is written $B \subseteq A$.

EXAMPLE 4   **(a)** If $A = \{1, 2, 3, 4\}$ and $B = \{1, 3\}$, then $B \subseteq A$.

(b) If $P = \{1, 2, 3, 4\}$ and $Q = \{4, 5\}$, then Q is not a subset of P and we write $Q \nsubseteq P$.

(c) Given $W = \{0, 1, 2, 3, 4, \ldots\}$ and $E = \{2, 4, 6, 8, \ldots\}$, then $E \subseteq W$ since every even whole number (set E) is a whole number (set W). As you can see, subsets are not restricted to finite sets.

(d) If $A = \{3, 4, 5\}$, then $A \subseteq A$ because every element of A is an element of A. Every set is a subset of itself.

(e) For consistency, the empty set is thought of as being a subset of every set. In other words, for any set A, we say that $\varnothing \subseteq A$.   ▮▮

Operations on Sets

There are two basic set operations, *union* and *intersection*.

> The **union** of two sets A and B, denoted $A \cup B$, is the set of all elements in A or in B or in both.

> The **intersection** of two sets A and B, denoted $A \cap B$, is the set of all elements common to both A and B.

EXAMPLE 5   Let $A = \{1, 2, 3, 4\}$, $B = \{2, 4, 6\}$, and $C = \{1, 3, 5\}$.

(a) $A \cup B = \{1, 2, 3, 4, 6\}$

(b) $A \cap B = \{2, 4\}$

(c) $A \cup \varnothing = \{1, 2, 3, 4\}$

(d) $B \cup C = \{1, 2, 3, 4, 5, 6\}$

(e) $B \cap C = \varnothing$: sets B and C have no elements in common and are said to be **disjoint.**

(f) $B \cap \varnothing = \varnothing$ ▮▮

1.1   Exercises

▮ ▮ ▮ ▮ ▮

▼ Write Exercises 1–10 using set notation.

1. The set of counting numbers between 5 and 10.

2. The set of letters in the word *genius*.

3. The set of states beginning with the word *New*.

4. The set of the names of the days of the week.

5. The set of even counting numbers larger than 5.

6. The set of odd counting numbers less than 10.

7. The set of months of the year beginning with the letter *H*.

8. The set of three-sided rectangles.

9. The set of letters in the last name of the first president of the United States.

10. The set of counting numbers that are evenly divisible by 3.

▼ For Exercises 11–16, indicate which sets are finite and which are infinite; in the case of finite sets, write their cardinal numbers.

11. $T = \{10, 20, 30, \ldots, 700\}$

12. The set of all people currently living in Atlanta, Georgia, who are over 12 feet tall.

13. The set of even counting numbers between 9 and 17.

14. $T = \{1000, 2000, 3000, \ldots\}$

15. The letters of the alphabet.

16. The set of counting numbers that can be evenly divided by 5.

▼ For Exercises 17–28, if set $A = \{2, 4, 6, \text{July}, *\}$, indicate whether the following statements are true or false.

17. $4 \in A$

18. June $\in A$

19. $\{*\} \subseteq A$

20. $A \subseteq A$

21. $\varnothing \subseteq A$

22. $\{2, \text{July}\} \subseteq A$

23. $\varnothing \in A$

24. $\{2, 4, 6\} \subseteq A$

25. $\{4, 6\} \in A$

26. $A \in A$

27. $\{\varnothing\} \subseteq A$

28. $\{\varnothing\} \in A$

▼ For Exercises 29–37, let $A = \{1, 2, 3, 4, 5\}$, $B = \{2, 4, 6, 8\}$, $C = \{1, 3, 5, 7, 9\}$, and $D = \{6, 7\}$. Find the following:

29. $A \cup C$

30. $A \cap B$

31. $A \cup D$

32. $B \cap A$

33. $A \cap D$

34. $C \cap A$

35. $D \cap D$

36. $D \cup D$

37. $\varnothing \cap A$

1.2   The Real Numbers

The real numbers are the building blocks of algebra, and in this section we will describe the set of real numbers in terms of its subsets and look at some of its properties.

The set of **counting numbers** is also called the **natural numbers**:

$$\{1, 2, 3, 4, \ldots\}$$

The set of **whole numbers** consists of the natural numbers and zero:

$$\{0, 1, 2, 3, 4, \ldots\}$$

HOW TO BE SUCCESSFUL AT MATHEMATICS

WHY DO I NEED ALGEBRA?

As long as people can add and subtract, they will probably get along quite well mathematically, as long as they limit their use of mathematics to the grocery store or to a checkbook. In the beginning, students will see very little use for algebra in their everyday lives. However, as you continue on into higher-level courses in mathematics, the importance of learning algebra will become more obvious.

   We live in an increasingly technical world and many occupations require that you function at a certain mathematical level. This is certainly true if you are interested in working in fields such as banking or finance, business, psychology, sociology, archaeology, computer science, data processing, physics, biology, chemistry, pharmacy, nursing and other health-related fields, engineering, environmental science, as well as many others. It is a harsh fact of life—those who cannot function at the required level of mathematics will be excluded from these jobs.

The positive and negative counting numbers along with zero make up the **signed numbers** or **integers:**

$$\{ \ldots, -3, -2, -1, 0, 1, 2, 3, \ldots \}$$

A convenient way of illustrating the set of integers is on a number line.

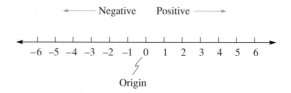

The Rational Numbers

• • • • •

> A **rational number** is any number that can be written as a quotient $\dfrac{a}{b}$, where a and b are integers and b does not equal 0.

Here are some examples of rational numbers:

$$\frac{-3}{4} \qquad \frac{1}{2} \qquad \frac{8}{5} \qquad \frac{7}{-4} \qquad \frac{6}{1}$$

Rational numbers written in the form $\dfrac{a}{b}$ are called **fractions,** where a is the **numerator** and b is the **denominator.**

An expression like $\dfrac{7}{0}$ is not defined, since 0 is not allowed as a denominator. A complete discussion involving division by 0 will be given in the next section.

All integers are rational numbers, since any integer can be written as a fraction with the given integer as the numerator and 1 as the denominator; for example, $-6 = \dfrac{-6}{1}$.

Terminating and Repeating Decimals

• • • • •

Since a fraction is a quotient, the numerator may be divided by the denominator. For example, $\dfrac{5}{8}$ can be written as the result obtained by dividing 5 by 8.

$$8\overline{)5.000} \quad .625$$

The decimal equivalent for $\dfrac{5}{8}$, is thus 0.625; it is called a **terminating** decimal since the sequence of digits comes to an end.

Similarly, $\dfrac{4}{11}$ can be expressed as a decimal:

$$11\overline{)4.0000\ldots} \quad .3636\ldots$$

If we continue the division, the 36 pattern continues to repeat. The decimal 0.363636. . . is called a **repeating** decimal. This is also written $0.\overline{36}$, where the bar is placed over the digits that repeat.

EXAMPLE 1   The following are rational numbers.

TERMINATING DECIMALS	REPEATING DECIMALS
(a) $\frac{1}{4} = 0.25$	**(d)** $\frac{1}{3} = 0.333\ldots$ or $0.\overline{3}$
(b) $\frac{3}{8} = 0.375$	**(e)** $\frac{3}{11} = 0.2727\ldots$ or $0.\overline{27}$
(c) $\frac{2}{5} = 0.4$	**(f)** $\frac{4}{7} = 0.571428571428\ldots$ or $0.\overline{571428}$ ▮▮

To summarize:

> Every rational number can be written as either a terminating or a repeating decimal.

Irrational Numbers

• • • • •

Consider the decimal

$$0.01001000100001000001\ldots$$

The three dots indicate that the sequence of digits does not end, so it is **nonterminating.** If you examine it carefully, however; you will see that the pattern does not repeat, so we call it a **nonrepeating** decimal. Such numbers are called **irrational numbers.** It can be shown that irrational numbers cannot be written as the quotient of two integers.

> An **irrational number** is a nonterminating, nonrepeating decimal. An irrational number cannot be written as the quotient of two integers.

EXAMPLE 2   The following are irrational numbers.

(a) 5.121121112. . .

(b) $\pi = 3.14159$. . .

(c) $\sqrt{2} = 1.4142$. . .

(d) Square roots of nonperfect squares, such as

$$\sqrt{2}, \sqrt{3}, \sqrt{5}, \sqrt{6}, \text{ and } \sqrt{7}$$

are all irrational numbers.   ▮▮

A common misconception is that $\pi = \frac{22}{7}$. Since it is the quotient of two integers, $\frac{22}{7}$ is a rational number. Dividing 22 by 7 gives $3.\overline{142857}$, a repeating decimal; so $\frac{22}{7}$ is a **rational** approximation to π, which is an **irrational** number.

The Real Numbers

• • • • •

> The set of **real numbers** is the union of all the rational numbers and all the irrational numbers. Each real number is represented by a point on the *real-number line.*

EXAMPLE 3   The following are real numbers.

(a) $\dfrac{9}{16}$ fraction (rational)

(b) $\sqrt{3}$ irrational

(c) $-\sqrt{5}$ irrational

(d) $0.\overline{231}$ repeating decimal (rational)

(e) -0.343343334. . .   nonrepeating decimal (irrational)

(f) 1.341 terminating decimal (rational)

We illustrate the position of these numbers on the real-number line.

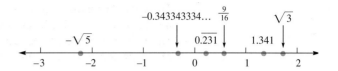

The following diagram illustrates the set relationship among the different sets of numbers that we have been considering. For example, we can see that the set of natural numbers is a subset of the set of whole numbers, of the set of integers, and of the set of rational numbers. Similarly, the whole numbers form a subset of the set of integers and the set of rational numbers; and the set of integers is a subset of the set of rational numbers. You can also see that the union of the sets of rational numbers and irrational numbers forms the set of real numbers.

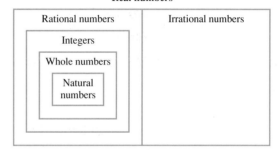

Prime Numbers

> A **prime number** is a counting number greater than 1 whose only divisors are itself and 1.

The first few primes are

$$2, 3, 5, 7, 11, 13, 17, 19, 23, 29, \ldots$$

A counting number greater than 1 that is not prime is called **composite**. All composite numbers can be written as the product of prime numbers.

EXAMPLE 4   Write each composite number as a product of prime numbers.

(a) $18 = 2 \cdot 9$
$ = 2 \cdot 3 \cdot 3$

(b) $90 = 2 \cdot 45$
$ = 2 \cdot 3 \cdot 15$
$ = 2 \cdot 3 \cdot 3 \cdot 5$

(c) $210 = 3 \cdot 70$
$ = 3 \cdot 7 \cdot 10$
$ = 3 \cdot 7 \cdot 2 \cdot 5$

1.2   Exercises

▮▮▮▮▮

▼ Put an X in the columns that apply to the number on the left.

	Number	Rational	Irrational	Real
1.	3	X		
2.	0.4242. . .	X		
3.	π		X	
4.	$\dfrac{22}{7}$	X		
5.	$\sqrt{9}$	X		
6.	$\sqrt{5}$		X	
7.	$-\sqrt{3}$		X	
8.	$\dfrac{-5}{4}$	X		
9.	0	X		
10.	0.555. . .	X		
11.	$0.\overline{62}$	X		
12.	$2\frac{2}{3}$	X		

Any # that repeat

because it's a fraction

▼ Tell whether the following statements are true or false.

13. 0 is a natural number.

14. 0 is an integer.

15. 0 is an irrational number.

16. 0 is a real number.

17. Every natural number is an integer.

18. Every integer is a natural number.

19. All whole numbers are also natural numbers.

20. All irrational numbers are also real numbers.

21. The intersection of the set of rational numbers with the set of irrational numbers is the empty set.

22. The union of the set of rational numbers with the set of irrational numbers is the set of real numbers.

23. π is a real number.

24. There is no largest natural number.

25. There is no smallest whole number.

26. There is no smallest integer.

27. Some integers are irrational numbers.

28. Some integers are natural numbers.

29. Some rational numbers are whole numbers.

30. Some rational numbers are irrational numbers.

▼ In Exercises 31–50, write each number as the product of prime numbers.

31. 14

32. 22

33. 12

34. 16

35. 32

36. 36

37. 47

38. 43

39. 85

40. 144

41. 100

42. 176

43. 225

44. 57

45. 126

46. 180

47. 221

48. 158

49. 192

50. 300

 **HOW TO BE SUCCESSFUL AT MATHEMATICS**

DEVELOP A POSITIVE ATTITUDE

"I've never been any good at math." "I can't even balance my checkbook." "I've forgotten all the math I ever knew." Do you ever find yourself making statements like these? *Math anxiety* is a popular term these days. Have you decided that you have it?

Regardless of the negative experiences you have had in the past, it's time to develop a more positive attitude for the future. Replace those negative thoughts with more helpful ones, like "I can do mathematics if I really try my best." "I'm going to do better this time than I've ever done

before." and "My third grade teacher who told me 'you can't do it' was wrong."

Acting as if you have total control over your success in your math course can make a big difference. If you behave in a successful way, your confidence will grow and your belief in yourself as a person who can do mathematics will evolve.

During the first few weeks of class, go overboard to dedicate yourself to the course. This extra work will pay off not only in increased knowledge but, equally important, in a more confident attitude toward your future success.

1.3   Operations With Signed Numbers

Absolute Value

Before adding signed numbers, we must review the concept of **absolute value.**

> The **absolute value** of a signed number x, denoted $|x|$, is the distance between x and zero on the number line.

EXAMPLE 1   **(a)** $+3$ is 3 units from zero on the number line so $|+3| = 3$.

(b) -3 is 3 units from zero on the number line so $|-3| = 3$.

(c) $|0| = 0$

(d) $\left|-\frac{3}{4}\right| = \frac{3}{4}$

Addition of Signed Numbers

> **To add signed numbers:**
>
> 1. If both numbers have the same sign, add their absolute values and keep the common sign for the answer.
>
> 2. If the numbers have different signs, find the difference in their absolute values and take the sign of the number with the larger absolute value for the answer.

E X A M P L E 2   Add the following:

(a) $-6 + (-5) = -11$

(b) $3 + (-8) = -5$

(c) $-4 + 1 = -3$

(d) $6 + 7 = 13$

(e) $-8 + 8 = 0$

(f) $-\frac{3}{5} + \frac{2}{5} = -\frac{1}{5}$ ▮▮

Subtraction of Signed Numbers

· · · · ·

The concept of the **negative** of a number suggests the **opposite** of that number. The opposite of moving 3 units to the right on the number line is moving 3 units to the left. Since $+3$ represents a move of 3 units to the right, the negative of $+3$, a move of 3 units to the left, is represented by -3. So the negative, or the opposite, of $+3$ is -3.

Following this same thinking, since -3 represents a move of 3 units to the left, the opposite, or negative, of -3 is a move of 3 units to the right. So the negative of -3 is $+3$, and we write this as $-(-3) = +3 = 3$.

E X A M P L E 3   **(a)** The negative of 7 is -7.

(b) The negative of -4 is $-(-4) = 4$.

(c) The negative of -6 is 6.   ▮▮

The rule for subtraction of signed numbers is stated in terms of the rule for addition of signed numbers, as follows:

> To subtract one signed number from another, add its negative.

E X A M P L E 4 $7 - 9 = 7 + (-9)$ To subtract 9, add its negative, -9.

$= -2$ ▮▮

E X A M P L E 5 $-4 - (-3) = -4 + 3$ To subtract -3, add its negative, $+3$.

$= -1$ ▮▮

E X A M P L E 6   Subtract:

(a) $8 - 3 = 8 + (-3) = 5$

(b) $-4 - 6 = -4 + (-6) = -10$

(c) $6 - (-1) = 6 + 1 = 7$

(d) $-11 - (-4) = -11 + 4 = -7$

(e) $\frac{2}{3} - \frac{5}{3} = \frac{2}{3} + \left(-\frac{5}{3}\right) = -\frac{3}{3} = -1$ ▮▮

Adding and Subtracting More than Two Signed Numbers

• • • • •

E X A M P L E 7

$$4 - 6 - 5 = 4 + (-6) + (-5)$$ To subtract +6 and +5, add their negatives.

$$= -2 + (-5)$$ Add from left to right.

$$= -7$$ ▮▮

E X A M P L E 8

$$6 - (-3) - 7 = 6 + 3 + (-7)$$ To subtract, add the negatives.

$$= 9 + (-7)$$ Add from left to right.

$$= 2$$ ▮▮

We frequently encounter problems in algebra like the following one.

E X A M P L E 9   Solve: $6 + 4 - 7 + 3 - (-8) - 4 - 2$.

S O L U T I O N   The easiest way to solve a long problem of this type is to change all subtractions to additions, add all of the positive values, add all of the negative values, and then combine these two results.

$$6 + 4 - 7 + 3 - (-8) - 4 - 2$$
$$= 6 + 4 + (-7) + 3 + 8 + (-4) + (-2)$$ Change subtractions to additions of the negatives.

$$6 + 4 + 3 + 8 = 21$$ Add the positive values.

$$(-7) + (-4) + (-2) = -13$$ Add the negative values.

$$21 + (-13) = 8$$ Combine the results from the preceding two steps.   ▮▮

Multiplication of Signed Numbers

• • • • •

> **To multiply two signed numbers:**
>
> 1. Multiply their absolute values.
>
> 2. The product is positive if their signs are alike.
>   The product is negative if their signs are different.

EXAMPLE 10 Multiply:

(a) $(-4)(6) = -24$ The signs are different

(b) $(-3)(-6) = 18$ The signs are alike.

(c) $4 \cdot 8 = 32$ The signs are alike.

(d) $7 \cdot (-6) = -42$ The signs are different.

(e) $-\dfrac{3}{4} \cdot \dfrac{1}{5} = -\dfrac{3}{20}$

(f) $(-1.6)(-3.5) = 5.6$

(g) $\left(-5\dfrac{1}{4}\right)\left(-2\dfrac{2}{7}\right) = \left(-\dfrac{21}{4}\right)\left(-\dfrac{16}{7}\right)$

$$= +\left(\dfrac{\overset{3}{\cancel{21}}}{\underset{1}{\cancel{4}}} \cdot \dfrac{\overset{4}{\cancel{16}}}{\underset{1}{\cancel{7}}}\right) = 12$$

Multiplying More than Two Signed Numbers

• • • • •

Let's extend the notion that we have just learned. Consider a problem in which there are more than two signed numbers to be multiplied. One way to solve the problem is to multiply the first two signed numbers together, multiply this result by the third signed number, multiply that result by the fourth signed number, and so on. To illustrate:

$(-2)(+1)(-3)(-2)(+4)$ Multiply $(-2)(+1) = -2$.

$= (-2)(-3)(-2)(+4)$ Multipy $(-2)(-3) = +6$.

$= (+6)(-2)(+4)$ Multiply $(+6)(-2) = -12$.

$= (-12)(+4)$ Multiply $(-12)(+4) = -48$.

$= (-48)$

There is an easier method for solving this type of problem. Each pair of negative signed numbers, when multiplied together, yields a positive answer. Therefore, if there is an even number of negative signs in the problem, the product will be positive. If there is an odd number of negative signs, each pair of negative factors yields a positive answer, but there will be one negative factor remaining, making the final product negative. The numerical part of the product will always be the product of the absolute values of all the factors in the problem.

To multiply more than two signed numbers:

1. Multiply the absolute values of the factors.

2. If there is an *even* number of negative signs, the product is *positive*. If there is an *odd* number of negative signs, the product is *negative*.

EXAMPLE 11   Multiply: $(-2)(+1)(-3)(-2)(+4)$.

$(2)(1)(3)(2)(4) = 48$ Multiply the absolute values.

Since there are three negative factors, an *odd* number, our answer is *negative,*

$(-2)(+1)(-3)(-2)(+4) = -48$ ▮▮

EXAMPLE 12   Multiply: $(-2)(-2)(-2)(-2)(-2)$.

$(2)(2)(2)(2)(2) = 32$ Multiply the absolute values.

Five negative signs is an odd number, so our answer is negative.

$(-2)(-2)(-2)(-2)(-2) = -32$ ▮▮

EXAMPLE 13   Multiply: $(-1)(2)(3)(-1)(4)$.

$(1)(2)(3)(1)(4) = 24$

Two negative factors, an even number, tells us our answer is positive, so

$(-1)(2)(3)(-1)(4) = +24 = 24$ ▮▮

Division of Signed Numbers

• • • • •

Division can be shown in three different ways:

$$\underset{\substack{\uparrow \qquad\qquad \nwarrow\ \ \nearrow \\ \text{dividend} \quad\ \text{divisor}}}{\overset{\overset{\text{quotient}}{\downarrow}}{15 \div 3 = 5}} \qquad \underset{\substack{\nearrow \quad\ \nwarrow \\ \text{divisor}\ \ \text{dividend}}}{\overset{\overset{\text{quotient}}{\overset{\downarrow}{\underline{5}}}}{3\overline{)15}}} \qquad \underset{\substack{\uparrow \\ \text{divisor}}}{\overset{\overset{\text{dividend}\quad\text{quotient}}{\ \ \nwarrow\qquad\downarrow}}{\frac{15}{3} = 5}}$$

In all cases, the number you divide by is the **divisor,** the number being divided is called the **dividend,** and the result is called the **quotient.**

To check a division problem, we multiply the divisor by the quotient to obtain the dividend.

$$15 \div 3 = 5 \qquad \text{because } 3 \times 5 = 15$$

Applying this principle to signed numbers gives us rules for dividing signed numbers:

$(3)(5) = 15$	because	$15 \div 3 = 5$
$(-3)(5) = -15$	because	$-15 \div -3 = 5$
$(3)(-5) = -15$	because	$-15 \div 3 = -5$
$(-3)(-5) = 15$	because	$15 \div -3 = -5$

These examples show that division of signed numbers follows the same rule of signs as multiplication of signed numbers. These rules are stated as follows:

To divide one signed number by another:

1. Divide their absolute values.

2. The quotient is positive if their signs are alike.
   The quotient is negative if their signs are different.

EXAMPLE 14   Divide:

(a) $-42 \div 6 = -7$ The signs are different.

(b) $\dfrac{-72}{-9} = 8$ The signs are alike,

(c) $-4\overline{)\,24}$ with quotient -6 The signs are different.

(d) $-28 \div (-4) = 7$ The signs are alike,

(e) $-\dfrac{9}{16} \div \dfrac{3}{8} = -\dfrac{9}{16} \cdot \dfrac{8}{3}$ Invert the divisor.

$$= -\dfrac{\cancel{9}^{3}}{\cancel{16}_{2}} \cdot \dfrac{\cancel{8}^{1}}{\cancel{3}_{1}} \qquad \text{Cancel and multiply.}$$

$$= -\dfrac{3}{2}$$

(f) Divide: $2.4\overline{)\,-1.44}$

$$2{\scriptstyle\,4.}\overline{)\,-1{\scriptstyle\,4.4}} \quad \text{quotient } -.6$$
$$\underline{1\ 4\ 4}$$
$$0$$

■ ■

Division Involving Zero

• • • • •

$$\dfrac{0}{3} = 0 \qquad \text{because} \qquad (0)(3) = 0$$

$$\dfrac{0}{-3} = 0 \qquad \text{because} \qquad (0)(-3) = 0$$

Now consider the equation

$$\dfrac{3}{0} = ?$$

What number can replace the question mark so that $? \times 0 = 3$? Since any number multiplied by zero is zero, there is no possible replacement for the question mark. We conclude, then, that $\frac{3}{0}$ is **undefined.**

Next consider the equation

$$\frac{0}{0} = ?$$

What number can replace the question mark so that $? \times 0 = 0$? Since any number times zero is zero, we can replace the question mark by any number we like. This gives us $\frac{0}{0} = 5$, -6, 14, or any other number. We say that $\frac{0}{0}$ is **indeterminate.**

Division Involving Zero:

$\dfrac{0}{a} = 0, \quad a \neq 0$	$\dfrac{a}{0}$ is undefined, $\quad a \neq 0$	$\dfrac{0}{0}$ is indeterminate

---

EXAMPLE 15   **(a)** $\dfrac{-74}{0}$ is undefined.

(b) $\dfrac{0}{-4} = 0$ because $(0)(-4) = 0$.

(c) $\dfrac{0}{0}$ is indeterminate.

1.3   Exercises

▼ Perform the indicated operations.

1. $4 + (-8)$

2. $-7 + 2$

3. $-4 + (-6)$

4. $-9 + 11$

5. $15 + (-3)$

6. $-16 + 15$

7. $4 - 9$

8. $-6 - 5$

9. $16 - 7$

10. $-11 - 8$

11. $-14 - (-6)$

12. $8 - (-9)$

13. $6 - 6$

14. $-14 - (-11)$

15. $-6 - (-12)$

16. $-11 - (-11)$

17. $\dfrac{3}{4} - \dfrac{5}{8}$

18. $\dfrac{5}{12} - \left(-\dfrac{2}{3}\right)$

19. $-\dfrac{2}{3} - \dfrac{3}{4}$

20. $-3 + 8 - 4$

21. $-6 - 2 + 5$

22. $-\dfrac{1}{2} + \dfrac{5}{8} - \dfrac{3}{4}$

23. $\dfrac{1}{8} - \dfrac{5}{4} - \left(-\dfrac{1}{2}\right)$

24. $-4.62 - (-2.13)$

25. $5.81 - 4.22 - 6.91$

26. $-8 + 9 - 4 + 1$

27. $-6 - 3 + 8 - 7$

28. $-6 - 3 + 4 - 5 + 6 + 9$

29. $8 - 4 + 6 + 7 - 9 - 14$

30. $-3 - (-8) + 5 - 6 - 2$

31. $-4 - (-4) - (-5) - 5$

32. $(-2)(-8)$

33. $6(-3)$

34. $(-7)(5)$

35. $(-11)(-4)$

36. $-7 \cdot 8$

37. $(-9)8$

38. $(6)(7)$

39. $(-4.13)(-2.2)$

40. $(6.15)(-8.3)$

41. $\left(-\dfrac{5}{16}\right)\left(\dfrac{4}{5}\right)$

42. $\left(-\dfrac{2}{3}\right)\left(-\dfrac{9}{8}\right)$

43. $(-4)(-2)(-2)$

44. $(3)(-4)(-5)$

45. $(-3)(2)(2)(-4)$

46. $(4)(-1)(-1)(-1)(-2)$

47. $(-8)(-2)(4)(-3)(-1)$

48. $(6)(-1)(-2)(5)(3)$

49. $\left(\dfrac{1}{2}\right)\left(-\dfrac{9}{16}\right)\left(-\dfrac{12}{5}\right)\left(\dfrac{15}{9}\right)$

50. $\left(-\dfrac{3}{16}\right)\left(-\dfrac{25}{9}\right)\left(\dfrac{8}{15}\right)\left(-\dfrac{9}{2}\right)$

51. $\dfrac{-8}{2}$

52. $\dfrac{-12}{-4}$

53. $\dfrac{-8}{-8}$

54. $\dfrac{32}{-4}$

55. $\dfrac{0}{8}$

56. $\dfrac{0}{-6}$

57. $\dfrac{-7}{0}$

58. $\dfrac{-4}{0}$

59. $\dfrac{0}{0}$

60. $-16 \div (-4)$

61. $-24 \div (-8)$

62. $-\dfrac{3}{8} \div \left(-\dfrac{5}{6}\right)$

63. $-\dfrac{3}{4} \div \left(-\dfrac{1}{2}\right)$

64. $-\dfrac{7}{8} \div 0$

HOW TO BE SUCCESSFUL AT MATHEMATICS

READING YOUR TEXTBOOK

Reading a mathematics textbook is not like reading a novel: it needs to be read slowly and carefully. Just reading one page can sometimes take half an hour or more. Always read your book with a pencil or pen in hand and don't be afraid to write in the book. A mathematics book without notes written throughout is one that has not been read correctly. Underline new words and important rules. Work through examples with your author step by step, filling in details that help you understand the problem. Pay attention to each word and to each mathematical statement.

Work out your homework exercises with thoroughness and persistence. Try to do them as soon after class as possible, while the concepts are fresh in your mind. If you have difficulty, refer to similar worked-out exercises that that the author has provided for you.

Remember, you can't rush this process—the extra time that you take in reading your text will pay off in the end.

1.4   Properties of Real Numbers

The properties discussed in this section are a summary of rules that we know to be true from past experience in working with real numbers. We will not try to prove them; we will merely organize them in some logical pattern.

Whenever you reverse the order of two real numbers in an addition problem, the sum does not change.

EXAMPLE 1   **(a)** $6 + 3 = 9$
$3 + 6 = 9$

(b) $(-4) + 5 = 1$
$5 + (-4) = 1$

This property is true in general and is called the **commutative property of addition.**

The commutative property of addition:
If a and b represent any real numbers, then $a + b = b + a$.

Similarly, if the order of the factors in a multiplication problem is reversed, the product remains the same.

EXAMPLE 2   **(a)** $6 \cdot 3 = 18$
$3 \cdot 6 = 18$

(b) $(-4)(5) = -20$
$(5)(-4) = -20$

This property is called the **commutative property of multiplication** and is stated formally as follows:

The commutative property of multiplication:

If a and b represent any real numbers, then $a \cdot b = b \cdot a$.

Note that in subtraction and division of real numbers, reversing the order *does* make a difference. You get a different result.

EXAMPLE 3   **(a)** $\left. \begin{array}{l} 7 - 5 = 2 \\ 5 - 7 = -2 \end{array} \right\}$ not the same

         (b) $\left. \begin{array}{l} 20 \div 4 = 5 \\ 4 \div 20 = \dfrac{4}{20} = \dfrac{1}{5} \end{array} \right\}$ not the same   ▮▮

Therefore neither subtraction nor division of real numbers is commutative.

When we add three numbers, we have a choice of which two to add first. Does it make any difference? We use parentheses to indicate which two numbers we are adding first.

EXAMPLE 4   Add: $3 + 2 + 6$.

         (a) $(3 + 2) + 6 = 5 + 6$ Add $3 + 2$ first.
                     $= 11$

         (b) $3 + (2 + 6) = 3 + 8$ Add $2 + 6$ first
                     $= 11$ ▮▮

Example 4 illustrates that when adding three real numbers, it makes no difference which two we add first. This property is always true and is called the **associative property of addition.**

The associative property of addition:

If a, b, and c are any real numbers, then $(a + b) + c = a + (b + c)$.

An example suggests that a similar property holds for multiplication of three real numbers.

EXAMPLE 5   Multiply: $-3 \cdot 4 \cdot 2$

(a) $(-3 \cdot 4) \cdot 2$ Multiply inside the parentheses first.
$= -12 \cdot 2$ $-3 \cdot 4 = -12$
$= -24$

(b) $-3 \cdot (4 \cdot 2)$ Multiply inside the parentheses first.
$= -3 \cdot 8$ $4 \cdot 2 = 8$
$= -24$

As you can see, we obtained the same result no matter which two numbers we multiplied first.   ▮▮

The associative property of multiplication:

If a, b, and c are any real numbers, then $(a \cdot b) \cdot c = a \cdot (b \cdot c)$.

As is the case with the commutative property, neither subtraction nor division is associative. This is illustrated in the following example.

EXAMPLE 6   **(a)** $(6 - 4) - 7 = 2 - 7$ $6 - (4 - 7) = 6 - (-3)$
$= -5$ $= 9$

—— Results differ ——

(b) $(-20 \div 4) \div 2 = -5 \div 2$ $-20 \div (4 \div 2) = -20 \div 2$
$= -\dfrac{5}{2}$ $= -10$

—— Results differ ——   ▮▮

These properties are quite intuitive, but they are stated here because they are used frequently as we continue on in algebra.

To summarize, we can change the *order* or the *grouping* in any addition or multiplication problem whenever it suits our purpose.

Study the next example carefully to discover yet another important property of real numbers.

EXAMPLE 7 Evaluate (a) $4(2 + 7)$ (b) $4 \cdot 2 + 4 \cdot 7$

SOLUTION **(a)** $4(2 + 7)$ Add $2 + 7$ inside the parentheses first.
      $= 4 \cdot 9$ Multiply.
      $= 36$

(b) $4 \cdot 2 + 4 \cdot 7$ Multiply first.
      $= 8 + 28$ Add.
      $= 36$ ▮▮

Since both results are the same, we see that $4(2 + 7) = 4 \cdot 2 + 4 \cdot 7$.

This property is true in general and is called the **distributive property of multiplication over addition.**

The distributive property of multiplication over addition:

If a, b, and c are real numbers, then $a \cdot (b + c) = a \cdot b + a \cdot c$.

EXAMPLE 8 Verify that the distributive property holds in each case.

(a) $3(2 + 6) = 3 \cdot 2 + 3 \cdot 6$

(b) $-7(4 + 2) = (-7)(4) + (-7)(2)$

(c) $12\left(\dfrac{1}{4} + \dfrac{1}{6}\right) = 12 \cdot \dfrac{1}{4} + 12 \cdot \dfrac{1}{6}$

SOLUTION

Left side	Right side
(a) $3(2 + 6) = 3(8)$ $\qquad\qquad = 24$	$3 \cdot 2 + 3 \cdot 6 = 6 + 18$ $\qquad\qquad\qquad = 24$
(b) $-7(4 + 2) = -7(6)$ $\qquad\qquad\quad = -42$	$(-7)(4) + (-7)(2) = (-28) + (-14)$ $\qquad\qquad\qquad\qquad = -42$
(c) $12\left(\dfrac{1}{4} + \dfrac{1}{6}\right) = 12\left(\dfrac{3}{12} + \dfrac{2}{12}\right)$ $\qquad\qquad\quad = \cancel{12}^{\,1} \cdot \dfrac{5}{\cancel{12}_{\,1}}$ $\qquad\qquad\quad = 5$	$12 \cdot \dfrac{1}{4} + 12 \cdot \dfrac{1}{6} = \cancel{12}^{\,3} \cdot \dfrac{1}{\cancel{4}_{\,1}} + \cancel{12}^{\,2} \cdot \dfrac{1}{\cancel{6}_{\,1}}$ $\qquad\qquad\qquad = 3 + 2$ $\qquad\qquad\qquad = 5$ ▮▮

The real numbers 0 and 1 have unique properties, which are summarized as follows:

Zero is the additive identity:

If a is a real number, then $a + 0 = a$ and $0 + a = a$.

E X A M P L E 9 $-6 + 0 = -6$ and $0 + (-6) = -6$.   ∎

---

1 is the multiplicative identity:

If a is a real number, then $a \cdot 1 = a$ and $1 \cdot a = a$.

E X A M P L E 1 0 $-6 \cdot 1 = -6$ and $1 \cdot (-6) = -6$.   ∎

The additive inverse property:

Every real number a has an additive inverse, $-a$, such that $a + (-a) = 0$ and $(-a) + a = 0$.

The additive inverse of a number is found simply by changing its sign.

E X A M P L E 1 1   **(a)** The additive inverse of 6 is -6, and

$$6 + (-6) = -6 + 6 = 0$$

(b) The additive inverse of -3 is $+3$, and

$$-3 + 3 = 3 + (-3) = 0$$ ∎

The multiplicative inverse property:

Every real number a (except zero) has a multiplicative inverse $\dfrac{1}{a}$ such that $a \cdot \dfrac{1}{a} = 1$ and $\dfrac{1}{a} \cdot a = 1$.

EXAMPLE 12   **(a)** The multiplicative inverse of 6 is $\frac{1}{6}$, and

$$6 \cdot \frac{1}{6} = \frac{1}{6} \cdot 6 = 1$$

(b) The multiplicative inverse of $\frac{2}{3}$ is its *reciprocal*, $\frac{3}{2}$, and is found by *inverting* the fraction:

$$\frac{2}{3} \cdot \frac{3}{2} = \frac{3}{2} \cdot \frac{2}{3} = 1$$

(c) The reciprocal of 0 would be $\frac{1}{0}$, which is undefined, so 0 has no multiplicative inverse.   ▮▮

1.4   Exercises

▮▮▮▮▮

▼ In Exercises 1–14, answer true or false.

1. Changing the order of addition does not change the result.

2. Changing the order of division does not change the result.

3. Changing the grouping in a subtraction problem does not change the difference.

4. Changing the order of multiplication does change the product.

5. Changing the grouping in an addition problem has no effect on the sum.

6. $6 - 4 = 4 - 6$ 7. $r(st) = (rs)t$

8. $\dfrac{a}{b} = \dfrac{b}{a}$

9. $(e \div f) \div g = e \div (f \div g)$

10. $(x + 3) + 4 = x + (3 + 4)$

11. $x - 15 = 15 - x$

12. $(7 - 5) - 2 = 7 - (5 - 2)$

13. $6 + m = m + 6$ 14. $m \cdot 6 = 6 \cdot m$

▼ In Exercises 15–20, verify the distributive property.

15. $6(3 + 5) = 6 \cdot 3 + 6 \cdot 5$

16. $-4(-5 + 7) = (-4)(-5) + (-4)(7)$

17. $-8(4 + 7) = (-8)(4) + (-8)(7)$

18. $\dfrac{1}{2}(7 + 5) = \dfrac{1}{2} \cdot 7 + \dfrac{1}{2} \cdot 5$

19. $4.3(1.2 + 0.3) = 4.3)(1.2) + (4.3)(0.3)$

20. $6.01(-2.1 + 0.05) = (6.01)(-2.1) + (6.01)(0.05)$

▼ In Exercises 21–26, give the property of real numbers that justifies each result.

21. $7\left(\dfrac{1}{7}\right) = 1$ **22.** $-6 + 0 = -6$

23. $11 \cdot 1 = 11$ **24.** $-9 + 9 = 0$

25. $\left(-\dfrac{5}{8}\right)\left(-\dfrac{8}{5}\right) = 1$ **26.** $-\dfrac{2}{3} + \dfrac{2}{3} = 0$

▼ Using the given property, complete each statement.

27. Commutative property of multiplication: $(6)(x) =$

28. Commutative property of addition: $6 + x =$

29. Distributive property: $3(x + 4) =$

30. Additive identity: $0 + 9 =$

31. Additive inverse property: $6 + (-6) =$

32. Associative property of addition: $(2 + x) + 5 =$

33. Multiplicative inverse property: $7 \cdot \dfrac{1}{7} =$

34. Distributive property: $4 \cdot x + 4 \cdot y =$

35. Multiplicative identity: $\dfrac{1}{8} \cdot 1 =$

36. Multiplicative inverse property: $\dfrac{5}{8} \cdot \dfrac{8}{5} =$

 37. Tell why subtraction is not commutative. Use an example.

 38. Explain the meaning of the commutative property of multiplication.

39. Why does zero have an additive inverse but not a multiplicative inverse?

1.5   Order of Operations and Grouping Symbols

Order of Operations

If we are given an expression containing more than one operation, we must be careful to perform the operations in the proper order. The following convention is used:

Order of operations:

1. Evaluate the expression inside any parentheses ( ), brackets [ ], or braces { }.

2. Powers and roots are performed.

3. Multiplications and divisions are done as they appear, from left to right.

4. Additions and subtractions are done as they appear, from left to right.

HOW TO BE SUCCESSFUL AT MATHEMATICS

PARTICIPATE IN CLASS

Your goal is to learn as much as you can from every class. Start by coming to class on time or even a little early. Choose your seat as close to the front and center of the room as possible. Not only will you be able to see and hear better, but if you tend to be shy or feel funny about asking questions, you won't have everybody turning around and looking at you every time you speak.

Do your best to get involved in the class activities. Volunteer to answer questions that you know the answer to. Not only does answering questions help your instructor and the rest of the class, but it improves your self-confidence and attitude, which in turn will help you to learn more easily and more thoroughly. Your contributions are valuable to the class and to yourself.

Don't be afraid to ask questions if you don't understand something. Chances are that other students may have the same concerns and will be happy that you voiced them. No one is going to think you are "stupid" for asking for an explanation. You will learn much more as an active participant than you will as an uninvolved spectator.

EXAMPLE 1 Evaluate: $27 \div (12 - 3) \cdot 2 + \sqrt{36}$.

SOLUTION

$27 \div (12 - 3) \cdot 2 + \sqrt{36}$ Evaluate inside the parentheses.

$= 27 \div 9 \cdot 2 + \sqrt{36}$ Take the square root.

$= 27 \div 9 \cdot 2 + 6$ Divide and multiply from left to right.

$= 3 \cdot 2 + 6$

$= 6 + 6$ Add.

$= 12$ ▮▮

EXAMPLE 2 Evaluate: $\sqrt{16} - 4(\sqrt{25} - 3) \div 4$.

SOLUTION

$\sqrt{16} - 4(\sqrt{25} - 3) \div 4$ Evaluate the expression inside the parentheses.

$= \sqrt{16} - 4(5 - 3) \div 4$

$= \sqrt{16} - 4(2) \div 4$ Take the root.

$= 4 - 4(2) \div 4$ Do multiplication and division from left to right.

$= 4 - 8 \div 4$

$= 4 - 2$ Subtract.

$= 2$ ▮▮

EXAMPLE 3 Evaluate: $3 \div 3^3 - 4\sqrt{64} + 3 \cdot 5$.

SOLUTION

$3 \cdot 3^3 - 4\sqrt{64} + 3 \cdot 5$ Do the powers and roots.

$= \underline{3 \cdot 27} - \underline{4 \cdot 8} + \underline{3 \cdot 5}$ Do multiplications.

$= 81 - 32 + 15$ Subtract.

$= 49 + 15$ Add.

$= 64$ ▮▮

Grouping Symbols

In the preceding discussion we introduced these grouping symbols:

- ( )   parentheses
- [ ]   brackets
- { }   braces

These symbols provide a means for changing the normal order of operations. For example, in the expression $6 - 4 \cdot 5$, using the correct order of operations, we multiply first and then subtract.

$$6 - 4 \cdot 5 = 6 - 20 = -14$$

If we want to subtract first and *then* multiply, we use parentheses to indicate this:

$$(6 - 4) \cdot 5 = 2 \cdot 5 = 10$$

Grouping symbols can also be used in more complicated expressions. The procedure when grouping symbols occur within other grouping symbols is to evaluate the innermost grouping first. The key phrase to remember is work from the "inside out."

EXAMPLE 4   Evaluate: $14 - [7 - (9 - 4)]$.

SOLUTION

$14 - [7 - (9 - 4)]$	Evaluate the inner parentheses.
$= 14 - [7 - 5]$	Evaluate the brackets.
$= 14 - 2$	Subtract.
$= 12$	▌▌

EXAMPLE 5   Evaluate: $7 + 2\{3 - [11 - (3 + 4)]\}$.

SOLUTION

$7 + 2\{3 - [11 - (3 + 4)]\}$	Evaluate $(3 + 4)$.
$= 7 + 2\{3 - [11 - 7]\}$	Evaluate $[11 - 7]$.
$= 7 + 2\{3 - 4\}$	Evaluate $\{3 - 4\}$.
$= 7 + 2(-1)$	Multiply.
$= 7 - 2$	Subtract.
$= 5$	▌▌

EXAMPLE 6   Evaluate: $6 - [3 - (4 - 8)]$.

SOLUTION

$6 - [3 - (4 - 8)]$	Evaluate $(4 - 8)$.
$= 6 - [3 - (-4)]$	Replace $-(-4)$ by $+4$.
$= 6 - [3 + 4]$	Evaluate $[3 + 4]$.
$= 6 - 7$	Evaluate $6 - 7$.
$= -1$	▌▌

EXAMPLE 7   Compare: (a) $3 - 6 \cdot 2$ (b) $(3 - 6) \cdot 2$

SOLUTION   **(a)** $3 - 6 \cdot 2 = 3 - 12$ Multiply first.

$= -9$

(b) $(3 - 6) \cdot 2 = -3 \cdot 2$ Evaluate the parentheses first.

$= -6$

1.5   Exercises

▼ Evaluate:

1. $9 \div 3 + 6$

2. $9 \div (3 + 6)$

3. $4 - 3 \cdot 5 + 7$

4. $7 - 4 \cdot 5 + 3$

5. $-4(3 - 12)$

6. $6(2 - 7)$

7. $13 - (4 - 6)$

8. $18 - (7 - 12)$

9. $3 + 44 \div 11 - 6$

10. $5 - 6 \div 2 + 6 - (2 + 2)$

11. $4 - [3 - (2 + 4)]$

12. $8 - [5 + 2(1 - 3)]$

13. $3^2 - 6 \div 3 + 4 \cdot 2$

14. $4^2 - 8 \div 4 + 9 \cdot 3$

15. $6 + [2 - (5 - 8)]$

16. $3 + 5[4 - (6 - 2)]$

17. $15 - 3[2 - (5 - 8)]$

18. $7 - \{-4 + [2 + (3 - 4)]\}$

19. $(-2 \cdot 5^2 + 8) \div (-6)$

20. $(3 \cdot 4^2 - 8) \div (-10)$

21. $(2\sqrt{121} + 3) \div 5$

22. $(3\sqrt{49} - 6) \div (-5)$

23. $4 + [3 - (4 + 6 \div 2)]$

24. $7 + [8 - (9 - 8 \div 2)]$

25. $14 - \{6 - [15 - (3 + 3^2)]\}$

27. $32 + 3\{4 - 3[2 - (5 - 8)]\}$

26. $15 - \{9 - [12 - (4^2 - 8)]\}$

28. $16 - 4\{5 - 2[4 - 3(7 - 5)]\}$

HOW TO BE SUCCESSFUL AT MATHEMATICS

GET ALL THE HELP YOU CAN

Even though you have worked hard, gone to class faithfully, attempted all of your homework, and generally done whatever you think is necessary to be successful, you may still be having trouble. If this is the case, there may be additional help that you can take advantage of. Your instructor should be your first source for information. Perhaps he or she may be able to offer extra help or know where it is readily available. Take advantage of one or more of the following resources that may be available at your school:

1. Mathematics laboratory
2. Computer software
3. Tutoring service
4. Video or audio tapes
5. Private tutoring

As soon as you notice you are having difficulty, seek help. If you put it off, you may find yourself hopelessly behind, and no one wants to be in that situation.

SUMMARY—CHAPTER 1

• • • • • • • • ▼

EXAMPLES

	EXAMPLES
A **set** is a group or collection of things.	Let $A = \{a, b, c, d\}$.
The symbol $\in$ is read "is an element of."	$b \in A, f \notin A$
The **cardinal number** of a set is the number of elements in the set.	$n(A) = 4$
A **finite set** has a whole number as its cardinal number.	A is a finite set, since $n(A) = 4$.
An **infinite set** does not have a whole number as a cardinal number.	$\{1, 3, 5, . . .\}$ is an infinite set.
The **empty set**, written $\varnothing$, is a set containing no elements.	The set of ten-pound pink elephants is equal to $\varnothing$
A set B is a **subset** of a set A, written $B \subseteq A$, if every element of the set B is an element of the set A.	If $A = \{a, b, c, d\}$, $B = \{a, c, e\}$, and $C = \{b, d\}$, then $C \subseteq A$ and $B \nsubseteq A$.
The **union** of two sets A and B, written $A \cup B$, is the set of all elements in A or B or in both sets.	$A \cup B = \{a, b, c, d, e\}$
The **intersection** of two sets A and B, written $A \cap B$, is the set of all elements common to A and B.	$A \cap B = \{a, c\}$
Disjoint sets have no elements in common. Their intersection is the empty set.	B and C are disjoint, since $B \cap C = \varnothing$.
The **integers** or **signed numbers** are the positive and negative counting numbers plus zero.	-3, 5 and 0 are examples of integers.
A **rational number** is one that can always be written in the form $\frac{a}{b}$, where a and b are integers and $b \neq 0$.	$\frac{3}{4}$ is rational.
A **rational number** can always be written either as a terminating or as a repeating decimal.	0.75 is a terminating decimal. $0.\overline{42}$ is a repeating decimal.
An **irrational number** can never be written in the form $\frac{a}{b}$, where a and b are both integers.	π and $\sqrt{3}$ are examples of irrational numbers.
An **irrational number** is always a nonterminating and nonrepeating decimal.	$0.353353335 . . .$ never terminates and it does not have a pattern that repeats, so it is irrational.
The **real numbers** consist of all the rational numbers together with the irrational numbers.	All of the rational and irrational numbers given above are examples of real numbers.
A **prime number** is a counting number greater than 1 whose only divisors are itself and 1.	$2, 3, 5, 7, 11, . . .$ are prime numbers.
A **composite** number has divisors other than itself and 1.	$6 = 2 \cdot 3$, so it is composite.
The **absolute value** of a real number x, denoted $|x|$, is the distance between x and 0 on the number line.	$|-7| = 7 \qquad |+7| = 7$ $|0| = 0$
To add signed numbers:	
1. If both numbers have the same sign, add their absolute values and keep the common sign for the answer.	$6 + 4 = 10$ $-6 + (-4) = -10$

2. If the numbers have different signs, find the difference in their absolute values and take the sign of the number with the larger absolute value.

$$6 + (-4) = 2$$
$$-6 + 4 = -2$$

To subtract one signed number from another, add its negative.

$$5 - 9 = 5 + (-9) = -4$$

To add or subtract more than two signed numbers:

1. Change all the subtractions to additions by adding the negatives.

$$-4 + 5 - (-7) - 3 + 4 - 1$$
$$= -4 + 5 + 7 + (-3) + 4 + (-1)$$

2. Add all the positive values.

$$5 + 7 + 4 = 16$$

3. Add all the negative values.

$$-4 + (-3) + (-1) = -8$$

4. Add the results from Steps 2 and 3.

Combine the two results: $16 + (-8) = 8$

To multiply two signed numbers:

1. Multiply their absolute values.

2. The product is positive if their signs are alike.

$$(-7)(-4) = 28$$
$$(7)(4) = 28$$

The product is negative if their signs are different.

$$(-7)(4) = -28$$
$$(7)(-4) = -28$$
$$(-7)(4) = -28$$
$$(7)(-4) = -28$$

To multiply more than two signed numbers:

1. Multiply the absolute values of the factors.

2. If there is an even number of negative signs, the product is positive.

$$(-2)(-3)(2)(-1)(2)(-3) = 72$$
Four negative signs is an even number.

3. If there is an odd number of negative signs, the product is negative.

$$(-5)(-1)(3)(2)(-2) = -60$$
Three negative signs is an odd number.

To divide one signed number by another:

1. Divide their absolute values.

2. The quotient is positive if their signs are alike.

$$\frac{-36}{-9} = 4$$

The quotient is negative if their signs are different.

$$\frac{14}{-7} = -2$$

$$\frac{-18}{6} = -3$$

$$\frac{0}{-6} = 0$$

Division by zero is not allowable.

$$\frac{-3}{0} \text{ is undefined,}$$

$$\frac{0}{0} \text{ is indeterminate.}$$

EXAMPLES

Properties of real numbers.

The commutative property of addition: If a and b represent any real numbers, then $a + b = b + a$.

$$-4 + 7 = 7 + (-4) = 3$$

The commutative property of multiplication: If a and b represent any real numbers, then $a \cdot b = b \cdot a$.

$$(-4)(6) = (6)(-4) = -24$$

The associative property of addition: If a, b, and c are any real numbers, then $(a + b) + c = a + (b + c)$.

$$(-6 + 3) + 2 = -6 + (3 + 2)$$
$$-3 + 2 = -6 + 5$$
$$-1 = -1$$

The associative property of multiplication: If a, b, and c are any real numbers, then $(a \cdot b) \cdot c = a \cdot (b \cdot c)$.

$$(-6 \cdot 3) \cdot 2 = -6 \cdot (3 \cdot 2)$$
$$-18 \cdot 2 = -6 \cdot 6$$
$$-36 = -36$$

The distributive property of multiplication over addition: If a, b, and c are any real numbers, then $a \cdot (b + c) = a \cdot b + a \cdot c$.

$$-2(3 + 5) = (-2)(3) + (-2)(5)$$
$$-2 \cdot 8 = -6 + (-10)$$
$$-16 = -16$$

The additive identity is zero:
$a + 0 = 0 + a = a$

$$-8 + 0 = 0 + (-8) = -8$$

The multiplicative identity is 1:
$a \cdot 1 = 1 \cdot a = a$

$$-8 \cdot 1 = 1 \cdot (-8) = -8$$

The additive inverse: Every real number a has an additive inverse $-a$ such that $a + (-a) = (-a) + a = 0$.

$$6 + (-6) = (-6) + 6 = 0$$
6 and -6 are additive inverses of each other.

The multiplicative inverse: Every real number a ($a \neq 0$) has a multiplicative inverse $\dfrac{1}{a}$ such that

$a \cdot \dfrac{1}{a} = \dfrac{1}{a} \cdot a = 1.$

$$4 \cdot \frac{1}{4} = \frac{1}{4} \cdot 4 = 1$$

The multiplicative inverse of a fraction $\dfrac{a}{b}$ is its

reciprocal, $\dfrac{b}{a}$.

$$\frac{3}{8} \cdot \frac{8}{3} = \frac{8}{3} \cdot \frac{3}{8} = 1$$

Order of operations:

1. Evaluate the expression inside any parentheses ( ), brackets [ ], or braces { }.

$$3 + 2(6 + 4) \div (-5)$$
$$= 3 + 2 \cdot 10 \div (-5)$$
$$= 3 + 20 \div (-5)$$
$$= 3 + (-4)$$
$$= -1$$

2. Powers and roots are performed.

3. Multiplications and divisions are done as they appear, from left to right.

4. Additions and subtractions are done as they appear, from left to right.

When **grouping symbols** occur within other grouping symbols, evaluate the innermost grouping first. Remember: work "inside out."

$13 - [4 - 2(3 - 6)]$
$= 13 - [4 - 2(-3)]$
$= 13 - [4 + 6]$
$= 13 - 10$
$= 3$

REVIEW EXERCISES—CHAPTER 1

▼ In Exercises 1 and 2, write out each set.

1. The set of even counting numbers between 9 and 17.

2. The set of odd counting numbers greater than 4.

3. What is the cardinal number of $H = \{t, a, p, e\}$?

4. Give an example of an infinite set.

5. Give an example of a finite set.

▼ In Exercises 6–14, let $S = \{m, n, p, q\}$, $T = \{n, p\}$, and $V = \{r, s\}$.

6. Find $S \cup V$ 7. Find $S \cap T$ 8. Find $T \cap V$

▼ Answer true or false.

9. $p \in T$ 10. $V \subseteq S$ 11. $T \subseteq S$

12. $\{r\} \in V$ 13. $T \in S$ 14. $\varnothing \subseteq S$

▼ In Exercises 15–18, classify each number as rational or irrational.

15. $\sqrt{8}$ 16. $6.2\overline{35}$

17. 0.4136 18. $7.464464446\ldots$

▼ Write each number as the product of prime numbers.

19. 45 20. 84

▼ Perform the indicated operations.

21. $-4 - 9$ 22. $6 - (-4)$

23. $-9 - 4 - (-12)$ 24. $-6 + 4 - 8 - 2 + 7$

25. $5 - (-8) + 12 - 14 - (-1)$

26. $(-6)(-11)$ 27. $(-4)(9)$

28. $8(-8)$ 29. $(-3)(-4)(-2)(-1)$

30. $(4)(6)(-2)(-1)(-1)$ 31. $\dfrac{-32}{-8}$

32. $\dfrac{-26}{13}$ 33. $\dfrac{0}{-9}$

34. $\dfrac{14}{0}$ 35. $\dfrac{0}{0}$

36. $-\dfrac{3}{8} \div \dfrac{3}{4}$

▼ Tell whether the following are true or false.

37. $a \div b = b \div a$

38. $6 + (4 + 5) = (6 + 4) + 5$

39. $(-3)(4 + 6) = (-3)4 + (-3)6$

40. $\left(-\dfrac{3}{2}\right)\left(-\dfrac{2}{3}\right) = 1$

▼ Evaluate:

41. $16 - (3 - 9)$ 42. $-6 + 12 \div 3 + 5 \cdot 3$

43. $-8 - \{4 - [3 - (2 - 6)]\}$

CRITICAL THINKING EXERCISES—CHAPTER 1

▼ There is an error in each of the following statements. Can you find it?

1. Since $\pi = \frac{22}{7}$, it is a rational number.

is an apprx   *$\pi \approx$ = axprox*

2. $\varnothing \in \varnothing$

an empty element dosn't have a number

3. Every whole number is also a natural number.

dosn't have "0"

4. Since 3.242242224 . . . is a repeating decimal, it must be a rational number.

dosn't repeat

5. $\frac{-5}{0} = 0$ *"0" can't be the denominator is undefined.*

6. $\frac{0}{8}$ is undefined.   *"0" can be the numberater*

7. $\sqrt{16}$ is an irrational number.   *rational - it has a squ. rat.*

8. $13 - 5 \div 2 = 4$ *=10.5 - it was sub 1st instead of division 1st*

9. Every whole number has a multiplicative inverse.   *"0" has no inverse.*

10. Every rational number has a multiplicative inverse.   *"0" has no inverse.*

ACHIEVEMENT TEST—CHAPTER 1

• • • • • • • • ▼

1. Write the set of counting numbers that are evenly divisible by 4.

2. Is the set in Problem 1 finite or infinite?

3. What is the cardinal number of $H = \{2, 4, 6, 8\}$?

▼ For Exercises 4–9, let $A = \{9, 10, 11, 12\}$ and $B = \{5, 8, 10\}$.

4. Find $A \cap B$.

5. Find $A \cup B$.

6. Is $\varnothing \in B$?

7. Is $\{10, 11\} \subseteq A$?

8. Is $B \subseteq A$?

9. Is $8 \in B$?

▼ Classify the following as rational or irrational.

10. 3.030030003. . .

11. 7.135135. . .

12. $\dfrac{13}{6}$

▼ Write each number as the product of prime numbers:

13. 36

14. 90

▼ Perform the indicated operations.

15. $(-7)(-8)$

16. $-42 \div 3$

17. $11 - (-6) + 5 - 3 - 9$

18. $3^2 + 5 \cdot 4 - 6 \div 2$

19. $4 - \{12 - [3(4 - 6)]\}$

1. _____

2. _____

3. _____

4. _____

5. _____

6. _____

7. _____

8. _____

9. _____

10. _____

11. _____

12. _____

13. _____

14. _____

15. _____

16. _____

17. _____

18. _____

19. _____

20. $\dfrac{0}{8}$

21. $\dfrac{-21}{0}$

22. $(-2)(-2)(-2)(-2)(-2)$

▼ Tell whether the following are true or false.

23. $(ab)c = a(bc)$

24. $24 \div 6 = 6 \div 24$

25. $-14 + 14 = 0$

20. _____

21. _____

22. _____

23. _____

24. _____

25. _____

Linear Equations and Inequalities

INTRODUCTION

We have already learned some preliminary concepts involving real numbers: what they are and how we work with them.

In this chapter we will use these skills to solve equations and inequalities; this is perhaps the most fundamental and important thing that we do in mathematics. Much of what follows in your future work in mathematics will be based on your knowledge of the concepts you learn in this chapter.

CHAPTER 2—NUMBER KNOWLEDGE

▮ ▮ ▮ ▮ ▮

NUMBER TRICKS INVOLVING 9

The Missing Number

Instructions	Example
1. Write any four-digit number.	3287
2. Add the digits.	$3 + 2 + 8 + 7 = 20$
3. Cross out any one of the digits in the original number.	32̸87 = 327
4. Subtract the sum in Step 2 from the remaining number in Step 3.	$\begin{array}{r} 327 \\ -\ 20 \\ \hline 307 \end{array}$
5. Add the digits from the result in Step 4.	$3 + 0 + 7 = 10$
6. Subtract the result in Step 5 from the closest multiple of **9** that is larger (18 in this case). This difference will be the number that you crossed out in Step 3.	$\begin{array}{r} 18 \\ -10 \\ \hline 8 \end{array}$

Try the same problem again, this time crossing out a different digit in Step 3. Then make up your own four-digit number and try it again.

Another Trick with 9s

Instructions	Example
1. From a push-button phone or calculator keyboard, choose any three-digit row, column, or diagonal. Multiply this number by any other three-digit row, column, or diagonal.	$456 \times 753 = 343{,}368$
2. Repeatedly add the digits in the product obtained in Step 1. The result will be **9**.	$3 + 4 + 3 + 3 + 6 + 8 = 27$ $2 + 7 = \mathbf{9}$

Still Another 9s Trick

Instructions	Example
1. Write down any six-digit number.	784,235
2. Mix up the digits in the number chosen in Step 1	427,358
3. Subtract the smaller number from the larger one.	$\begin{array}{r} 784{,}235 \\ -427{,}358 \\ \hline 356{,}877 \end{array}$
4. Add the digits of the difference together. The sum will be **9** or a number greater than **9**. If it is larger than **9**, repeatedly add the digits. The result will be **9.**	$3 + 5 + 6 + 8 + 7 + 7 = 36$ $3 + 6 = \mathbf{9}$

One Last Trick with 9's

Instructions	Example
1. Write any three-digit number.	318
2. Reverse the digits.	813
3. Subtract the smaller number from the larger one.	$\begin{array}{r} 813 \\ -318 \\ \hline 495 \end{array}$
4. The middle digit will always be 9, and the sum of the first and third digits will always be 9.	$4 + 5 = \mathbf{9}$

2.1   Solving Linear Equations

Equations are perhaps the most useful and important concept you will learn in algebra. The main reason for studying algebra is to learn how to solve problems in the real world, and many of these problems are solved using equations.

Two algebraic expressions separated by an equal sign is called an **equation.** An example of a **linear equation** is

$$6x + 2 = 3x - 7$$

In this equation $6x$ and $3x$ are called **variable terms** and 2 and -7 are **constant terms.**

Solution to an Equation

A **solution** to an equation is a value that, when substituted for the variable, makes the equation a true statement.

In the preceding example, if we substitute -3 for x, we get

$$6x + 2 = 3x - 7$$
$$6(\ -3\ ) + 2 \overset{?}{=} 3(\ -3\ ) - 7$$
$$-18 + 2 \overset{?}{=} -9 - 7$$
$$-16 = -16$$

which is a true statement, and we say that -3 is a solution to the equation $6x + 2 = 3x - 7$, or that -3 **satisfies** the equation.

EXAMPLE 1   Is 4 a solution to the equation $3y - 8 = y$?

SOLUTION
$$3y - 8 = y$$
$$3 \cdot \boxed{4} - 8 \overset{?}{=} \boxed{4} \qquad \text{Substitute 4 for } y \text{ wherever it appears.}$$
$$12 - 8 \overset{?}{=} 4 \qquad \text{Evaluate each side of the equation.}$$
$$4 = 4 \qquad \text{Left side} = \text{right side}$$

Therefore 4 is a solution.   ∎∎

Solving Equations

● ● ● ● ●

Now that we know what the solution to an equation is, our next step will be to discover how to find one.

The left and right sides of an equation are two ways of representing the same quantity. In this way we can think of an equation as a balance, with the left member on one side and the right member on the other. For example, the equation $x - 5 = -1$ can be represented by the following diagram:

If we add the same quantity, say 7, to both sides of the balance, it should stay in equilibrium, since both sides still carry the same amount.

Algebraically, we are adding 7 to both sides of the original equation:

$$
\begin{array}{r}
x - 5 = -1 \\
\underline{+7 \quad +7} \\
x + 2 = 6
\end{array}
$$

If you check the equations $x - 5 = -1$ and $x + 2 = 6$, you will find that $x = 4$ is the solution to both.

$$x - 5 = -1 \qquad\qquad x + 2 = 6$$
$$\boxed{4} - 5 = -1 \qquad\qquad \boxed{4} + 2 = 6$$
$$-1 = -1 \qquad\qquad\quad 6 = 6$$
$$\uparrow \qquad\qquad\qquad\quad \uparrow$$

4 is a solution to both equations.

Two equations that have the same solution are called **equivalent equations.** It is apparent from the preceding example that we can add the same number to both sides of any equation and obtain a new equation that is equivalent to the original one; that is, they both have the same solution.

In a similar way, if we subtract the same quantity from both sides of an equation, the result is an equivalent equation. Actually, this is the same as adding a negative quantity to both sides of the equation. This principle is basic to the solution of equations and is stated as follows.

The addition principle:

The *same* number can be added to (or subtracted from) each side of any equation and the solution will remain the same.

It seems reasonable, also, that we should be able to multiply or divide both sides of an equation by the *same* number and produce an equivalent equation. This is the second basic principle that we will use in solving equations.

The multiplication principle:

If both sides of an equation are multiplied (or divided) by the *same* number (except zero), an equivalent equation results.

Let's now apply these principles to finding the solutions to equations. Our objective is always the same, to isolate the variable on one side of the equation. To do this, we first get all the terms containing the variable on one side of the equation and the constant terms on the other side of the equation.

To solve an equation using both the addition and multiplication principles:

1. Put all the terms containing the variable on one side of the equation by using the addition principle.

2. Put all constant terms on the *other* side of the equation by using the addition principle again.

3. Divide both sides of the equation by the coefficient of the variable.

EXAMPLE 2   Solve: $5x + 1 = 3x + 7$.

SOLUTION

$5x + 1 = 3x + 7$	Get all x terms on the left side.
$5x + (-3x) + 1 = 3x + (-3x) + 7$	Add $-3x$ to both sides.
$2x + 1 = 7$	All x terms are now on the left side.
$2x + 1 - 1 = 7 - 1$	Subtract 1 from both sides to get all the constant terms on the right.
$2x = 6$	
$\dfrac{2x}{2} = \dfrac{6}{2}$	Divide both sides by 2, the coefficient of x.
$x = 3$	The solution is 3.

Check

$$5x + 1 = 3x + 7$$
$$5 \cdot 3 + 1 \overset{?}{=} 3 \cdot 3 + 7$$
$$15 + 1 \overset{?}{=} 9 + 7$$
$$16 = 16 \quad ✔ \qquad\qquad ▮▮$$

Do we have to put the x terms on the *left* and the constant terms on the *right?* Or can we do it the other way around? Let's try it and see.

EXAMPLE 3   Solve: $5x + 1 = 3x + 7$.

ALTERNATE SOLUTION

$5x + 1 = 3x + 7$	Get all x terms on the right side.
$5x + (-5x) + 1 = 3x + (-5x) + 7$	Add $-5x$ to both sides.
$1 = -2x + 7$	Get all constant terms on the left.
$1 + (-7) = -2x + 7 + (-7)$	Add -7 to both sides.
$\dfrac{-6}{-2} = \dfrac{-2x}{-2}$	Divide both sides by -2.
$3 = x$ or $x = 3$	The solution is the same. ▮▮

EXAMPLE 4   Solve: $3y - 2 = 13$.

SOLUTION

$3y - 2 = 13$	All y terms are already on the left.
$3y - 2 + 2 = 13 + 2$	Eliminate -2 on the left.
$3y = 15$	All constant terms are now on the right.
$\dfrac{3y}{3} = \dfrac{15}{3}$	Divide both sides by 3.
$y = 5$	The solution is 5.

Check $3 \cdot 5 - 2 \overset{?}{=} 13$

$$15 - 2 \overset{?}{=} 13$$

$$13 = 13 \quad \text{✓}$$

E X A M P L E 5   Solve: $-22 - 3x = 5x$.

S O L U T I O N   Since the only constant is already on the left side of the equation, there is some advantage to putting all the variables on the right:

$$-22 - 3x = 5x$$

$$-22 - 3x + \boxed{3x} = 5x + \boxed{3x} \qquad \text{Add } 3x \text{ to both sides.}$$

$$-22 = 8x$$

$$\frac{-22}{8} = \frac{8x}{8} \qquad \text{Divide both sides by 8.}$$

$$\frac{-11}{4} = x \qquad \text{The solution is } \frac{-11}{4}.$$

Check $-22 - 3 \cdot \left(\dfrac{-11}{4} \right) \overset{?}{=} 5 \cdot \left(\dfrac{-11}{4} \right)$

$$\frac{-88}{4} + \frac{33}{4} \overset{?}{=} \frac{-55}{4}$$

$$\frac{-55}{4} = \frac{-55}{4} \quad \text{✓}$$

E X A M P L E 6   Solve: $5 + \dfrac{y}{6} = 8$.

S O L U T I O N $5 + \dfrac{y}{6} = 8$

$$5 + \boxed{(-5)} + \frac{y}{6} = 8 + \boxed{(-5)} \qquad \text{Add } -5 \text{ to both sides.}$$

$$\frac{y}{6} = 3$$

$$\boxed{6} \cdot \frac{y}{6} = \boxed{6} \cdot 3 \qquad \text{Multiply both sides by 6.}$$

$$y = 18 \qquad \text{The solution is 18.}$$

Check $5 + \dfrac{18}{6} \overset{?}{=} 8$

$$5 + 3 \overset{?}{=} 8$$

$$8 = 8 \quad \text{✓}$$

2.1   Exercises

▮▮▮▮▮

▼   Solve each equation and check the solution.

1. $7x = 35 + 2x$

2. $3y - 2 = 13$

3. $4x - 6 = 3x + 2$

4. $6a - 3 = 2a + 6$

5. $6a + 9 = 2a - 7$

6. $6y - 26 = 3y + 1$

7. $-2y + 1 = 13$

8. $5 - 7z = 3z + 15$

9. $-6t - 1 = -5t + 6$

10. $9y + 1 = 4y$

11. $18 - 4y = 5y$

12. $-36 = 24x + 12$

13. $\frac{1}{2}x + 3 = 7$

14. $m = 5m - 8$

15. $0.6n + 0.5 = -3.7$

16. $3x - 5.2 = 4.1$

17. $6 + \dfrac{p}{5} = 8$

18. $\dfrac{h}{3} - 5 = 2$

19. $\dfrac{x}{4} + 3 = -5$

20. $-8x + 2 = 3x - 4$

21. $5x + 6 = 3x + 5$

22. $7y + 4 = 8y + 7$

23. $7x - 5 = 2x + 5$

24. $-2x - 4 = 3x - 4$

25. $11x - 7 = 3x - 7$

26. $-2x + 1 = -7x + 5$

27. $-8x - 3 = -4x + 5$

28. $14x - 3 = 10x + 7$

29. $7x + 5 = -2x + 7$

30. $8x - 4 = 2x - 4$

31. $-4x + 17 = 11x + 5$

32. $23x - 4 = 11x + 17$

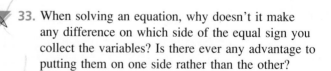

33. When solving an equation, why doesn't it make any difference on which side of the equal sign you collect the variables? Is there ever any advantage to putting them on one side rather than the other?

HOW TO BE SUCCESSFUL AT MATHEMATICS

USE THE FEW MINUTES BEFORE CLASS EFFECTIVELY

Frequently there are a few minutes available to you before your class begins. Don't waste this time, even if it isn't very much. Over the course of a semester, it can amount to two or three hours of extra time. Do as many of the following things as you can during these few moments:

1. Go over the previous class notes one more time.

2. Read through the new material to be presented in today's class.

3. Prepare any questions you wish to ask in class.

4. Rework one of the homework problems assigned that day, or try one that wasn't assigned.

2.2 Solving Equations Containing Parentheses, Fractions, and Decimals

Equations Containing Parentheses

When an equation contains parentheses, we must simplify it before solving.

EXAMPLE 1 Solve: $4(x + 3) = 8$.

SOLUTION Using the distributive property, we multiply both x and 3 by 4 on the left side of the equation:

$$4(x + 3) = 8 \qquad \text{Use the distributive property.}$$

$$4x + 12 = 8 \qquad \text{Multiply through the parentheses by 4.}$$

$$4x + 12 - 12 = 8 - 12 \qquad \text{Subtract 12 from both sides.}$$

$$4x = -4$$

$$\frac{4x}{4} = \frac{-4}{4} \qquad \text{Divide each side by 4.}$$

$$x = -1$$

Check

$$4(x + 3) = 8$$

$$4(-1 + 3) \overset{?}{=} 8$$

$$4(2) \overset{?}{=} 8$$

$$8 = 8 \quad ✓$$

STOP

A negative sign written in front of an expression in parentheses means the same as multiplying the entire expression by -1:

$$-(3x - 4) \qquad \text{means the same as} \qquad -1(3x - 4)$$

Applying the distributive property gives us

$$-(3x - 4) = -1(3x - 4)$$
$$= (-1)(3x) - 1(-4)$$
$$= -3x + 4$$

You can see, then, that the net result of parentheses preceded by a negative sign is to change the sign of each term within the parentheses. For example:

$$-(6x + 5) = -6x - 5$$
$$-(4 - x) = -4 + x \quad \text{or} \quad x - 4$$

EXAMPLE 2   Solve: $3x - 2(x + 4) = 6 - (x + 4)$.

SOLUTION

$3x - 2(x + 4) = 6 - (x + 4)$	Remove parentheses from both sides.
$3x - 2x - 8 = 6 - x - 4$	Combine like terms on each side.
$x - 8 = -x + 2$	Add x to each side.
$x + x - 8 = -x + x + 2$	
$2x - 8 = 2$	
$2x - 8 + 8 = 2 + 8$	Add 8 to each side.
$2x = 10$	Divide each side by 2.
$\dfrac{2x}{2} = \dfrac{10}{2}$	
$x = 5$	The solution is 5.

Check

$$3x - 2(x + 4) = 6 - (x + 4)$$
$$3 \cdot 5 - 2(5 + 4) \stackrel{?}{=} 6 - (5 + 4)$$
$$15 - 2(9) \stackrel{?}{=} 6 - 9$$
$$15 - 18 \stackrel{?}{=} -3$$
$$-3 = -3 \quad \text{☑} \qquad\qquad \blacksquare$$

Equations Containing Fractions

Equations containing fractions are handled quite easily. We find the least common denominator (LCD) of all the fractions in the equation, multiply both sides of the equation by this LCD, and all the fractions will disappear. Stated formally:

To eliminate fractions from an equation:
Multiply both sides of the equation by the LCD of all the fractions that appear in the equation.

EXAMPLE 3   Solve: $\dfrac{1}{2}x + \dfrac{1}{5} = \dfrac{2}{5}x - \dfrac{1}{2}$.

SOLUTION

$$\frac{1}{2}x + \frac{1}{5} = \frac{2}{5}x - \frac{1}{2}$$

$$10 \cdot \left(\frac{1}{2}x + \frac{1}{5}\right) = 10 \cdot \left(\frac{2}{5}x - \frac{1}{2}\right)$$ Multiply both sides by 10, the LCD.

$$10 \cdot \frac{1}{2}x + 10 \cdot \frac{1}{5} = 10 \cdot \frac{2}{5}x - 10 \cdot \frac{1}{2}$$ Apply the distributive rule.

$$\overset{5}{10} \cdot \frac{1}{\underset{}{2}}x + \overset{2}{10} \cdot \frac{1}{\underset{}{5}} = \overset{2}{10} \cdot \frac{2}{\underset{}{5}}x - \overset{5}{10} \cdot \frac{1}{\underset{}{2}}$$ Cancel.

$$5x + 2 = 4x - 5$$ The fractions are gone!

$$5x + (-4x) + 2 + (-2) = 4x + (-4x) - 5 + (-2)$$ Add $-4x$ and -2 to each side.

$$x = -7$$ The solution is -7.

Check

$$\frac{1}{2}x + \frac{1}{5} = \frac{2}{5}x - \frac{1}{2}$$

$$\frac{1}{2} \cdot -7 + \frac{1}{5} \overset{?}{=} \frac{2}{5} \cdot -7 - \frac{1}{2}$$

$$\frac{-7}{2} + \frac{1}{5} \overset{?}{=} \frac{-14}{5} - \frac{1}{2}$$

$$\frac{-35}{10} + \frac{2}{10} \overset{?}{=} \frac{-28}{10} - \frac{5}{10}$$

$$\frac{-33}{10} = \frac{-33}{10}$$

    As you can see, with complicated equations such as these, checking the solution is sometimes more work than finding it in the first place. So the recommendation is to work carefully and in an orderly way in order to eliminate errors.

EXAMPLE 4   Solve: $\dfrac{2}{3}x + 1 = \dfrac{1}{2}x + \dfrac{3}{4}$.

SOLUTION

$$\dfrac{2}{3}x + 1 = \dfrac{1}{2}x + \dfrac{3}{4}$$

$$12 \cdot \left(\dfrac{2}{3}x + 1 \right) = 12 \cdot \left(\dfrac{1}{2}x + \dfrac{3}{4} \right)$$ Multiply each side by 12, the LCD.

$$12 \cdot \dfrac{2}{3}x + 12 \cdot 1 = 12 \cdot \dfrac{1}{2}x + 12 \cdot \dfrac{3}{4}$$ Apply the distributive rule.

$$\overset{4}{\cancel{12}} \cdot \dfrac{2}{\cancel{3}}x + 12 \cdot 1 = \overset{6}{\cancel{12}} \cdot \dfrac{1}{\cancel{2}}x + \overset{3}{\cancel{12}} \cdot \dfrac{3}{\cancel{4}}$$ Cancel.

$$8x + 12 = 6x + 9$$ Add $-6x$ and -12 to each side.

$$2x = -3$$ Divide each side by 2.

$$\dfrac{2x}{2} = \dfrac{-3}{2}$$

$$x = \dfrac{-3}{2}$$ The solution is $\dfrac{-3}{2}$

Check

$$\dfrac{2}{3}x + 1 = \dfrac{1}{2}x + \dfrac{3}{4}$$

$$\dfrac{2}{3} \cdot \dfrac{-3}{2} + 1 \overset{?}{=} \dfrac{1}{2} \cdot \dfrac{-3}{2} + \dfrac{3}{4}$$

$$-1 + 1 \overset{?}{=} \dfrac{-3}{4} + \dfrac{3}{4}$$

$$0 = 0$$ ✔

Equations Containing Decimals

• • • • •

It is usually more difficult to solve an equation that contains decimals than to solve one that doesn't contain decimals. We "clear" equations of decimals by using the following rules:

To clear an equation of decimals:

1. Count the greatest number of decimal places that appear in any of the decimals in the equation.

2. Multiply each side of the equation by 1 followed by that number of zeros.

E X A M P L E 5 Solve: $0.04x + 3.1 = 0.39x + 3.2$.

S O L U T I O N The greatest number of decimal places in any of the decimals is *two* (in both $0.04x$ and $0.39x$), so we must multiply each side of the equation by 100 (1 followed by *two* zeros). Multiplying by 100 will move the decimal point *two* places to the right.

$$0.04x + 3.1 = 0.39x + 3.2$$ Multiply each side by 100.

$$100(0.04x) + 100(3.1) = 100(0.39x) + 100(3.2)$$

$$0.04x + 3.10 = 0.39x + 3.20$$ The decimal point moves two places to the right.

$$4x + 310 = 39x + 320$$ Add $-39x$ and -310 to both sides.

$$-35x = 10$$ Divide both sides by -35.

$$x = \frac{10}{-35} = -\frac{2}{7}$$

or $x \approx -0.286$ Rounded to the nearest thousandth (The symbol $\approx$ means "approximately equal to".)

2.2 Exercises

▼ Solve each equation and check the solution.

1. $3(x + 1) = 2x + 7$

2. $7(x - 2) = 6x - 5$

3. $7x = 2(x + 20)$

4. $12x = 2(x + 30)$

5. $y - 2(y + 4) = -1$

6. $3(x - 1) + 7 = 2x$

7. $3(x - 4) - 2(x + 6) = -20$

8. $x - 5(x + 2) = 4 - 2(x - 1) + 6$

9. $3(x - 3) + 2(x - 2) = 5(3 - x)$

10. $3(n + 5) = 2(n - 1)$

11. $-2(t - 5) = 2(3t + 5)$

12. $0.2x = 2.4$

13. $0.05x = 7.5$

14. $6.2x = 13.02$

15. $4.1x + 7.8 = 8.62$

16. $7.1x - 4.2 = 0.1x + 0.7$

17. $\dfrac{3}{8}x - \dfrac{1}{3} = \dfrac{1}{4}x + \dfrac{5}{12}$

28. $\dfrac{5}{16}y - \dfrac{1}{4} = \dfrac{1}{2}y - \dfrac{7}{16}$

18. $\dfrac{1}{15}x + \dfrac{2}{5} = \dfrac{1}{3}$

29. $0.5(w + 1) = 3.5w - 7.5$

19. $\dfrac{1}{3}x + \dfrac{1}{2}x = 5$ 20. $\dfrac{1}{2}y - \dfrac{1}{5}y = 6$

30. $5\left(y - \dfrac{1}{2}\right) = 6 + \dfrac{1}{5}\left(2y - \dfrac{1}{3}\right)$

21. $\dfrac{1}{3}y - \dfrac{1}{7}y = 4$ 22. $\dfrac{2}{3}x + 1 = \dfrac{1}{2}x - \dfrac{3}{4}$

31. $3\left(\dfrac{1}{2}x - 4\right) = 2 + \dfrac{1}{2}(x + 1)$

32. $-(x - 4) = \dfrac{2}{3}(x - 5)$

23. $\dfrac{2}{3}y + 9 = y$ 24. $\dfrac{1}{3}x + \dfrac{1}{2} = \dfrac{3}{2}$

25. $\dfrac{3}{4}t + \dfrac{1}{2} = \dfrac{1}{8}t - 2$ 26. $\dfrac{1}{2}y + \dfrac{3}{2} = \dfrac{2}{3}y - 1$

33. In solving the equation $\dfrac{1}{4}x + \dfrac{2}{3} = x + \dfrac{1}{2}$, we multiply by 12, the LCD, to eliminate the fractions. Why should the x on the right side be multiplied by the LCD even though it's not a fraction?

27. $\dfrac{1}{3}(x - 4) = \dfrac{7}{8}(x - 1)$

 **HOW TO BE SUCCESSFUL AT MATHEMATICS**

LEARN TO TAKE GOOD NOTES

Many students take *too many* notes. They try frantically to copy down everything that the instructor says and every detail of every problem that is put on the board. When they get home they may well find that their notes are missing some important steps and they were so busy trying to copy every word that they missed some of the necessary parts of what was being explained in the lesson.

Taking good notes requires that you be a good *listener*, not just a good copier. Follow the explanations carefully and write down only the important ideas. Looking over the new material before coming to class can help you to know what the important ideas are. If there are any useful suggestions given by the insructor, be sure to write those down in your notes. If two examples are done for you on the board, copy only one of them. Follow the other example carefully, focusing on how the instructor takes you through the problem, step by step. The right combination of careful, attentive listening and selective note-taking should be your goal.

2.3 Conditional Equations, Identities, and Contradictions

A **conditional equation** is one that is true for only certain values of the variable.

For example, when we solve $2x - 1 = 5$, we find that it has only one solution, namely, $x = 3$. The two sides are equal only under the *condition* that $x = 3$. Any other value for x will not make the equation a true statement, so it is a **conditional equation.** All of the equations we have considered so far have been conditional equations.

An **identity** (or **identical equation**) is one that is true for all values of the variable.

For example, the equation

$$3x - 1 = 3x - 1$$

is true no matter what number is substituted for the variable. If you try to get all the x terms on one side of the equation and all the constant terms on the other side, you eventually get $0 = 0$.

If you get $0 = 0$ when trying to solve an equation, then the original equation was an identity.

EXAMPLE 1   Solve: $3(x - 4) + 7 = 3x - 5$.

SOLUTION

$$3(x - 4) + 7 = 3x - 5 \qquad \text{Remove the parentheses.}$$
$$3x - 12 + 7 = 3x - 5 \qquad \text{Combine } -12 \text{ and } 7.$$
$$3x - 5 = 3x - 5 \qquad \text{Add } -3x \text{ and } 5 \text{ to both sides.}$$
$$0 = 0$$

Since our result is $0 = 0$, we conclude that the given equation is an **identity** and *every* numerical replacement for the variable is a solution.   ▮ ▮

> A **contradiction** is an equation that has no solutions.

For example, the equation

$$x = x + 1$$

can't possibly have a solution, since if we replace x by *any* number, the right side of the equation will always be 1 larger than the left. If we try to get all of the x terms on the left, we obtain a strange result:

$$x = x + 1 \qquad \text{Subtract } x \text{ from each side.}$$
$$x + (-x) = x + (-x) + 1$$
$$0 = 1?? \qquad \text{There is no solution.}$$

If the solution of any equation results in the two sides being unequal, like $0 = 1$, $5 = 7$, etc., then the given equation is a contradiction and there is no solution.

EXAMPLE 2   Solve: $6x - 9 = 3(2x - 5)$.

SOLUTION

$$6x - 9 = 3(2x - 5) \qquad \text{Remove the parentheses.}$$
$$6x - 9 = 6x - 15 \qquad \text{Add } -6x \text{ and } 9 \text{ to both sides.}$$
$$0 = -6$$

Since $0 \neq -6$, we conclude that there is no solution and the original equation is a **contradiction.**   ▮ ▮

To summarize:

> **There are three possibilities when attempting to solve an equation:**
>
> 1. Conditional—when a solution results.
>
> 2. Identity—when the *solved* equation results in $0 = 0$.
>
> 3. Contradiction—when the two sides of the *solved* equation are unequal (such as $0 = 6$, etc.).

2.3   Exercises

▼ Classify each of the following equations as conditional, an identity, or a contradiction. Solve the conditional equations.

1. $3x = 3(x + 9) - 2$

2. $4x = 5(x - 2) + 7$

3. $3x - 7 = 3(x - 2) - 1$

4. $7x = 4(2 - x) + 5$

5. $6y + 5 = 3(4 + 2y)$

6. $3n = 2 - 3(2 - n)$

7. $8z - 4(3 + 2z) + 5 = -7$

8. $13y + 7(1 - 2y) = -(2 + y)$

9. $6z - 3(4 - 2z) = 0$

10. $4 - 6(t + 2) = 3(2t - 1)$

11. $2(x - 4) + 2x = 4x - 8$

12. $3(2 - 3x) = -[5x - (4 - 4x)]$

2.4   Literal Equations

A **literal equation** is one that contains more than one letter. It will usually contain a variable and constants that are represented by letters instead of numbers. Formulas are good examples of literal equations.

Some examples of literal equations are:

$A = \pi r^2$ Area of a circle

$F = \dfrac{9}{5}C + 32$ Celsius to Fahrenheit

$A = l \cdot w$ Area of a rectangle

$P = 2l + 2w$ Perimeter of a rectangle

$y = mx + b$ Equation of a line

To solve a literal equation for a particular letter, we isolate that letter on one side of the equation. Since letters have the same meanings as numbers in literal equations, we use the same equation-solving techniques applied throughout this chapter.

HOW TO BE SUCCESSFUL AT MATHEMATICS

ATTEND EVERY CLASS

Mathematics, by its very nature, is perhaps more cumulative than some other courses that you may take. That is, what you do in class today is dependent on what you learned in the previous class. Students often underestimate the harm they do themselves when they miss even one class. Questions about homework from the class that you missed won't make sense to you and you will begin to feel lost. The new material presented in class will be less familiar to you and your morale will begin to fade. Somehow, from that day on, you will always feel like you're playing "catch up."

Every time you miss a class, you miss important information that you may not be able to find in the textbook. Your instructor may point out certain things that are especially important or show the class some time-saving technique for solving a problem. Someone else's notes may be a poor substitute for actually being there.

Make it your goal to attend every class and be there on time.

EXAMPLE 1

Solve $y = mx + b$ for x.

SOLUTION

If the equation read $7 = 3x + 4$, we would know exactly what to do. To show the similarity let's look at them side by side.

$$7 = 3x + 4 \qquad\qquad y = mx + b$$

$$7 + \boxed{(-4)} = 3x + 4 + \boxed{(-4)} \quad \text{Add } -4. \qquad y + \boxed{(-b)} = mx + b + \boxed{(-b)} \quad \text{Add } -b$$

$$3 = 3x \qquad\qquad y - b = mx$$

$$\frac{3}{3} = \frac{3x}{3} \quad \text{Divide by 3.} \qquad \frac{y - b}{m} = \frac{mx}{m} \quad \text{Divide by } m.$$

$$1 = x \qquad\qquad \frac{y - b}{m} = x$$

The solution is

$$x = 1$$

The solution is

$$x = \frac{y - b}{m}$$

▮▮

As you can see, literal equations are solved using exactly the same methods that we have been using all along.

EXAMPLE 2

Solve $P = 2l + 2w$ for w.

SOLUTION

We isolate w on one side of the equation:

$$P = 2l + 2w$$

$$P + \boxed{(-2l)} = 2l + \boxed{(-2l)} + 2w \qquad \text{Add } -2l \text{ to each side.}$$

$$P - 2l = 2w$$

$$\frac{P - 2l}{2} = \frac{2w}{2} \qquad \text{Divide each side by 2.}$$

$$w = \frac{P - 2l}{2} \qquad \text{This is the solution.}$$

▮▮

E X A M P L E 3   Solve $d = a + (b - 1)c$ for b.

SOLUTION

$$d = a + (b - 1)c$$

$$d = a + bc - c \qquad \text{Eliminate the parentheses.}$$

$$d + \boxed{(-a)} + \boxed{c} = a + \boxed{(-a)} + bc - c + \boxed{c} \qquad \text{Isolate the } b \text{ term.}$$

$$d - a + c = bc$$

$$\frac{d - a + c}{c} = \frac{b\cancel{c}}{\cancel{c}} \qquad \text{Divide each side by } c.$$

$$b = \frac{d - a + c}{c} \qquad \text{This is the solution.} \quad \blacksquare$$

E X A M P L E 4   Solve $\dfrac{2}{3}x + \dfrac{3}{4}y = \dfrac{1}{2}z$ for y.

SOLUTION

$$\frac{2}{3}x + \frac{3}{4}y = \frac{1}{2}z \qquad \text{Clear of fractions.}$$

$$\overset{4}{\cancel{12}} \cdot \frac{2}{\underset{1}{\cancel{3}}}x + \overset{3}{\cancel{12}} \cdot \frac{3}{\underset{1}{\cancel{4}}}y = \overset{6}{\cancel{12}} \cdot \frac{1}{\underset{1}{\cancel{2}}}z \qquad \text{Multiply by 12, the LCD.}$$

$$8x + 9y = 6z$$

$$8x + \boxed{(-8x)} + 9y = 6z + \boxed{(-8x)} \qquad \text{Add } -8x \text{ to each side.}$$

$$9y = 6z - 8x$$

$$\frac{\cancel{9}y}{\cancel{9}} = \frac{6z - 8x}{9} \qquad \text{Divide each side by 9.}$$

$$y = \frac{6z - 8x}{9} \qquad \text{This is the solution.} \quad \blacksquare$$

E X A M P L E 5   Solve $r = \dfrac{d}{t}$ for t.

SOLUTION

$$r = \frac{d}{t} \qquad \text{Clear of fractions.}$$

$$r \cdot \boxed{t} = \frac{d}{t} \cdot \boxed{t} \qquad \text{Multiply each side by } t, \text{ the LCD.}$$

$$rt = d$$

$$\frac{\cancel{r}t}{\cancel{r}} = \frac{d}{r} \qquad \text{Divide each side by } r.$$

$$t = \frac{d}{r} \qquad \text{This is the solution.} \quad \blacksquare$$

2.4   Exercises

I I I I I

▼ Solve each literal equation for the indicated letter.

1. $a + y = b$; for y

2. $x - y = z$; for x

3. $x - y = z$; for y

4. $ax = b$; for x

5. $x + ay = z$; for y

6. $I = prt$; for r

7. $P = a + b + c$; for b

8. $E = mc^2$; for m

9. $5(x + y) = z$; for y

10. $a(x + y) = z$; for y

11. $A = \frac{1}{2}bh$; for b

12. $h = 3(p + 2k)$; for k

13. $t = 3t(2a + b)$; for b

14. $6s - 4t = 5(s + t)$; for t

15. $7x + 5y = 3(x - y)$; for x

16. $I = \dfrac{E}{R}$; for R

17. $\dfrac{PV}{T} = C$; for P

18. $s = \frac{1}{2}gt^2$; for g

19. $\frac{1}{3}x + \frac{1}{2}y = \frac{2}{3}z$; for y

20. $I = \dfrac{E}{R + r}$; for R

21. $V = lwh$; for w

22. $V = \pi r^2 h$; for h

23. $y - y_1 = m(x - x_1)$; for m

24. $C = 2\pi r$; for π

25. $I = P + Prt$; for r

26. $F = \frac{9}{5}C + 32$; for C

 28. Write down the steps you would use to solve the chemistry formula $P = \dfrac{nRT}{V}$ for R.

 27. Explain, step by step, the procedure you would use to solve the formula $V = \frac{1}{3}\pi r^2 h$ for h.

2.5 Applications of Linear Equations

Problems in the real world rarely are stated in algebraic equations. The equations usually come well camouflaged within the English language in the form of application problems. Our first job will be to translate English statements into mathematical expressions and equations.

Certain words and phrases appear over and over again in application problems that tell us which mathematical operations we should use. The following table of key words and phrases tells us when we should add, subtract, multiply, or divide and also which words mean "equal to."

Add (a + b)	Subtract (a − b)	Multiply (a · b)	Divide (a ÷ b or a/b)	Equals (a = b)
a plus *b*	*a* minus *b*	*a* times *b*	*a* divided by *b*	*a* equals *b*
the sum of *a* and *b*	the difference of *a* and *b*	the product of *a* and *b*	the quotient of *a* and *b*	*a* is the same as *b*
add *a* and *b*	subtract *b* from *a*	multiply *a* and *b*	divide *a* by *b*	*a* is equivalent to *b*
b more than *a*	*b* less than *a*	of		*a* is identical to *b*
a increased by *b*	*a* less *b*			is, are, was
	a decreased by *b*			

Solving application problems has traditionally been one of the most difficult parts of algebra for students. The following "battle plan" has helped many students learn to solve application problems. If you are willing to follow the steps of this strategy carefully through the examples and exercises, your efforts will be rewarded and, indeed, you *will* learn to solve application problems.

HOW TO BE SUCCESSFUL AT MATHEMATICS

PREPARE FOR THE NEXT CLASS

Look over the section in your book that will be covered *before* you attend your next class. The material that is scheduled to be covered next is listed in your course syllabus or can be obtained from your instructor.

You do not need to study this material in detail; rather, go over it briefly, looking for key ideas and some general notions of what the section is about. As your instructor presents the new material, you will have some idea of what it is about and you will find that the concepts seem more familiar to you.

Give this technique a try a few times. You may be surprised at how effective it is.

A strategy for solving application problems:

1. Read the problem several times, making note of any key words. Draw a picture if applicable.

2. Write down what you are supposed to find and what is given.

3. Represent one of the unknown quantities by a variable. Express any other unknown quantities in terms of that *same* variable.

4. Find out which expressions are the same (equal) and write an equation. Sometimes a simple sketch helps to determine which expressions are equal.

5. Solve the equation.

6. Check your solution in the *orginal word statement* to see if it really meets the conditions of the problem. This can catch an error that might have been made in writing the equation.

Number-Relation Problems

EXAMPLE 1   Eight times a number decreased by 5 is equal to 43. Find the number.

SOLUTION   Let x represent the unknown number. Then we write

8	times	a number	decreased by	5	is equal to	43
↓	↓	↓	↓	↓	↓	↓
8	·	x	−	5	=	43

or

$$8x - 5 = 43$$

$$8x = 48$$

$$x = 6 \qquad \text{The unknown number is 6.}$$

Now we put the solution $x = 6$ back in the original word statement to see if it meets the conditions of the problem.

Check

8	times	a number	decreased by	5	is equal to	43
↓	↓	↓	↓	↓	↓	↓
8	·	6	−	5	=	43

$$8 \cdot 6 - 5 \overset{?}{=} 43$$

$$48 - 5 \overset{?}{=} 43$$

$$43 = 43 \quad ☑$$

EXAMPLE 2   Three times the difference of a number and 7 is 12. Find the number.

SOLUTION   Let x represent the desired number.

3	times	the difference of a number and 7	is	12
↓	↓	↓	↓	↓
3	·	$(x - 7)$	=	12

or

$$3(x - 7) = 12$$

$$3x - 21 = 12$$

$$3x = 33$$

$$x = 11 \qquad \text{The desired number is 11.}$$

Check $x = 11$ checks, since three times the difference of 11 and 7 is $3 \cdot 4$, which is 12.   ☑

EXAMPLE 3   The sum of two numbers is 84. If one number is one-half the other, find the numbers.

SOLUTION   Let $x =$ one of the numbers.

Then $\dfrac{1}{2}x =$ the other number.

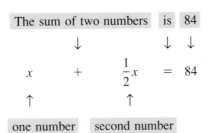

The sum of two numbers		is	84
↓		↓	↓
x	$+$ $\dfrac{1}{2}x$	=	84
↑	↑		
one number	second number		

$$x + \frac{1}{2}x = 84$$

$$2x + 2\left(\frac{1}{2}\right)x = 2(84) \qquad \text{Multiply by 2, the LCD.}$$

$$2x + x = 168$$

$$3x = 168$$

$$x = 56 \qquad \text{The first number}$$

$$\frac{1}{2}x = \frac{1}{2}(56) = 28 \qquad \text{The second number}$$

So the two numbers are 56 and 28.

Check   28 is one-half of 56 and $28 + 56 = 84$.   ✔   ■ ■

EXAMPLE 4   The total cost of an English book and an economics book is $81. If the economics book costs $5 more than the English book, how much does each book cost?

SOLUTION   Let x = the cost of the English book.
Then $x + 5$ = the cost of the economics book (since it costs $5 more).

Cost of English book	plus	cost of economics book	is	$81
↓	↓	↓	↓	↓
x	$+$	$x + 5$	$=$	81

$$x + x + 5 = 81$$

$$2x + 5 = 81$$

$$2x = 76$$

$$x = \$38 \qquad \text{The cost of the English book}$$

$$x + 5 = \$43 \qquad \text{The cost of the economics book}$$

Check   The economics book costs $5 more than the English book ($43 is $5 more than $38), and since $43 + \$38 = \$81$, the answers check.   ✔   ■ ■

Consecutive-Integer Problems

• • • • •

Consecutive integers are numbers that follow one right after another. For example,

$$5, 6, 7, 8, 9, \ldots \qquad \text{and} \qquad -14, -13, -12, -11, \ldots$$

If x is some integer, the next consecutive integer is one larger, or $x + 1$. The next one is one larger than that, or $x + 2$, and so on.

Consecutive integers are represented by:

$$x, \; x + 1, \; x + 2, \; x + 3, \ldots$$

Consecutive even integers are even numbers that follow one another without interruption. For example,

$$14, \; 16, \; 18, \; 20, \ldots \quad \text{and} \quad -10, \; -8, \; -6, \ldots$$

Each integer is two greater than the preceding number, so if an even integer is represented by x, the next one is $x + 2$, the next $x + 4$, and so on.

Consecutive even integers are represented by:

$$x, \; x + 2, \; x + 4, \; x + 6, \ldots$$

Consecutive odd integers are odd numbers that follow one another without interruption. For example,

$$11, \; 13, \; 15, \; 17, \ldots \quad \text{and} \quad -5, \; -3, \; -1, \; 1, \ldots$$

Each integer is two greater than the last number. So if x represents some odd integer, the next consecutive odd integer is two greater, or $x + 2$; the next will be $x + 4$, and so on.

Consecutive odd integers are represented by:

$$x, \; x + 2, \; x + 4, \; x + 6, \ldots$$

Notice that *both* consecutive even and consecutive odd integers are represented by $x, \; x + 2, \; x + 4, \; x + 6, \ldots$. This is because both consecutive even and consecutive odd integers differ from the preceding integer by two.

E X A M P L E 5   The sum of three consecutive integers is 57. Find the numbers.

S O L U T I O N   Let x = the first integer.
Then $x + 1$ = the second integer
and $x + 2$ = the third integer.

$$\boxed{\text{The sum of three consecutive integers}} \quad \boxed{\text{is}} \quad \boxed{57}$$

$$x + (x + 1) + (x + 2) \qquad\qquad = \quad 57$$

$$x + x + 1 + x + 2 = 57$$

$$3x + 3 = 57$$

$$3x = 54$$

$$x = 18 \qquad \text{The first integer}$$

$$x + 1 = 19 \qquad \text{The second integer}$$

$$x + 2 = 20 \qquad \text{The third integer}$$

Check   18, 19, 20 are consecutive integers and $18 + 19 + 20 = 57$.

EXAMPLE 6   Find three consecutive even integers such that the sum of the first and third integers is equal to twice the middle integer.

SOLUTION   Let $x =$ the first even integer,
$x + 2 =$ the second even integer,
$x + 4 =$ the third even integer.

$$\boxed{\text{Sum of 1st and 3rd integers}} \quad \boxed{\text{equals}} \quad \boxed{\text{twice}} \quad \boxed{\text{the middle integer}}$$

$$x + (x + 4) \qquad\qquad = \qquad 2 \cdot \qquad (x + 2)$$

$$x + (x + 4) = 2(x + 2)$$

$$x + x + 4 = 2x + 4$$

$$2x + 4 = 2x + 4$$

$$0 = 0 \qquad \text{This is an identity.}$$

In an identity, any value of the variable will be a solution. This means that if we take any three consecutive even integers, the sum of the first and the third integers will equal twice the middle one. Try a couple of examples and see for yourself!

EXAMPLE 7   Find three consecutive odd integers such that twice the smallest added to the largest is 67.

SOLUTION   Let $x =$ the first odd integer,
$x + 2 =$ the second odd integer,
$x + 4 =$ the third odd integer.

$$\boxed{\text{Twice the smallest}} \quad \boxed{\text{added to}} \quad \boxed{\text{the largest}} \quad \boxed{\text{is}} \quad \boxed{67}$$

$$2 \cdot x \qquad\qquad + \qquad (x + 4) \qquad = \quad 67$$

$$2x + x + 4 = 67$$
$$3x + 4 = 67$$
$$3x = 63$$
$$x = 21 \qquad \text{The first integer}$$
$$x + 2 = 23 \qquad \text{The second integer}$$
$$x + 4 = 25 \qquad \text{The third integer}$$

Check Twice the first integer is $2(21) = 42$, added to the third, 25, is $42 + 25 = 67$, and it checks.

Problems Involving Averages

EXAMPLE 8   If you receive a grade of 52 on a history test, what grade must you earn on the second test to bring your average to 70?

SOLUTION   Let x be the grade on the second test.

Average of the 2 exams	is equal to	70
↓	↓	↓
$\dfrac{52 + x}{2}$	=	70

$$\frac{52 + x}{2} = 70$$
$$52 + x = 140$$
$$x = 88 \qquad \text{The grade on the second test}$$

Check The average of the two grades is $\dfrac{52 + 88}{2} = \dfrac{140}{2} = 70$, which checks.

EXAMPLE 9   If you receive grades of 78 and 75 on your first two exams in Spanish class, what grade must you earn on your third exam to have an overall average of 80?

SOLUTION   Let x denote your grade on the third exam. Then the average of the three exam scores is the sum divided by 3:

Sum divided by 3	equals	average
↓	↓	↓
$\dfrac{78 + 75 + x}{3}$	=	80

$$(\cancel{3}) \cdot \frac{78 + 75 + x}{\cancel{3}} = (3) \cdot 80 \qquad \text{Multiply both sides by 3.}$$
$$153 + x = 240$$
$$x = 87$$

Check   If you receive an 87 on the third exam, your average will be
$$\frac{78 + 75 + 87}{3} = \frac{240}{3} = 80.$$

2.5   Exercises

▼ Solve and check each of the following problems.

NUMBER-RELATION PROBLEMS

1. When 4 is added to seven times a number, the result is 32. Find the number.

2. When a number is subtracted from 21, the result is twice the number. Find the number.

3. If 16 is subtracted from two-thirds of a number, the result is 8. Find the number.

4. Eight plus an unknown number is the same as twice the difference of the number and 2. What is the number?

5. Four more than a number is 3 less than twice that number. What is the number?

$4 + x = 2x - 3$

6. When a number is decreased by 18, the result is one-half of the number. Find the number.

7. Three-fourths of a number added to twice the number is equal to 44. What is the number?

8. One-third of a number subtracted from one-half of the number is 6. Find the number.

9. Separate 63 into two parts such that twice the first part is equal to the second part.

$2x = 63 - x$

10. A yardstick is cut into two pieces such that one piece is 7 in. shorter than the other. How long is each piece?

11. A 40-ft rope is cut into two pieces in such a way that one piece is 4 ft longer than twice the other. How long is each piece?

12. A repairman charges $40 for a service call plus a rate of $25 per hour after the first hour. If his bill was $115, how many hours did he spend on the service call altogether?

13. One-third the sum of a number and 6 is equal to three-fourths of the number. Find the number.

14. Paul is three times as old as his son Danny. Find the age of each if the sum of their ages is 48.

15. In a math class of 30 students, there are 4 more males than there are females. How many of each sex are in the class?

16. The perimeter of a rectangle is 40. Find the length and width of the rectangle if the length is 4 more than the width. Use $P = 2l + 2w$.

17. Mr. Nyberg wants to build a sandbox for his daughter. If he wants the length to be 1 ft more than the width and he has 26 ft of lumber, find the length and width of the sandbox.

18. When three times the sum of 6 and a number are added to five times the number, the result is equal to nine times the number. What is the number?

19. A mother is 20 years older than her daughter. Ten years ago the mother was twice as old as the daughter. How old are they now?

20. The total cost for a set consisting of a table and six chairs is $727. How much do the table and chairs each cost if eight times the cost of a chair is $1 more than the cost of the table?

21. If you spend half of your money on item A and one-third of your money on item B and you have $30 left, how much money did you have at the beginning?

22. The length of a rectangle is 3 less than three times the width. If the perimeter is equal to seven times the width, find the length and width of the rectangle.

23. Mrs. McGowan plans to fence in her rectangular garden. What will the dimensions of the garden be if the length is 2 ft more than twice the width, and she has 64 ft of fencing available?

CONSECUTIVE-INTEGER PROBLEMS

24. Find three consecutive numbers with a sum of 63.

25. Find three consecutive even integers that add to 90.

26. Find three consecutive odd integers such that the sum of the first two is 3 greater than the third.

27. Find two consecutive even integers such that four times the first is 2 greater than three times the second.

28. Find three consecutive odd integers such that the first plus twice the second plus three times the third is equal to 70.

29. Find three consecutive odd integers such that the sum of the first two is 27 greater than the third.

30. Find two consecutive even integers such that four times the first is 2 greater than three times the second.

31. Find three consecutive even integers such that twice the smallest is 4 larger than the largest.

32. Find three consecutive odd integers such that three times the sum of the first and second is 7 more than five times the third.

33. The sum of the first two of three consecutive odd integers added to the sum of the last two is 116. Find the integers.

34. Find four consecutive integers such that twice the sum of the first three is 2 less than five times the fourth.

35. Find three consecutive even integers such that four times the first is 12 less than twice the sum of the second and third.

36. Find three consecutive odd integers such that the sum of the smallest and the largest is equal to twice the middle integer.

37. Find three consecutive even integers with a sum of 90.

38. Find three consecutive even integers such that the sum of the smallest and the largest is equal to five times the middle integer.

39. Find three consecutive even integers such that the sum of the second two is 3 less than twice the first.

40. Find three consecutive odd integers such that one-fourth of the sum of the first two is equal to the third.

41. Find five consecutive even integers such that the sum of the first three integers equals the sum of the last two integers.

42. Find three consecutive integers that add to −105.

43. Find three consecutive odd integers with a sum of −111.

44. Find four consecutive even integers totaling −84.

PROBLEMS INVOLVING AVERAGES

45. If you receive a grade of 52 on a history test, what grade must you earn on the second test to bring your average up to 70?

46. If you receive grades of 72 and 76 on the first two exams of the semester, what grade do you have to get on the third exam in order to bring your average up to a 90? If the highest grade possible is 100, is this possible?

47. So far this semester, Brenda Frost has received exam grades of 86, 96, and 91 in psychology. What is the lowest grade she can receive on her fourth exam and still maintain an average of 90?

48. Suppose that you received grades of 82, 87, and 84 on the first three exams in your math course. (a) What grade must you get on your next exam to average 90 for all four exams? (b) Assuming that you need an average of 90 to get an A in the course and the highest grade you can earn on your fourth exam is 100, is it possible to get an A in the course? (c) What is the lowest grade you can get on the fourth exam and still get a B in the course (an average of 80)?

49. Mr. Quinn is a car salesman. In the first week of October he sold 7 cars; in the second and third weeks, he sold 3 cars each week. His goal is sell an average of 5 cars per week for the four-week period in October. How many cars must he sell in the fourth week to reach his goal?

50. During the first five weeks of his diet, Steve Coccia lost 4 lb, 2 lb, 2 lb, 1 lb, and 3 lb, respectively. How many pounds must he lose in the sixth week in order to have an average weight loss of $2\frac{1}{2}$ lb per week?

51. In preparation for running in the New York City Marathon, Bill Messner wants to average 8 miles per day. For the first five days of the week, he ran 6 miles, 4 miles, 10 miles, 10 miles, and 12 miles, respectively. How far must he run in order to meet his goal of averaging 8 miles per day for the six-day period?

52. The Middletown Public Works Department currently has four cars, which average 22, 31, 18, and 14 mpg. The Commissioner of Public Works has received permission from the City Council to purchase a fifth vehicle, but she has been instructed that the average number of miles per gallon for all five cars must be 30 mpg. How many miles per gallon must the new car get in order to comply with the City Council's mandate? Do you think they will be able to find such a vehicle?

53. Three college friends are driving from Philadelphia to Los Angeles, a distance of 2,980 miles. What is the average number of miles per day that they will have to travel in order to make the trip in 5 days? If they drive 480 miles the first day and 610 miles the second day, how many miles per day must they average for the remaining 3 days?

54. Darlene has scored 76, 88, 84, and 74 on her first four Spanish exams. She is preparing to take her final, which will count twice as much as the regular exams (200 points rather than 100 points). What grade must she score on the final exam in order to get an average of 80 for the semester?

2.6   Linear Inequalities

■ ■ ■ ■ ■

As was stated earlier; an **equation** is a statement that two algebraic expressions are equal. An **inequality** is a statement that one algebraic expression is either greater than or less than another one.

$$x - 3 < 6$$
$$2x - 5 > 4x + 7$$
$$5x < 2x - 6$$

are all examples of inequalities.

Fortunately, with only one exception, the rules for solving inequalities are the same ones we have learned for solving equations.

To solve an inequality, we isolate the variable on one side of the inequality sign and put everything else on the other side of the inequality sign.

Starting with the true inequality $4 < 6$, the following examples will show how the methods are the same or different for equations and inequalities.

a. $4 < 6$ *Add* the same number, 3, to both sides of the inequality.

$4 + \boxed{3} < 6 + \boxed{3}$

$7 < 9$ A true inequality results.

 **HOW TO BE SUCCESSFUL AT MATHEMATICS**

WHY SHOULD YOU DO HOMEWORK?

You may have heard the expression, "Mathematics is not a spectator sport." This is certainly true. Mathematics is best learned by active involvement rather than by staying on the sidelines and watching. Like learning a sport, learning mathematics requires practice. Watching someone do mathematics is like watching someone play tennis or play the piano. It helps, but to learn to do it yourself you must practice frequently and on a regular basis. That old saying, "Practice makes perfect," applies to learning mathematics.

Your instructor will carefully select the exercises that he or she feels will best help you to learn what is required. Find a comfortable place to do your homework, free of interruption from other people, and don't try to do your homework with the television on. Many students report, however, that having music in the background—as long as it's not too loud or too lively,—actually helps them do their homework better.

A rule of thumb is to spend a minimum of two hours of work outside of class for each hour spent in class. Of course, many students require even more time than this.

b.
$$4 < 6$$
$$4 + (-3) < 6 + (-3)$$
$$1 < 3$$

Subtract the same number, 3, from both sides of the inequality.

A true inequality results.

c.
$$4 < 6$$
$$3 \cdot (4) < 3 \cdot (6)$$
$$12 < 18$$

Multiply both sides of the inequality by the same positive number, 3.

A true inequality results.

d.
$$4 < 6$$
$$\frac{4}{2} < \frac{6}{2}$$
$$2 < 3$$

Divide both sides of the inequality by the same positive number, 2.

A true inequality results.

e.
$$4 < 6$$
$$-3 \cdot 4 \; ? \; -3 \cdot 6$$
$$-12 > -18$$

Multiply both sides of the inequality by the same negative number, −3.

The *direction* or *sense* of the inequality sign must be *changed* to obtain a true inequality.

f.
$$4 < 6$$
$$\frac{4}{-2} \; ? \; \frac{6}{-2}$$
$$-2 > -3$$

Divide both sides of the inequality by the same *negative* number, −2.

The direction or sense of the inequality sign must be changed to obtain a true inequality.

Examples (a) and (b) illustrate that we can add or subtract the same number on both sides of the inequality.

Examples (c) and (d) illustrate that we can multiply or divide both sides of an inequality by the same *positive* number.

Examples (e) and (f) illustrate that if both sides of an inequality are multiplied or divided by the same *negative* number, the direction of the inequality must be reversed.

These procedures for solving an inequality are summarized as follows:

To solve an inequality:

Use all the same rules and procedures as those for solving equations except, when multiplying or dividing both sides of an inequality by a *negative* number, the direction of the inequality symbol must be reversed.

E X A M P L E   1   Solve: $4x - 1 > 2x + 7$.

SOLUTION

$$4x - 1 > 2x + 7$$ Isolate the x terms on the left side.

$$4x + \boxed{(-2x)} - 1 > 2x + \boxed{(-2x)} + 7$$ Subtract $2x$ from each side.

$$2x - 1 > 7$$

$$2x - 1 + \boxed{1} > 7 + \boxed{1}$$ Add 1 to each side.

$$2x > 8$$

$$\frac{2x}{2} > \frac{8}{2}$$ Divide each side by 2.

$$x > 4$$ This is the solution.

To say "$x > 4$ is the solution" means that if x is replaced by any number greater than 4 in the original inequality, a true statement will result. Graphically, this looks like the following:

The open circle at 4 means that 4 *is not* part of the solution.   ▮▮

The expression $x \le y$ is read "x is less than or equal to y" and means that if either $x < y$ or $x = y$, then $x \le y$ is true.

$$2 \le 7 \quad \text{because} \quad 2 < 7$$
$$2 \le 2 \quad \text{because} \quad 2 = 2$$

The expression $x \ge y$ is read "x is greater than or equal to y" and means that if either $x > y$ or $x = y$, then $x \ge y$ is true.

$$9 \ge 5 \quad \text{bcause} \quad 9 > 5$$
$$6 \ge 6 \quad \text{because} \quad 6 = 6$$

To solve inequalities with the symbols $\leq$ and $\geq$, we follow the same rules that we used for "<" and ">."

EXAMPLE 2   Solve: $2x + 3 \leq 5(x - 3)$.

SOLUTION

$$2x + 3 \leq 5(x - 3) \qquad \text{Clear of parentheses.}$$

$$2x + 3 \leq 5x - 15$$

$$2x + (-5x) + 3 + (-3) \leq 5x + (-5x) - 15 + (-3) \qquad \text{Add } -5x \text{ and } -3 \text{ to both sides.}$$

$$-3x \leq -18$$

$$\frac{-3x}{-3} \geq \frac{-18}{-3} \qquad \text{Divide each side by } \textit{negative } 3, \text{ changing the direction of the inequality symbol.}$$

$$x \geq 6 \qquad \text{This is the solution.}$$

The graph of the solution is

The solid circle at 6 means that 6 *is* part of the solution.   ▐▐

EXAMPLE 3   Solve: $\dfrac{2x}{3} < \dfrac{3x}{4} + 3$.

SOLUTION

$$\frac{2x}{3} < \frac{3x}{4} + 3$$

$$\overset{4}{\cancel{12}} \cdot \frac{2x}{\cancel{3}} < \overset{3}{\cancel{12}} \cdot \frac{3x}{\cancel{4}} + 12 \cdot 3 \qquad \text{Multiply both sides by 12, the LCD.}$$

$$8x < 9x + 36$$

$$8x + (-9x) < 9x + (-9x) + 36 \qquad \text{Add } -9x \text{ to both sides.}$$

$$-x < 36 \qquad -x \text{ is the same as } -1 \cdot x.$$

$$-1x < 36$$

$$\frac{-1x}{-1} > \frac{36}{-1} \qquad \text{Divide each step by } -1, \text{ reversing the direction of the inequality.}$$

$$x > -36 \qquad \text{This is the solution.}$$

The graph of the solution is

┼───┼───⊕───┼───┼───┼→
−38 −37 −36 −35 −34 −33

▐▐

2.6   Exercises

▼ Solve the following inequalities and graph the solution.

1. $x + 4 > 2$

2. $x - 3 < 1$

3. $6x + 2 < 14$

4. $2y - 5 \geq 15$

5. $3x < 5x + 6$

6. $6x + 3 \geq 8x - 5$

7. $5 - 2x < 7$

8. $4 - 3x \leq -8$

9. $8x + 5 > 10x - 5$

10. $9x - 4 \geq 12x + 8$

11. $6(x - 1) < 5x + 7$

12. $3(x + 4) > 5(x - 2)$

13. $4(x + 1) > 3x + 7$

14. $6(2 - x) \leq 4x - (x - 3)$

15. $7(b - 2) \leq 3(b + 2) + 4$

16. $3(3 + 2y) - (3y + 2) < 12$

17. $-\dfrac{5}{8}y > -25$

<---------------------->

20. $\dfrac{h}{-4} < 7$

<---------------------->

18. $-\dfrac{2}{3}t \geq 6$

<---------------------->

21. $\dfrac{2}{3}x + 2 > \dfrac{1}{2}x - \dfrac{3}{4}$

<---------------------->

19. $\dfrac{x}{3} - 1 \geq \dfrac{x}{2}$

<---------------------->

22. $\dfrac{2x}{5} - \dfrac{1}{4} < \dfrac{3}{4}(x + 2)$

<---------------------->

2.7   Absolute Value Equations

Previously we defined the absolute value of x, written $|x|$, to be the distance between x and 0 on the number line.

EXAMPLE 1   Solve: $|x| = 4$.

SOLUTION   According to the definition, the distance between x and 0 on the number line is 4 units. Therefore, x can be at either 4 or -4, since both of these numbers are 4 units from 0.

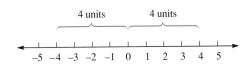

This gives us two related equations:

$$x = 4 \qquad \text{or} \qquad x = -4$$

The solutions are 4 or -4.

HOW TO BE SUCCESSFUL AT MATHEMATICS

STUDY WITH SOMEONE FROM YOUR CLASS

Try to find a member of your class to study with. Get his or her telephone number, and make arrangements to study together on a regular basis. Comparing and going over your notes, asking each other questions, and working through problems together can be beneficial to both of you. Studying math is much more enjoyable when you have a fellow student who gives you support.

If one of you *must* miss a class, you can get the notes from the other person.

You must be certain not to copy the other person's homework or to lean on that person too heavily; this should be a joint venture. Each of you can learn more by working together than you can by working individually.

EXAMPLE 2   Solve: $|2x + 1| = 3$.

SOLUTION $2x + 1$ is 3 units from 0, so $2x + 1$ is equal to either 3 or -3. We must solve both equations:

$$2x + 1 = 3 \quad \text{or} \quad 2x + 1 = -3$$
$$2x = 2 \quad \text{or} \quad 2x = -4$$
$$x = 1 \quad \text{or} \quad x = -2$$

The solutions are 1 or -2.

Check

$$
\begin{array}{c|c}
x = 1 & x = -2 \\
|2x + 1| \overset{?}{=} 3 & |2x + 1| \overset{?}{=} 3 \\
|2(1) + 1| \overset{?}{=} 3 & |2(-2) + 1| \overset{?}{=} 3 \\
|2 + 1| \overset{?}{=} 3 & |-4 + 1| \overset{?}{=} 3 \\
|3| \overset{?}{=} 3 & |-3| \overset{?}{=} 3 \\
3 = 3 \quad ✓ & 3 = 3 \quad ✓
\end{array}
$$

As you can see, *both* solutions check.   ▮ ▮

EXAMPLE 3   Solve: $2|2x - 5| - 1 = 17$.

SOLUTION   The secret to solving an absolute value equation of this type is to isolate the absolute value on one side of the equation.

$$2|2x - 5| - 1 = 17 \qquad \text{Add 1 to each side.}$$
$$2|2x - 5| = 18 \qquad \text{Divide each side by 2.}$$
$$|2x - 5| = 9$$

Now it looks similar to the equation we solved before, and we solve it in the same way: $|2x - 5| = 9$ means $2x - 5$ is equal to 9 or -9.

$$2x - 5 = 9 \quad \text{or} \quad 2x - 5 = -9$$
$$2x = 14 \quad \text{or} \quad 2x = -4$$
$$x = 7 \quad \text{or} \quad x = -2$$

The solutions are 7 or -2. ▮▮

E X A M P L E 4   Solve: $|5x + 1| = -3$.

S O L U T I O N   The left side of the equation *is never negative,* since an absolute value is always greater than or equal to zero. Since the right side of the equation *is negative,* we conclude that there can be no solution to the inequality. Try substituting some values for x and you will see that you can never get -3. The correct answer is *no solution.* ▮▮

STOP   An alternate algebraic definition of the absolute value is

$$|x| = \begin{cases} x & \text{if} \quad x \geq 0 \\ -x & \text{if} \quad x < 0 \end{cases}$$

Students sometimes find the second step of this definition confusing, because they see $-x$ and yet they know that the absolute value is never negative. It is important to realize that $-x$ is not necessarily negative. Remember that $|x| = -x$ only when $x < 0$. For example, if $x = -3$, then $-x = -(-3) = 3$, which is positive. If x is negative, then $-x$ is indeed positive.

2.7   Exercises

▮ ▮ ▮ ▮ ▮

▼   Solve:

7. $|x - 4| = 7$ **8.** $|y - 5| = 1$

1. $|x| = 6$ **2.** $|y| = 7$

9. $|2y - 5| = 7$ **10.** $|5x + 3| = 18$

3. $|x| = -5$ **4.** $|y| = -4$

11. $|4x - 1| = -3$ **12.** $|4 - x| = 4$

5. $|x + 1| = 7$ **6.** $|y - 3| = 8$

13. $|-6x - 3| = 39$ **14.** $|7y + 1| - 3 = 5$ **19.** $2|5y - 3| + 2 = 6$ **20.** $-3|4x + 2| - 2 = -14$

15. $|3x - 2| + 5 = 2$ **16.** $4|2x - 5| = 28$

   21. Explain why the equation $|2x - 3| = -5$ cannot possibly have a solution.

17. $|3x - 4| + 1 = 8$ **18.** $4|2x + 5| - 1 = 27$

2.8   Absolute Value Inequalities

▮ ▮ ▮ ▮ ▮

Recall that the absolute value of x, written $|x|$, is the distance from x to 0 on the number line. Let's now look at three absolute value expressions and what they each mean in terms of this definition.

1. $|x| = 3$ means x is precisely 3 units from 0 on the number line.

Interpretation: $x = -3$ or $x = 3$.

2. $|x| < 3$ means x is *less than* 3 units from 0 on the number line.

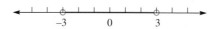

Interpretation: $x > -3$ and $x < 3$, which can also be written $-3 < x < 3$.

3. $|x| \geq 3$ means that x is *greater than or equal to* 3 units from 0 on the number line.

Interpretation: $x \leq -3$ or $x \geq 3$.

HOW TO BE SUCCESSFUL AT MATHEMATICS

PREPARE FOR EXAMS

The following steps will maximize your chances for success on any mathematics exam.

1. Do your work on a reglar basis so that you won't find yourself cramming or unprepared when exam time comes.

2. Your goal should be to get 100% on the exam.

3. Make a list of all of the topics that will be covered on the exam.

4. Make sure that you can solve the problems involved in each of these topics. Vary the order of these topics when solving the problems.

5. Be sure that you can correctly solve all of the problems in the Chapter Reviews and on the Achievement Tests from the applicable chapters in the text.

6. Begin studying for your exam as soon as possible to avoid last-minute cramming. Only a brief summary should be necessary on the day of the exam.

7. Get a good night's sleep. If you have followed the Steps 1–6, you should have no reason to stay up all night studying.

EXAMPLE 1   Solve $|2x - 1| > 5$ and sketch the graph of the solution.

SOLUTION   This is similar to Case 3, since $|2x - 1| > 5$ means that $2x - 1$ is more than 5 units from 0 on the number line. Graphically:

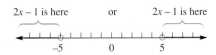

Our two inequalities to be solved are

$$2x - 1 < -5 \quad \text{or} \quad 2x - 1 > 5$$
$$2x < -4 \quad \text{or} \quad 2x > 6$$
$$x < -2 \quad \text{or} \quad x > 3$$

The solution is $x < -2$ or $x > 3$, the graph of which is

Any values of x that are either less than -2 or greater than 3 will satisfy the original inequality $|2x - 1| > 5$. No other values will work. (Try some and see.)

EXAMPLE 2   Solve $|x - 3| \geq 4$ and sketch the graph of the solution.

SOLUTION $|x - 3| \geq 4$ means that $x - 3$ is greater than or equal to 4 units from 0. This gives us two inequalities:

$$x - 3 \leq -4 \qquad \text{or} \qquad x - 3 \geq 4$$
$$x \leq -1 \qquad \text{or} \qquad x \geq 7$$

The solution is $x \leq -1$ or $x \geq 7$, the graph of which is   ∎ ∎

EXAMPLE 3   Solve and graph $|3x - 2| < 4$.

SOLUTION $|3x - 2| < 4$ means that $3x - 2$ is less than 4 units from 0 on the number line or, graphically,

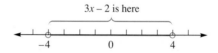

We say that $3x - 2$ is *between* -4 and 4 or equivalently; that $3x - 2 > -4$ and $3x - 2 < 4$. This is usually written as a continued inequality, $-4 < 3x - 2 < 4$, which is solved as follows:

$$-4 < 3x - 2 < 4 \qquad \text{Add 2 to all three members.}$$
$$-2 < 3x < 6 \qquad \text{Divide each member by 3.}$$
$$-\frac{2}{3} < x < 2 \qquad \text{This is our solution.}$$

The graph of the solution is   ∎ ∎

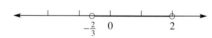

STOP   The continued inequality $1 < x < 4$ is read from the variable, from the inside out in both directions, as "x is greater than 1 and less that 4." This notation always means **and** and it never means or.
      The statement $x < 1$ or $x > 4$ *cannot* be written using a continued inequality.

EXAMPLE 4   Solve: $|5x + 3| < -3$.

SOLUTION   Before jumping right in, let's examine the situation first. The left side of the inequality is never negative because it is an absolute value. The right side is negative. How can we have a positive quantity less than a negative quantity? The answer is, of course, that we can't, and we conclude that there is no solution. An absolute value can never be less than a negative number.   ∎ ∎

E X A M P L E   5 Solve: $|2y - 3| > -1$.

S O L U T I O N The left side of the inequality is never negative so it must always be greater than -1, a negative quantity. Therefore the statement $|2y - 3| > -1$ is true no matter what number we substitute for y. The solution is the set of all real numbers. ▮▮

2.8 Exercises

▮▮▮▮▮

▼ Solve each inequality and graph the solution.

1. $|x| > 3$

2. $|x| \geq 4$

3. $|x| < 3$

4. $|x| \leq 5$

5. $|x + 3| < 4$

6. $|y - 2| > 7$

7. $|y - 7| \geq 5$

8. $|x + 6| \leq 3$

9. $|x| + 3 < 5$

10. $|y| - 4 > 2$

11. $|2x + 5| \leq 7$

12. $|3x - 2| \geq 7$

13. $|x - 4| < -5$

14. $|3y + 9| < 3$

15. $|x + 1| > -2$

16. $|2x + 1| < 9$

17. $|3x + 1| - 2 \geq 3$

18. $|2y - 2| + 2 \leq 4$

19. $|4x + 1| + 5 < 3$

20. $2|3y - 1| + 4 < 14$

21. $\left| \dfrac{a - 2}{2} \right| < 5$

22. $\left| \dfrac{a - 2}{2} \right| > 5$

23. $2|2x - 5| - 1 \geq 1$

24. $3|4x + 1| - 7 \leq -1$

 25. Explain why the inequality $|4x - 3| < -5$ cannot have a solution

 26. Explain why any number will work as a solution to the inequality $|3x + 5| > -2$

▮▮▮▮ ▰ ●●●● HOW TO BE SUCCESSFUL AT MATHEMATICS ──

TAKING AN EXAM

Here are some hints to help you maximize your chances for success on a math exam:

1. Arrive at the exam rested and a few minutes early.

2. Bring all necessary materials—pencils or pens, an eraser, a ruler if necessary, and a watch.

3. Read the instructions and skim the entire exam.

4. Do the easiest problems first.

5. If you don't understand something, ask your instructor for clarification.

6. Show all of your work.

7. Don't leave any answers blank.

8. Don't leave early. Check your work as much as time allows.

SUMMARY—CHAPTER 2

EXAMPLE

The **addition principle:** The same number may be added to (or subtracted from) both sides of any equation and the solution will remain the same.

Solve: $x + 5 = -3$.
$$x + 5 + (-5) = -3 + (-5)$$
$$x = -8$$

The **multiplication principle:** If both sides of an equation are multiplied (or divided) by the same number (except zero), the new equation is equivalent to the original equation.

Solve: $\frac{2}{3}x = 12$.

Multiply both sides by $\frac{3}{2}$, the reciprocal of $\frac{2}{3}$.

The **reciprocal** of a fraction $\frac{a}{b}$ is $\frac{b}{a}$.

$$\frac{\cancel{3}}{\cancel{2}} \cdot \frac{\cancel{2}}{\cancel{3}}x = \frac{3}{\cancel{2}} \cdot \overset{6}{\cancel{12}}$$
$$x = 18$$

To solve an equation using both the addition and multiplication principles:

1. Put all the terms containing the variable on one side of the equation by using the addition principle.

2. Put all the constant terms on the other side of the equation by again using the addition principle.

3. Multiply both sides of the equation by the reciprocal of the coefficient of the variable using the multiplication principle.

Solve: $3x + 1 = 5x - 11$.
$$3x + (-5x) + 1 = 5x + (-5x) - 11$$
$$-2x + 1 = -11$$
$$-2x + 1 + (-1) = -11 + (-1)$$
$$-2x = -12$$
$$\frac{-2x}{-2} = \frac{-12}{-2}$$
$$x = 6$$

To eliminate fractions from an equation, multiply both sides of the equation by the LCD (**least common denominator**) of all the fractions that appear in the equation.

Solve: $\frac{1}{2}x + \frac{3}{4} = \frac{1}{3}$.

The LCD is 12.

$$\overset{6}{\cancel{12}} \cdot \frac{1}{\cancel{2}}x + \overset{3}{\cancel{12}} \cdot \frac{3}{\cancel{4}} = \overset{4}{\cancel{12}} \cdot \frac{1}{\cancel{3}}$$
$$6x + 9 = 4$$
$$6x = -5$$
$$x = -\frac{5}{6}$$

To clear an equation of decimals:

1. Count the greatest number of decimal places that appear in any of the decimals in the equation.

2. Multiply each side of the equation by 1 followed by that number of zeros.

Solve: $0.03x + 0.21 = -0.06$.

Multiply by 100:
$$100 \cdot (0.03x) + 100 \cdot (0.21) = 100 \cdot (-0.06)$$
$$3x + 21 = -6$$
$$3x = -27$$
$$x = -9$$

A **conditional equation** is one that is true for only certain values of the variable.

An **identity** or an **identical equation** is one that is true no matter what value is substituted for the variable.

A **contradiction** is an equation that has no solutions.

When solving an equation, it is:

1. *conditional* if a solution results;

2. *an identity* if $0 = 0$ results;

3. *a contradiction* if an unequal expression, such as, $0 = 4$, results.

A **literal equation** is one that contains a variable and constants represented by letters.

A strategy for solving word problems:

1. Read the problem several times, making note of any key words.

2. Write down what is given and what you are supposed to find.

3. Represent one of the unknown quantities by a variable. Express any other unknown quantities in terms of that same variable.

4. Find out which expressions are the same and write an equation. Sometimes a simple sketch helps to determine which expressions are equal.

5. Solve the equation.

6. Check your solution in the original word statement to see if it really meets the conditions of the problem.

Consecutive integers are integers that follow one right after another, represented by $x, x + 1, x + 2, \ldots$

Consecutive even integers are even integers that follow one another, represented by $x, x + 2, x + 4, \ldots$

Consecutive odd integers are odd integers that follow one another, represented by $x, x + 2, x + 4, \ldots$

An **inequality** is a statement that one algebraic expression is either greater than or less than another algebraic expression.

To solve an inequality, use the same rules and procedures as those for solving equations, except when multiplying or dividing both sides of an inequality by a negative number, the direction of the inequality symbol must be reversed.

The absolute value of x, written $|x|$, is the distance between x and 0 on the number line.

EXAMPLE

Solve $3a + 2x = b$ for x.
$$3a + (-3a) + 2x = b + (-3a)$$
$$2x = b - 3a$$
$$x = \frac{b - 3a}{2}$$

Find three consecutive odd integers such that the sum of the first two added to twice the third is 110.

$x = $ first integer
$x + 2 = $ second integer
$x + 4 = $ third integer
$$x + x + 2 + 2(x + 4) = 110$$
$$2x + 2 + 2x + 8 = 110$$
$$4x + 10 = 110$$
$$4x = 100$$
$$x = 25$$

The 1st integer is 25.
The 2nd integer is $x + 2 = 27$.
The 3rd integer is $x + 4 = 29$.

Solve: $3x - 7 < 8$.
$$3x - 7 + 7 < 8 + 7$$
$$3x < 15$$
$$x < 5$$

$|x| = 4$ means $x = -4$ or $x = 4$.
$|x| > 4$ means $x < -4$ or $x > 4$.
$|x| < 4$ means $-4 < x < 4$.

REVIEW EXERCISES—CHAPTER 2

▼ **Solve each equation:**

1. $2x + 6 = 6x - 18$

2. $7x - 6 = -4x - 28$

3. $3(x - 2) = 12$

4. $2x - 3(x - 1) = 2(x + 1)$

5. $0.41x + 0.3 = 0.39x - 0.42$

6. $0.2x + 3.1 = 1.2(x - 3)$

7. $\frac{1}{2}x + 2 = \frac{1}{3}x - 3$

8. $\frac{2}{3}x - 2 = \frac{1}{2}x + \frac{3}{4}$

▼ **Classify each of the following equations as conditional, an identity, or a contradiction.**

9. $4(x - 1) = 2x - 2(3 - x)$

10. $-2(x + 3) = 4x - 6(x + 1)$

▼ **Solve each equation for x:**

11. $3y + 4x = 5z$

12. $\frac{1}{2}(x - 2y) = \frac{1}{3}(x + 2)$

▼ **Solve and check:**

13. When 4 is subtracted from three times a number, the result is 17. Find the number.

14. Find three consecutive odd integers such that the sum of the first two is 9 more than the third.

15. When two-thirds of a number is added to one-half of the same number, the result is 1 more than the number. Find that number.

16. What grade must you earn on your third exam so that your average will be 90 if you scored 94 and 78 on your first two exams?

▼ **Solve the inequalities:**

17. $-2x - 1 \geq 7$

18. $5x + 6 < 2x - 3$

19. $\frac{1}{2}x - 5 > \frac{2}{3}x + 1$

20. $3(x - 1) \leq 5(2 - x)$

▼ **Solve each equation:**

21. $|x + 4| = 9$ **22.** $|3x + 1| = -4$

23. $|2x - 7| + 4 = 13$ **24.** $3|x + 2| - 4 = 11$

▼ **Solve the inequality and graph the solution:**

25. $|x| \leq 6$

←——————————→

26. $|x - 3| > 4$

←——————————→

27. $|3x + 1| < -4$

←——————————→

28. $2|3x + 5| + 3 \geq 13$

←——————————→

29. $|2x + 5| > -4$

←——————————→

30. $|x + 2| \leq -8$

←——————————→

CRITICAL THINKING EXERCISES—CHAPTER 2

▼ There is an error in each of Exercises 1–7. Can you find it?

1. Solve for x: $-7(x - 4) = 3$ Distribute the -7.
$$-7x - 4 = 3$$
$$-7x = 7$$ Divide by -7.
$$x = -1$$

2. Solve for x: $8x - 5 = -5$ Add 5 to both sides.
$$8x = 0$$ Divide both sides by 8.
$$\frac{8x}{8} = \frac{0}{8}$$
$$x = 8$$

3. Solve for x: $8x = 24$ Subtract 8 from both sides.
$$8x - 8 = 24 - 8$$
$$x = 16$$

4. Solve for x: $\dfrac{1}{2}x + 3 = \dfrac{1}{3}$ Multiply by 6, the LCD.
$$\overset{3}{\cancel{6}}\left(\frac{1}{\cancel{2}}x\right) + 3 = \overset{2}{\cancel{6}}\left(\frac{1}{\cancel{3}}\right)$$
$$3x + 3 = 2$$ Subtract 3 from both sides.
$$3x = -1$$ Divide each side by 3.
$$x = -\frac{1}{3}$$

5. Solve for x: $-6x \leq -30$ Divide both sides by -6.
$$\frac{-6x}{-6} \leq \frac{-30}{-6}$$
$$x \leq 5$$

6. Solve for x: $|2x - 1| = -5$.

$$2x - 1 = -5 \quad \text{or} \quad 2x - 1 = 5$$
$$2x = -4 \quad \text{or} \quad 2x = 6$$
$$x = -2 \quad \text{or} \quad x = 3$$

The solutions are $x = -2$ or $x = 3$.

7. There is no solution to the inequality $|4x - 1| > -3$.

8. Suppose we are given that $a > b$, and we know that $2 > 1$. Criticize the following "proof":

$$2 > 1 \qquad \text{Given.}$$
$$2(b - a) > 1(b - a) \qquad \text{Multiply both sides by } (b - a).$$
$$2b - 2a > b - a \qquad \text{Distribute the 2 on the left.}$$
$$2b - b > 2a - a \qquad \text{Add } 2a \text{ and } -b \text{ to both sides.}$$
$$b > a$$

However, it was given that $a > b$, so how can $b > a$?

9. What can you conclude about $-x$ if $-2 < x < 3$?

10. What can you conclude about x if $5 < x < 2$?

NAME _____   **▮ ▮ ▮ ▮ ▮ CLASS**

ACHIEVEMENT TEST—CHAPTER 2

▼ Solve for *x:*

1. $6x - 3 = 8x + 9$ 1. _____

2. $2x - (x + 4) = 5 - 3(x - 1)$ 2. _____

3. $x + \frac{1}{2} = \frac{1}{2}x - \frac{1}{3}$ 3. _____

4. $0.7x - 4 = 0.6x - 3.2$ 4. _____

5. $ax + b = c - 2d$ 5. _____

6. $-3x + 4 \geq 13$ 6. _____

7. $\frac{1}{3}x - 1 < \frac{1}{4}(x - 3)$ 7. _____

8. $|2x + 3| + 3 = 10$

8. _____

9. $|6x - 1| = -3$

9. _____

▼ Classify as conditional, an identity, or a contradiction. Solve the conditional equations.

10. $-2(1 - 3x) = 3(2x + 1) - 5$

10. _____

11. $4x + 3 = -2(3 - 2x) + 1$

11. _____

▼ Solve each inequality and graph the solution.

12. $|x - 4| < 3$

12. _____

⟵—————————⟶

13. $2|4x - 1| + 5 \geq 11$

13. _____

⟵—————————⟶

14. $|3x - 1| < -2$

14. _____

⟵—————————⟶

15. When two-thirds of a number is added to 12, the sum is 24. Find the number.

15. _____

16. Find three consecutive even integers such that three times the first is equal to 10 more than the sum of the second two integers.

16. _____

17. If you receive grades of 80 and 77 on your first two exams, what grade must you receive on the third exam to average 85 for the three exams?

17. _____

Polynomials

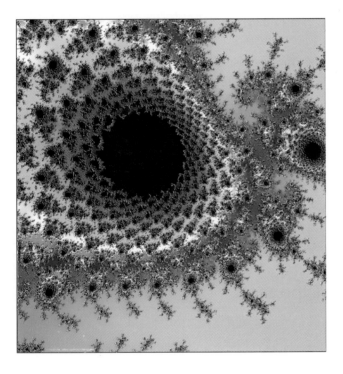

INTRODUCTION

In Chapter 3, we introduce an algebraic expression known as a *polynomial*. We will investigate how to combine them using the basic operations of addition, subtraction, multiplication, and division. Also we will spend considerable time rewriting them as products, that is, *factoring* them. Some of the factoring may be a review and some kinds will be new to you. Regardless, a thorough knowledge of factoring is necessary for the work we will do as we move on.

CHAPTER 3—NUMBER KNOWLEDGE

Why are Chain Letters Illegal?

It is a violation of federal law to mail chain letters using the U.S. Postal Service.

Suppose you receive a **chain letter** containing a list of six names and addresses. (It can have either more or less names.) You are instructed to send some money or a recipe or whatever to the top name on the list, cross off that name, add your name to the bottom, and send a copy of the letter to six of your friends. As the process continues, your name eventually moves to the top of the list and you will receive the money or recipes or whatever from 46,656 people. Chain letters illustrate a mathematical concept known as a **geometric progression** and involve **exponential growth.** The following chart shows how many people will receive a copy of the letter, assuming that no one breaks the chain and no one receives more than one copy of the letter, both unlikely events.

1st mailing	6^1	6
2nd mailing	6^2	36
3rd mailing	6^3	216
4th mailing	6^4	1,296

5th mailing	6^5	7,776
6th mailing	6^6	46,656
7th mailing	6^7	279,936
8th mailing	6^8	1,679,616
9th mailing	6^9	10,077,696
10th mailing	6^{10}	60,466,176
11th mailing	6^{11}	362,797,056
12th mailing	6^{12}	2,176,782,336
13th mailing	6^{13}	13,060,694,016
14th mailing	6^{14}	78,364,164,096

The eighth mailing involves about twice the population of the entire state of Montana; the tenth mailing involves more than twice the population of California; the eleventh mailing involves more people than live in the United States; and the fourteenth mailing would result in more letters than there are people in the entire world! This would keep the Postal Service too busy to do anything else.

3.1　Integral Exponents

Exponents are used whenever a number or a variable is to be multiplied by itself:

$$\underbrace{a \cdot a \cdot a \cdot a \cdot a}_{5 \text{ factors}} = a^{\overset{\text{exponent or power}}{5}}_{\text{base}}$$

In the **exponential expression** a^5, we call a the **base** and 5 the **exponent** or **power.**

EXAMPLE 1　**(a)** $x \cdot x = x^2$　　Read "x squared."

　(b) $6 \cdot 6 \cdot 6 = 6^3$　　Read "6 cubed."

　(c) $(-2)(-2)(-2)(-2) = (-2)^4$　　Read "negative two to the fourth power."

　(d) $7^1 = 7$　　The first power of any number is just the number itself.

(e) When we multiply a variable a times itself n times, we write

$$a^n = \underbrace{a \cdot a \cdot a \cdot a \ldots a}$$

a is used as a factor n times

■■

E X A M P L E   2   Write the following powers without using exponents.

(a) $2^4 = 2 \cdot 2 \cdot 2 \cdot 2 = 16$

(b) $3x^2 = 3 \cdot x \cdot x$

The exponent, 2, does not apply to the 3.

(c) $(3x)^2 = (3x) \cdot (3x)$

Here the exponent applies to everything in the parentheses.

(d) $6^1 = 6$

(e) $1^6 = 1 \cdot 1 \cdot 1 \cdot 1 \cdot 1 \cdot 1 = 1$

1 raised to any power equals 1.

(f) $(-2)^4 = (-2) \cdot (-2) \cdot (-2) \cdot (-2) = 16$

The exponent applies only to the 2, unless there are parentheses.

(g) $-2^4 = -(2 \cdot 2 \cdot 2 \cdot 2) = -16$

(h) $\left(\dfrac{2}{3}\right)^3 = \dfrac{2}{3} \cdot \dfrac{2}{3} \cdot \dfrac{2}{3} = \dfrac{8}{27}$

■■

Consider the product $a^3 \cdot a^4$. Using our definition,

$$a^3 \cdot a^4 = \underbrace{(a \cdot a \cdot a)} \cdot \underbrace{(a \cdot a \cdot a \cdot a)} = \underbrace{a \cdot a \cdot a \cdot a \cdot a \cdot a \cdot a} = a^7$$

3 factors + 4 factors   =   7 factors

So we have $a^3 \cdot a^4 = a^{3+4} = a^7$. This is true in general, leading us to our first rule of exponents: to multiply two quantities with the same base, keep that base and **add** the exponents. Written in symbols, we have the following:

Exponent Rule I:

$$x^a \cdot x^b = x^{a+b}$$

E X A M P L E   3   **(a)** $h^2 \cdot h^3 = h^{2+3} = h^5$

(b) $2^2 \cdot 2^4 = 2^{2+4} = 2^6 = 64$

(c) $x^3 \cdot x = x^3 \cdot x^1 = x^{3+1} = x^4$ Recall that $x = x^1$.

(d) $10^5 \cdot 10^4 = 10^{5+4} = 10^9 = 1,000,000,000$

(e) $x^4 \cdot y^5$ These cannot be combined since x and y are different bases.

(f) $x^3 \cdot x^7 \cdot x^4 = x^{14}$ The rule can be generalized to include more than two factors.

(g) $x^2 \cdot y^4 \cdot x^3 \cdot y^3 = x^5 y^7$ Combine only the factors with like bases.   ■■

A common error is to multiply the bases together and add the exponents. Note that

$$2^5 \cdot 2^4 \neq 4^9$$

Remember to keep the *same base* and add the exponents.

$$2^5 \cdot 2^4 = 2^9$$

Now let's consider expressions where we raise a power to a power, $(x^5)^2$:

$$(x^5)^2 = x^5 \cdot x^5 = x^{5+5} = x^{10}$$

We see that $(x^5)^2 = x^{5 \cdot 2} = x^{10}$. When a quantity raised to a power is again raised to a power, we **multiply** exponents. This is stated as follows:

Exponent Rule II:

$$(x^a)^b = x^{a \cdot b}$$

EXAMPLE 4   **(a)** $(x^6)^4 = x^{6 \cdot 4} = x^{24}$

(b) $(b^2)^5 = b^{2 \cdot 5} = b^{10}$

(c) $(2^3)^3 = 2^{3 \cdot 3} = 2^9$

(d) $(10^m)^n = 10^{m \cdot n} = 10^{mn}$

(e) $(y^4)^5 = y^{4 \cdot 5} = y^{20}$

(f) $(x^4)^y = x^{4 \cdot y} = x^{4y}$ ▮▮

Next we consider *division* of two quantities that have the same base raised to powers, for example, $\dfrac{x^5}{x^2}$:

$$\frac{x^5}{x^2} = \frac{x \cdot x \cdot x \cdot x \cdot x}{x \cdot x} = \boxed{\frac{x \cdot x}{x \cdot x}} \cdot x \cdot x \cdot x = x \cdot x \cdot x = x^3$$

↑
This is equal to 1.

We have $\dfrac{x^5}{x^2} = x^{5-2} = x^3$.

This is true in general (providing $x \neq 0$) and is stated as follows:

Exponent Rule III:

$$\frac{x^a}{x^b} = x^{a-b}$$

E X A M P L E   5 **(a)** $\dfrac{x^7}{x^3} = x^{7-3} = x^4$

(b) $\dfrac{x^5}{x^4} = x^{5-4} = x^1 = x$

(c) $\dfrac{2^9}{2^7} = 2^{9-7} = 2^2 = 4$

(d) $\dfrac{10^m}{10^n} = 10^{m-n}$

(e) $\dfrac{x^6}{y^2}$ Cannot be simplified since the bases are different.

(f) $\dfrac{x^{2n+1}}{x^{2n}} = x^{2n+1-2n} = x^1 = x$ ▌▐

You probably noticed that the examples were chosen so that the exponent in the numerator was greater than the exponent in the denominator. This is not always so, and in the next section we will consider that situation.

If we consider a *product* raised to a power, for example, $(xy)^3$, we find that the following happens:

$$(xy)^3 = xy \cdot xy \cdot xy$$
$$= x \cdot x \cdot x \cdot y \cdot y \cdot y$$
$$= x^3 \cdot y^3$$

Therefore

$$(xy)^3 = x^3 y^3$$

What this really means is that each factor in the parentheses is raised to the power. This gives us our next rule.

Exponent Rule IV:

$$(xy)^a = x^a y^a$$

E X A M P L E   6 **(a)** $(mn)^7 = m^7 n^7$

(b) $(6x)^3 = 6^3 \cdot x^3 = 216x^3$

(c) $(xyz)^8 = x^8 y^8 z^8$ The rule applies to more than two factors.   ▌▐

A *quotient* raised to a power is handled in a similar way:

$$\left(\frac{x}{y}\right)^3 = \frac{x}{y} \cdot \frac{x}{y} \cdot \frac{x}{y}$$
$$= \frac{x \cdot x \cdot x}{y \cdot y \cdot y}$$
$$= \frac{x^3}{y^3}$$

Therefore

$$\left(\frac{x}{y}\right)^3 = \frac{x^3}{y^3}$$

As you can see, when a fraction is raised to a power, both the numerator and denominator are raised to the power. This is stated symbolically as follows:

Exponent Rule V:

$$\left(\frac{x}{y}\right)^a = \frac{x^a}{y^a}$$

3.1 Exercises

▮ ▮ ▮ ▮ ▮

▼ Simplify using the rules of exponents:

1. $x^3 \cdot x^5$

2. $m^2 \cdot m^8$

3. $y^4 \cdot y^4$

4. $x^9 \cdot x^6$

5. $x^4 \cdot x^5 \cdot x^3$

6. $y^3 \cdot y^8 \cdot y^3$

7. $x \cdot x^5$

8. $y^4 \cdot y$

9. $2^4 \cdot 2$

10. $3 \cdot 3^2$

11. $x^3 \cdot y^4$

12. $x^3 \cdot y^2 \cdot z^4$

13. $2^3 \cdot 3^2$

14. $\left(x^5\right)^3$

15. $\left(y^2\right)^7$

16. $\left(x^3\right)^5$

17. $\left(y^7\right)^2$

18. $\left(2^3\right)^2$

19. $\left(3^2\right)^2$

20. $\dfrac{x^5}{x^2}$

21. $\dfrac{y^{15}}{y^{13}}$

22. $\dfrac{x^7}{x}$

23. $\dfrac{y^9}{y}$

24. $\dfrac{x^4}{y^3}$

41. $x^a \cdot y^{4b}$

42. $x^5 \cdot x^h$

25. $\dfrac{y^3}{x}$

26. $(2x)^3$

43. $x^h \cdot x^5$

44. $\dfrac{x^5}{x^h}$

27. $(4y)^2$

28. $(xy)^9$

45. $\dfrac{x^h}{x^5}$

46. $(x^5)^h$

29. $(xyz)^5$

30. $\left(\dfrac{x}{y}\right)^4$

47. $(5x)^h$

48. $\left(\dfrac{x}{5}\right)^h$

31. $\left(\dfrac{x}{3}\right)^2$

32. $\dfrac{2^{48}}{2^{46}}$

49. $\dfrac{2^a}{2^b}$

50. $6^a \cdot 6^b$

33. $\dfrac{3^{50}}{3^{49}}$

34. $(2xy)^4$

51. $5^4 \cdot 5^5$

52. $\dfrac{x^{12n}}{x^{4n}}$

35. $(-2xy)^4$

36. $x^a \cdot x^{2a}$

53. $\dfrac{y^{7a}}{y^{3a}}$

54. $(x^y)^z$

37. $y^{3b} \cdot y^b$

38. $x^{2a} \cdot x^{2b}$

55. $(6x)^y$

56. $(x^x)^x$

39. $x^{3a} \cdot x^{5b}$

40. $x^{3a} \cdot y^{2a}$

57. $(3^x)^x$ **58.** $(3^a)^3$ **69.** $(3x)^2$ **70.** $(6xy)^2$

MENTAL MATHEMATICS

▼ Solve each of the following problems mentally without the use of pencil or paper.

71. $x^3 \cdot y^4$ **72.** $\dfrac{x^{11}}{y^4}$

59. $x^4 \cdot x^5$ **60.** $7^5 \cdot 7^6$

61. $x^5 \cdot x$ **62.** $(x^3)^9$

73. $\dfrac{x^3}{y}$ **74.** $\left(\dfrac{x}{y}\right)^5$

63. $(6^5)^3$ **64.** $(y^3)^3$

75. $\left(\dfrac{2}{3}\right)^2$ **76.** $\left(\dfrac{3}{4}\right)^3$

65. $\dfrac{x^6}{x^5}$ **66.** $\dfrac{x^9}{x}$

77. Explain why $6^5 \cdot 6^7 \neq 36^{12}$.

67. $\dfrac{y^{15}}{y}$ **68.** $(xyz)^8$

3.2   Negative and Zero Exponents

▮ ▮ ▮ ▮ ▮

Negative Exponents

• • • • •

In our discussion of Rule III, the exponent in the numerator was always larger than the exponent in the denominator. If that is not the case, we have an expression like the following:

$$\frac{x^2}{x^5} = \frac{x \cdot x}{x \cdot x \cdot x \cdot x \cdot x} = \frac{x \cdot x}{x \cdot x} \cdot \frac{1}{x \cdot x \cdot x} = \frac{1}{x \cdot x \cdot x} = \frac{1}{x^3}$$

↑
This is equal to 1.

If we apply our third rule of exponents to the same problem, we subtract exponents, giving us

$$\frac{x^2}{x^5} = x^{2-5} = x^{-3}$$

So we have

$$\frac{x^2}{x^5} = \frac{1}{x^3}$$

and

Both quantities equal $\dfrac{x^2}{x^5}$, so they are equal to each other.

$$\frac{x^2}{x^5} = x^{-3}$$

We conclude that $x^{-3} = \dfrac{1}{x^3}$.

Generalizing gives us our next rule:

Exponent Rule VI:
$x^{-a} = \dfrac{1}{x^a}$

E X A M P L E   1   **(a)** $x^{-8} = \dfrac{1}{x^8}$

(b) $b^{-5} = \dfrac{1}{b^5}$

(c) $2^{-5} = \dfrac{1}{2^5} = \dfrac{1}{32}$

(d) $(-3)^{-2} = \dfrac{1}{(-3)^2} = \dfrac{1}{9}$

(e) $-3^{-2} = \dfrac{-1}{3^2} = -\dfrac{1}{9}$ The exponent applies only to the 3, not the

negative sign.   ▌▐

STOP   A negative exponent should not be confused with the sign of the base.

a. $4^{-2} \neq \dfrac{1}{-4^2}$ The sign of the 4 does not change.

b. $4^{-2} \neq -4^2$

c. $4^{-2} \neq (4)(-2)$

d. $4^{-2} = \dfrac{1}{4^2} = \dfrac{1}{16}$ Only the sign of the exponent changes.

What does $\dfrac{1}{x^{-3}}$ equal? Applying Rule VI to the denominator, x^{-3}, yields the

following:

$$\frac{1}{x^{-3}} = \frac{1}{\dfrac{1}{x^3}} \qquad \text{Rule VI}$$

$$= 1 \div \frac{1}{x^3} \qquad \text{Definition of division}$$

$$= \frac{1}{1} \cdot \frac{x^3}{1} \qquad \text{Invert and multiply.}$$

$$= x^3$$

Therefore $\dfrac{1}{x^{-3}} = x^3$.

This, too, is true in general, leading us to our next rule for exponents.

> **Exponent Rule VII:**
>
> $$\frac{1}{x^{-a}} = x^a$$

EXAMPLE 2 **(a)** $\dfrac{1}{x^{-8}} = x^8$

(b) $\dfrac{1}{b^{-5}} = b^5$

(c) $\dfrac{1}{2^{-5}} = 2^5 = 32$

(d) $\dfrac{1}{(-3)^{-2}} = (-3)^2 = 9$ (-3) is the base here.

(e) $\dfrac{1}{-3^{-2}} = \dfrac{3^2}{-1} = \dfrac{9}{-1} = -9$ Only the 3 is the base here; the negative sign is not in parentheses.   ▮▮

Combining Rules VI and VII gives us an important method for removing negative exponents.

> A **factor** may be moved from numerator to denominator, or vice versa, by changing the sign of the exponent.

EXAMPLE 3

(a) $\dfrac{x^3}{x^{-4}} = x^3 \cdot x^4 = x^7$ Change x^{-4} to x^{+4} and move from denominator to numerator.

Notice that we could also do this problem using Rule III:

$$\dfrac{x^3}{x^{-4}} = x^{3-(-4)} = x^{3+4} = x^7 \qquad \text{Subtract exponents.}$$

(b) $\dfrac{x^{-3}}{y^2} = \dfrac{1}{x^3 \cdot y^2} = \dfrac{1}{x^3 y^2}$ Change x^{-3} to x^{+3} and move from numerator to denominator.

(c) $x^{-5}y^3 = \dfrac{x^{-5} \cdot y^3}{1} = \dfrac{y^3}{x^5}$

(d) $x^{-2}y^3 z^{-4} = \dfrac{y^3}{x^2 z^4}$

(e) $\dfrac{x^{-3}}{y^{-4}} = \dfrac{y^4}{x^3}$

(f) $\dfrac{1}{2^{-4}} = \dfrac{2^4}{1} = 16$

(g) $\dfrac{x^2}{y^{-2}} = x^2 y^2$

(h) $2^{-2} + 4^{-1} = \dfrac{1}{2^2} + \dfrac{1}{4^1}$

$\qquad\qquad\quad = \dfrac{1}{4} + \dfrac{1}{4}$

$\qquad\qquad\quad = \dfrac{2}{4} = \dfrac{1}{2}$

(i) $2^{-4} + 4^{-3} = \dfrac{1}{2^4} + \dfrac{1}{4^3}$

$\qquad\qquad\quad = \dfrac{1}{16} + \dfrac{1}{64}$

$\qquad\qquad\quad = \dfrac{4}{64} + \dfrac{1}{64}$

$\qquad\qquad\quad = \dfrac{5}{64}$

E X A M P L E 4   Simplify: $\dfrac{x^{-3} + x^{-4}}{x^3}$.

S O L U T I O N

$\dfrac{x^{-3} + x^{-4}}{x^3} = \dfrac{\dfrac{1}{x^3} + \dfrac{1}{x^4}}{x^3}$ 　　Neither x^{-3} nor x^{-4} are *factors* of the numerator.

$\qquad\qquad = \dfrac{\dfrac{x}{x^4} + \dfrac{1}{x^4}}{x^3}$ 　　x^4 is the LCD in the numerator.

$\qquad\qquad = \dfrac{\dfrac{x+1}{x^4}}{\dfrac{x^3}{1}}$ 　　Add the fractions in the numerator.

$\qquad\qquad = \dfrac{x+1}{x^4} \cdot \dfrac{1}{x^3}$ 　　Multiply by the reciprocal of $\dfrac{x^3}{1}$.

$\qquad\qquad = \dfrac{x+1}{x^7}$ 　　Multiply the numerators together and multiply the denominators together. ▌▐

The next illustration provides a shortcut to help us simplify a fraction raised to a negative power.

R U L E   U S E D

$\left(\dfrac{x}{y}\right)^{-3} = \dfrac{x^{-3}}{y^{-3}}$ 　　$\left(\dfrac{x}{y}\right)^a = \dfrac{x^a}{y^a}$

$\qquad\quad = \dfrac{y^3}{x^3}$ 　　$x^{-a} = \dfrac{1}{x^a}$ and $\dfrac{1}{x^{-a}} = x^a$

$\qquad\quad = \left(\dfrac{y}{x}\right)^3$ 　　$\dfrac{x^a}{y^a} = \left(\dfrac{x}{y}\right)^a$

You can see that we simplified the fraction raised to a negative power by inverting the fraction (writing its reciprocal) and changing the sign of the exponent. This occurs frequently enough that it deserves to be stated formally:

Exponent Rule VIII:

$$\left(\frac{x}{y}\right)^{-a} = \left(\frac{y}{x}\right)^{a}$$

EXAMPLE 5   **(a)** $\left(\dfrac{x}{y}\right)^{-5} = \left(\dfrac{y}{x}\right)^{5} = \dfrac{y^5}{x^5}$

             (b) $\left(\dfrac{t}{s}\right)^{-n} = \left(\dfrac{s}{t}\right)^{n} = \dfrac{s^n}{t^n}$

             (c) $\left(\dfrac{2}{3}\right)^{-2} = \left(\dfrac{3}{2}\right)^{2} = \dfrac{3^2}{2^2} = \dfrac{9}{4}$ ▮▮

So to eliminate a negative exponent on a fraction, we simply take the reciprocal and then raise it to the appropriate power.

Zero Exponents

Applying Rule III to the expression $\dfrac{x^3}{x^3}$ means that we subtract exponents, giving us

$$\frac{x^3}{x^3} = x^{3-3} = x^0$$

However, we also know that

$$\frac{x^3}{x^3} = 1 \qquad \text{Any quantity (except zero) divided by itself equals 1}$$

So we have

$$\frac{x^3}{x^3} = x^0$$

and   Both quantities equal $\dfrac{x^3}{x^3}$, so they equal each other.

$$\frac{x^3}{x^3} = 1$$

Therefore we conclude that $x^0 = 1$, which gives us our final rule for exponents.

Exponent Rule IX:

$$x^0 = 1 \qquad (x \neq 0)$$

We must make the restriction that $x \neq 0$ here since if we work an expression like $\dfrac{0^2}{0^2}$ two different ways, we obtain

$$\frac{0^2}{0^2} = \frac{0}{0} \qquad \text{(since } 0^2 = 0\text{)}$$

and   Both quantities equal $\dfrac{0^2}{0^2}$, so they

equal each other.

$$\frac{0^2}{0^2} = 0^{2-2} = 0^0 \qquad \text{Subtract exponents}$$

yielding 0^0 and $\dfrac{0}{0}$, and since $\dfrac{0}{0}$ is indeterminate, we must also say that 0^0 is indeterminate. From now on, we will assume that when any variable is raised to the zero power, the variable is not equal to 0.

EXAMPLE 6

(a) $7^0 = 1$

(b) $a^0 = 1$ Remember $a \neq 0$.

(c) $3x^0 = 3 \cdot 1 = 3$ The exponent applies only to x.

(d) $(3x)^0 = 1$ The exponent applies to the entire expression.   ∎

STOP   Remember, x^0 is **not** equal to 0, rather, $x^0 = 1$.

3.2   Exercises

▼ Simplify, leaving answers with only positive exponents:

1. x^{-7}

2. y^{-4}

3. 2^{-3}

4. 5^{-2}

5. $\dfrac{1}{x^{-4}}$

6. $\dfrac{1}{y^{-1}}$

7. $\dfrac{1}{2^{-3}}$

8. $\dfrac{1}{6^{-2}}$

9. 10^{-1}

10. 10^{-3}

11. $\dfrac{1}{10^{-2}}$

12. $\dfrac{1}{10^{-3}}$

13. $x^{-2}y^2$

14. xy^{-3}

31. $\left(\dfrac{1}{5}\right)^{-3}$

32. $\dfrac{1}{4^{-2}}$

15. $\dfrac{x^{-2}}{y^2}$

16. $\dfrac{x^5}{y^{-2}}$

33. $\dfrac{5^0}{6^0}$

34. -4^0

17. $\dfrac{y^{-4}}{x^4}$

18. $\dfrac{y^7}{x^{-3}}$

35. $(-4)^0$

36. -7^0

19. $\dfrac{1}{x^{-2}y^2}$

20. $\dfrac{1}{x^4y^{-7}}$

37. $(-7)^0$

38. $a^{-5}b^{-3}c^2$

21. $6x^{-3}$

22. $-5x^{-4}$

39. $x^{-5}y^{-1}z^4$

40. $\left(\dfrac{3}{4}\right)^{-2}$

23. $3y^0$

24. $(3y)^0$

41. $\left(\dfrac{2}{3}\right)^{-1}$

42. $\left(\dfrac{a}{b}\right)^0$

25. 14^0

26. 7^0

43. $\dfrac{a^0}{b^{-2}}$

44. $\left(\dfrac{5}{2}\right)^{-3}$

27. $\left(\dfrac{x}{y}\right)^{-5}$

28. $\left(\dfrac{a}{b}\right)^{-7}$

45. $\dfrac{6^3}{6}$

46. $\dfrac{8^5}{8^4}$

29. $\left(\dfrac{1}{4}\right)^{-2}$

30. $\left(\dfrac{2}{3}\right)^{-3}$

47. $\left(\dfrac{8}{3}\right)^{-2}$

48. $\dfrac{1}{3^{-3}}$

49. $\dfrac{7^2}{7^0}$

50. $\dfrac{10^3}{10^0}$

51. $2^{-3} + 3^{-2}$

52. $4^{-2} + 2^{-3}$

53. $3^2 + 3^{-2}$

54. $5^3 + 5^{-3}$

65. $\dfrac{1}{x^{-1}}$

66. $\dfrac{1}{18^{-1}}$

67. $7x^{-6}$

68. $9x^{-3}$

69. $-2x^{-5}$

70. 23^0

71. y^0

72. $7y^0$

73. $(7y)^0$

74. $-y^0$

75. $(-y)^0$

76. $\left(\dfrac{x}{y}\right)^{-1}$

77. $\left(\dfrac{5}{3}\right)^{-1}$

78. $\left(\dfrac{2}{3}\right)^0$

79. $\left(\dfrac{1}{4}\right)^{-1}$

80. $\left(\dfrac{3}{4}\right)^{-2}$

81. $\dfrac{x^2}{x^3}$

82. $\dfrac{y^5}{y^7}$

83. $\dfrac{x^8}{x^4}$

84. $\dfrac{y^{12}}{y^4}$

MENTAL MATHEMATICS

▼ Perform the following exercises mentally without using a pencil or paper. Do **not** leave answers with negative exponents.

55. x^{-5}

56. y^{-4}

57. 2^{-11}

58. 6^{-5}

59. $\dfrac{1}{x^{-6}}$

60. $\dfrac{1}{y^{-8}}$

61. $\dfrac{1}{7^{-8}}$

62. $\dfrac{1}{9^{-11}}$

63. y^{-1}

64. 15^{-1}

 85. Explain why $4x^0$ is equal to 4 and not zero.

 86. Are $(x + y)^{-1}$ and $x^{-1} + y^{-1}$ the same or are they different? Try substituting numbers for x and y and evaluate each expression.

 87. Is $(x^{-1} + y^{-1})^{-1}$ the same as $x + y$? Substitute numbers for x and y and evaluate each expression.

3.3 Combining Rules of Exponents

Expressions containing exponents can be quite complicated and frequently require the application of more than one rule to simplify them. In most cases there is more than one order in which the rules may be applied.

EXAMPLE 1   RULE USED

(a) $\dfrac{x^2y^4}{xy^5} = x^{2-1}y^{4-5}$ $\dfrac{x^a}{x^b} = x^{a-b}$

$\qquad = x^1 \cdot y^{-1}$

$\qquad = \dfrac{x}{y}$ $x^{-a} = \dfrac{1}{x^a}$

                                                       RULE USED

(b) $(3x^2)^3 = 3^3(x^2)^3$ $(xy)^a = x^ay^a$

$\qquad = 27x^6$ $(x^a)^b = x^{ab}$

                                                       RULE USED

(c) $(3x^2)^{-3} = 3^{-3}(x^2)^{-3}$ $(xy)^a = x^ay^a$

$\qquad = 3^{-3}x^{-6}$ $(x^a)^b = x^{ab}$

$\qquad = \dfrac{1}{3^3x^6}$ $x^{-a} = \dfrac{1}{x^a}$

$\qquad = \dfrac{1}{27x^6}$

We could have changed the exponent -3 to $+3$ right away by moving the entire quantity $3x^2$ into the denominator and changing the sign of the exponent:

                                                       RULE USED

$(3x^2)^{-3} = \dfrac{1}{(3x^2)^3}$ $x^{-a} = \dfrac{1}{x^a}$

$\qquad\quad = \dfrac{1}{3^3(x^2)^3}$ $(xy)^a = x^ay^a$

$\qquad\quad = \dfrac{1}{27x^6}$ $(x^a)^b = x^{ab}$

(d) $\left(\dfrac{x^2y^{-3}}{x^1y^{-8}}\right)^0 = 1$ This simplifies immediately to 1, since it is an expression raised to the zero power.

                                                       RULE USED

(e) $\left(\dfrac{x^{-2}y^4}{x^{-5}y^{-3}}\right)^3 = (x^{-2} \cdot x^5 \cdot y^4 \cdot y^3)^3$ $\dfrac{1}{x^{-a}} = x^a$

$\qquad\qquad = (x^{-2+5} \cdot y^{4+3})^3$ $x^a \cdot x^b = x^{a+b}$

$\qquad\qquad = (x^3y^7)^3$

$\qquad\qquad = x^9y^{21}$ $(x^a)^b = x^{ab}$

RULE USED

(f) $-3x^{-2} = \dfrac{-3}{x^2}$ $\qquad\qquad x^{-a} = \dfrac{1}{x^a}$

Note that the exponent -2 applies only to the x and not to -3. Also, the sign of -3 is not affected by changing the sign of the exponent.

RULE USED

(g) $(-3x)^{-2} = \dfrac{1}{(-3x)^2}$ $\qquad x^{-a} = \dfrac{1}{x^a}$: The exponent -2 applies to the entire quantity $(-3x)$.

$\qquad\qquad\ = \dfrac{1}{(-3)^2 x^2}$ $\qquad (xy)^a = x^a y^a$

$\qquad\qquad\ = \dfrac{1}{9x^2}$

RULE USED

(h) $(3x^0 y^4)^2 = (3y^4)^2$ $\qquad x^0 = 1$

$\qquad\qquad\ = 3^2(y^4)^2$ $\qquad (xy)^a = x^a y^a$

$\qquad\qquad\ = 9y^8$ $\qquad (x^a)^b = x^{ab}$

RULE USED

(i) $\left(\dfrac{2x}{y}\right)^{-3} = \left(\dfrac{y}{2x}\right)^3$ $\qquad \left(\dfrac{x}{y}\right)^{-a} = \left(\dfrac{y}{x}\right)^a$

$\qquad\qquad\ = \dfrac{y^3}{(2x)^3}$ $\qquad \left(\dfrac{x}{y}\right)^a = \dfrac{x^a}{y^a}$

$\qquad\qquad\ = \dfrac{y^3}{2^3 x^3}$ $\qquad (xy)^a = x^a y^a$

$\qquad\qquad\ = \dfrac{y^3}{8x^3}$

RULE USED

(j) $(2x^{-3})^{-4} = 2^{-4}(x^{-3})^{-4}$ $\qquad (xy)^a = x^a y^a$

$\qquad\qquad\ = 2^{-4} x^{12}$ $\qquad (x^a)^b = x^{ab}$

$\qquad\qquad\ = \dfrac{x^{12}}{2^4}$ $\qquad x^{-a} = \dfrac{1}{x^a}$

$\qquad\qquad\ = \dfrac{x^{12}}{16}$

3.3   Exercises

▮ ▮ ▮ ▮ ▮

▼ Simplify, leaving your answers with only positive exponents. Try to do some of the problems more than one way by applying the rules of exponents in a different order.

15. $\dfrac{x^{-3}y^{-5}}{x^{-2}y^{-1}}$

16. $\dfrac{xy^4}{x^{-2}y}$

1. $\dfrac{x^2y^3}{x^4y}$

2. $\dfrac{x^6y^3}{xy^5}$

17. $\dfrac{x^0y^{-2}}{x^{-4}y^0}$

18. $\dfrac{x^{-2}y^{-4}}{x^0y^0}$

3. $(5x^2)^3$

4. $(-5x^2)^3$

19. $\dfrac{x^3y^{-4}}{x^3y^{-4}}$

20. $\dfrac{x^{-2}y^5}{x^{-2}y^{-3}}$

5. $(5x^2)^{-3}$

6. $(-5x^2)^{-3}$

21. $\left(\dfrac{2x^2}{y}\right)^3$

22. $\left(\dfrac{-4x^3}{y^2}\right)^2$

7. $6x^0y^4$

8. $(8x^0y^3)^2$

23. $\left(\dfrac{3x}{y^2}\right)^{-2}$

24. $\left(\dfrac{2x^3}{y^4}\right)^{-3}$

9. $\dfrac{x^{-2}y^3}{x^{-5}}$

10. $\dfrac{x^{-3}y^{-4}}{x^{-5}y^{-7}}$

25. $(x^2y)^{-3}$

26. $(xy^3)^{-2}$

11. $\dfrac{x^5y^{-5}}{x^3y^{-3}}$

12. $\dfrac{x^3y^{-4}}{x^{-3}y^4}$

27. $(2x^3y^2)^{-3}$

28. $(6x^2y^{-2})^{-2}$

13. $\dfrac{x^3y^4}{x^{-6}y^{-2}}$

14. $\dfrac{x^{-6}y^{-2}}{x^4y^5}$

29. $(-2x^3y^{-5})^{-3}$ 30. $(-2x^{-3})^{-2}$ 35. $(y^4)^0 \cdot x^{-1}$ 36. $\left(\dfrac{x^2y^{-2}}{x^3y^{-1}}\right)^{-2}$

31. $(3x^{-2})^{-4}$ 32. $(4x^3y^7)^0$ 37. $\left(\dfrac{x^{-3}y^4}{xy^{-1}}\right)^{-3}$ 38. $\left(\dfrac{2x^3y^{-4}}{-x^{-4}y^{-4}}\right)^{-2}$

33. $[(-4x^0)^2]^0$ 34. $\dfrac{5xy^2}{x^{-1}y^{-2}}$

3.4   Polynomials: Their Sum and Difference

Basic Principles

Before proceeding, we need to discuss some basic definitions:

EXAMPLES

Variable: A letter used to represent a number

$x,\ y,\ t$

Term of a polynomial: A number or the product of a number and one or more variables raised to whole-number powers

$3x^5,\ 8,\ -6x^2y^3$

Coefficient: The numerical factor of a term

$3\,x^5,\ \ -4\,x^3y,\ \ -x$

↑

coefficient is −1

Like terms: Terms that differ only in their coefficients

$-2\,x^2$ and $4\,x^2$

Unlike terms: Terms that have different variable parts

$3\,x^2$ and $4\,x$

Polynomial: A term or some combination of sums and/or differences of terms. Polynomials never have a negative exponent on any variable and no variables appear in any denominator.

   A **polynomial in one variable** contains only one variable.

$3x^3 - 2x^2 - 8x + 7$ is a polynomial in x.

A **polynomial in several variables** contains more than one variable	$3x^2y - 8x + 7y + 2$ is a polynomial in x and y.
Monomial: A polynomial with one term	$6x^3$, $-4x^3y$, -7
Binomial: A polynomial with two terms	$6x^2 - 5$, $3x + 15$
Trinomial: A polynomial with three terms	$x^2 - 2x + 1$, $3x + 2y^4 - 5$
The **degree of a term in one variable** is the exponent on the variable.	The degree of $2x^4$ is 4.
The **degree of a term in several variables** is the sum of the exponents of all of the variables.	The degree of $2x^3y^4$ is $3 + 4 = 7$.
The **degree of a polynomial** is the largest degree of any one term in the polynomial.	The degree of $3x^4 - 8x - 5$ is 4.
The usual way to write a polynomial in a single variable is **in order of descending powers.** The exponents on the variable decrease from left to right.	$6x^5 - 8x^3 + 2x^2 - 4$

EXAMPLE 1 Find the degree of each polynomial.

(a) $6x^3 + 2x^2 - 8x - 5$

The largest exponent of the single variable, x, is 5, so the degree of the polynomial is 5.

(b) $7x$

The degree is 1, since $7x = 7x^1$.

(c) 6

The degree of 6 is 0, since we can write 6 as $6x^0$.

(d) $6x^3y^2 - 8x^2y^4 + 2xy^3 - 5$

The degrees of the terms are 5, 6, 4, and 0; respectively. Therefore the degree of the polynomial is 6.

(e) 0

It is agreed that the polynomial 0 has *no* degree, since we can write 0 as $0^1 = 0^2 = 0^3$ and so on. ▮▮

Most of the polynomials that we will consider in this book are polynomials in a single variable.

Adding Polynomials

• • • • •

To add polynomials, we combine like terms by adding their numerical coefficients and keeping the same variable parts.

EXAMPLE 2 **(a)** $3\,x^2y + 5\,x^2y = 8\,x^2y$

(b) $-7\,x^4 + 5\,x^4 = -2\,x^4$

(c) $-4\,xy^3 - 3\,xy^3 = -7\,xy^3$ ▮▮

> **To add polynomials horizontally:**
>
> 1. Write the polynomials in descending order.
> 2. Combine all the like terms occurring in the polynomials being added.

EXAMPLE 3   Add: $3x^3 + 5x^2 - 2x + 1$ and $2x^3 + 6x^2 + 5x - 8$.

SOLUTION

$(3x^3 + 5x^2 - 2x + 1) + (2x^3 + 6x^2 + 5x - 8)$

$= \underbrace{3x^3 + 2x^3} + \underbrace{5x^2 + 6x^2} \underbrace{- 2x + 5x} + \underbrace{1 - 8}$ Group like terms.

$= \quad 5x^3 \quad + \quad 11x^2 \quad + \quad 3x \quad - \quad 7$ Combine like terms.

$= 5x^3 + 11x^2 + 3x - 7$ ∎

EXAMPLE 4   Add: $-4x^3 - 6x^4 + 2x^2 - 11$ and $-x^2 + 6 + 2x^4$.

SOLUTION

$(-6x^4 - 4x^3 + 2x^2 - 11) + (2x^4 - x^2 + 6)$ Write in descending order.

$= \ -4x^4 - 4x^3 + x^2 - 5$ Combine like terms.   ∎

Polynomials can also be added *vertically:*

> **To add polynomials vertically:**
>
> 1. Write the polynomials in descending order with like terms under one another.
> 2. Combine like terms.

EXAMPLE 5   Add: $4x^3 + 6x^4 - 3x^2 + 1$ and $-4 + 3x^3 - 2x^2 + 3x$.

SOLUTION

$$\begin{array}{r} 6x^4 + 4x^3 - 3x^2 \qquad + 1 \\ 3x^3 - 2x^2 + 3x - 4 \\ \hline 6x^4 + 7x^3 - 5x^2 + 3x - 3 \end{array}$$

Write like terms under one another, leaving space for missing terms. Combine like terms.   ∎

Subtracting Polynomials

• • • • •

Previously, we removed parentheses preceded by a negative sign by changing the sign of each term in the parentheses. We will make use of this procedure in subtracting polynomials.

> **To subtract polynomials:**
>
> 1. Write each polynomial in descending order.
> 2. Change the sign of each term of the polynomial being subtracted.
> 3. Combine like terms.

EXAMPLE 6   Subtract $6x^2 - 3x^3 + 2x - 1$ from $7x^3 - 3x + 1 - 4x^2$.

SOLUTION $(7x^3 - 4x^2 - 3x + 1) - (-3x^3 + 6x^2 + 2x - 1)$ Write in descending order.

$= 7x^3 - 4x^2 - 3x + 1 \boxed{+}\ 3x^3 \boxed{-}\ 6x^2 \boxed{-}\ 2x \boxed{+}\ 1$ Change the sign of each term in the polynomial being subtracted.

$= 10x^3 - 10x^2 - 5x + 2$ Combine like terms.   ▮▮

To do the same subtraction problem vertically, change the sign of each term in the polynomial being subtracted, place the like terms under one another, and combine.

EXAMPLE 7   Subtract $6x^2 - 3x^3 + 2x - 1$ from $7x^3 - 3x + 1 - 4x^2$.

SOLUTION
$$
\begin{array}{r}
7x^3 - 4x^2 - 3x + 1 \\
+\ 3x^3 - 6x^2 - 2x + 1 \\
\hline
10x^3 - 10x^2 - 5x + 2
\end{array}
$$
Write in descending order.
Change the sign of each term.
Combine like terms to obtain the answer.   ▮▮

EXAMPLE 8   Subtract $3x + 4x^3 - 2x^2 + 6$ from $7 - 5x + 5x^3$ **(a)** horizontally and **(b)** vertically.

SOLUTION   **(a)** $5x^3 - 5x + 7 - (4x^3 - 2x^2 + 3x + 6)$ Write in descending order.

$= 5x^3 - 5x + 7 \boxed{-}\ 4x^3 \boxed{+}\ 2x^2 \boxed{-}\ 3x \boxed{-}\ 6$ Change the sign of each term being subtracted.

$= x^3 + 2x^2 - 8x + 1$ Combine like terms.

(b)   Leave space for missing term.
$$\downarrow$$
$$
\begin{array}{r}
5x^3 \qquad - 5x + 7 \\
-\ 4x^3 + 2x^2 - 3x - 6 \\
\hline
x^3 + 2x^2 - 8x + 1
\end{array}
$$
Write in descending order.
Change the sign of each term subtracted.
Combine like terms.   ▮▮

3.4   Exercises

▮▮▮▮▮

▼ Find the degree of each polynomial.

1. $4x^3 - 8x^2 + 5x - 1$ 2. $6x^4 + 6x^3 - 2$

3. $6x^4 + 1$ 4. $2x^3 - 5$

5. $12y^5$ 6. $-8y^4$

7. $4y$ 8. $-3y$

9. 15

10. -6

11. $4x^3y^5$

12. $-2xy^4$

13. $6x^3y - 8xy + 7$

14. $7x^5y - 3x^2y^7 + 2$

15. $14x^3yz^4 + 5xy^5 - 2z$ 16. $8p^4qr^2 - 5p^3r + 7qr^7$

17. $4p^5q^4r + 5p^4q^5 - 2p^3r$ **18.** 0

 19. The exponent in the term 3^4 is 4. Is its degree 4?
Explain why or why not.

 20. Explain why 0 is considered to have no degree.

 21. Write a polynomial that expresses the surface area of
a square box with an open top and with dimensions
as given.

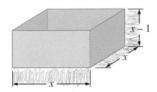

▼ In Exercises 22–33, add the polynomials. Write your
answer in descending order.

22. $6x^2 + 3x - 1$ and $4x^2 - 6x - 5$

23. $-3x + 4x^3 - 9 + x^2$ and $5x - 3x^2 + 2x^3 + 7$

24. $6y + 7y^2 - 5$ and $5y + 6y^2 + 10$

25. $7n^5 + 2n^3 - 3n$ and $7n - 5n^2 + 2n^3 + 14n^4$

26. $-3a^3 - 4a^4 + 17$ and $6a^4 - 24a - 15$

27. $-7t^3 + 8t^2 - 6 + 6t$, $14t + 5t^3 + 8$, and
$8t^3 + 12t - 4t^2$

28. $5y + 14y^5 + 12y^3$, $3y^2 + 6y - 14$, and
$-7y^5 - 3y^2 + 7$

29. $-14w^3 + 3w^7 - 8w$, $-w + w^3$, and
$7w^4 + 3w - 8w^3$

30. $6x^3 + 2x^2 - 8x - 4$
$\underline{7x^3 + 5x^2 \qquad + 8}$

31. $7y^3 - 8y^2 \qquad + 2$
$\underline{-7y^3 \qquad + 8y - 9}$

32. $-24a^4 + 7a^3 \qquad - a - 4$
$ 4a^3 + 2a^2 + a + 5$
$\underline{ a^4 - 6a^3 \qquad - 7}$

33. $7t^4 - 8t^3 - 6t^2$
$ 8t^3 + 6t^2 + 2t + 6$
$\underline{5t^4 \qquad\qquad\qquad -6}$

34. Subtract $3x^2 + 4x - 6$ from $9x^2 - 7x + 5$.

35. Subtract $8a + 7a^2 + 9$ from $4a + 6 - 8a^2$

36. Subtract $y^3 - 12y^2 - 8 + 4y$ from $6y + 9y^3$.

37. Subtract $-8x - 14x^2 + 2x^3 - 9$ from $8x - 2x^3 + 5$.

38. Subtract $7x^4 - 3x + 5$ from $2x^3 + 5x - 2x^2$.

39. Subtract $-4y^9 + 7y - 6y^5$ from $12 + 7y$.

▼ Subtract the polynomials.

40. $(-6a^3 + 7a^2 - 4a + 2) - (3a^3 - 6a^2 + 5a - 4)$

41. $(-2a^4 + 3a^2 - 7a) - (4a^3 + 2a^2 + 7a + 5)$

42. $(-3y - 17y^2 - 5) - (-6 + 2y^3 + 5y^2)$

43. $(6x - 4 + 2x^5 + 3x^2) - (-7 + 5x^4 + 2x^3)$

44. $-7x^3 + 5x^2 - 7x + 5$
      $- (4x^3 + 7x^2 - 8x + 9)$
      _____

45. $-8x^3 + 7x^2 \quad\quad + 5$
      $- (-4x^3 \quad\quad + 5x - 9)$
      _____

46. Subtract $-6x^2 + 2x - 4$ from the sum of
$-7x^2 + 2x - 8$ and $5x^2 + 8x - 1$.

47. Subtract $-8y^3 + 2y^2 + 6y - 8$ from the sum of
$-3y^3 - 8y^2 + 2y - 8$ and $6y^3 + 5y - 1$.

48. Subtract $-4x^4 + 2x^2 - 4$ from the sum of $5x^3 + 2x^2$
and $3x^4 + 9$.

49. Subtract $6 + 3x^2 - 4x^4$ from the sum of
$6x - 5x^4 + 2$ and $3x^2 - 9x^4 + 1$.

▼ Perform the indicated additions and subtractions
and simplify.

50. $(6x^3 + 12x^2 - 4x + 1) + (2x^3 - 5x^2 + 2x - 1) -$
$(6x^3 + 9x^2 - 3)$

51. $(-14 - 8x^3 + 2x) + (x^2 - 9x^3 + 1) - (x + 6x^3 + 9)$

52. $(-9y + 4) + (-7y - 9y^4 + 6y^2) - (5y^2 + 14 + 7y^3)$

53. $(6 - 3a^2 + 9a + 4a^3) - (6a^2 - 3 + 5a) +$
$(-7 + 6a - 5a^3)$

54. Given polynomials $-3x^2 - 4x^3 + 8 - 5x$,
$-x^3 + 4x - 6x^2$, and $17x - 15x^3 + 7 + x^2$, subtract
the sum of the last two from the first one.

55. Given polynomials $a^4 - a + 3a^2 - 14$, $9 + 6a^4 - a^3 - 7a$, $a^3 + 3a^4 + 3a^2$, and $2 - 4a^3 + 6a^4$, subtract the sum of the first two from the sum of the last two.

56. Given polynomials $y - 4y^3 + 2y^2 - 6$, $y^3 + 1$, $6y + 4y^3 - 8$, and $7y + 3y^2 - 6$, subtract the sum of the first two from the sum of the last two.

3.5 Products of Polynomials

▮ ▮ ▮ ▮ ▮

Multiplying a Monomial by a Monomial

• • • • •

> **To multiply monomials:**
>
> **1.** Multiply the numerical coefficients.
>
> **2.** Multiply the variables using the rule $x^a \cdot x^b = x^{a+b}$.

E X A M P L E 1   Multiply the given monomials.

(a) $(-6x^3)(5x^2) = (-6)(5)x^3 \cdot x^2$
$= -30x^5$

(b) $(7x^4)(-6x^2) = (7)(-6)x^4 \cdot x^2$
$= -42x^6$

(c) $(6mn^2)(3m^3n) = (6)(3)m \cdot m^3 \cdot n^2 \cdot n$
$= 18m^4n^3$ ▮▮

STOP   Be careful to distinguish between *adding* and *multiplying* monomials.

1. $3x^2y + 5x^2y = 8x^2y$:   Add only the coefficients. The variables do not change.

2. $3x^2y + 5x^3y^2$ cannot be combined because the terms are not like terms.

3. $(3x^2y)(5x^3y^2) = 15x^5y^3$:   Multiply the coefficients and multiply the variables by adding their exponents.

Multiplying a Polynomial by a Monomial

• • • • •

Multiplying a polynomial by a monomial is an application of the distributive property:

$$a(b + c + d + e + \ldots) = ab + ac + ad + ae + \ldots$$

and

$$(b + c + d + e + \ldots)a = ba + ca + da + ea + \ldots$$
$$= ab + ac + ad + ae + \ldots$$

It is conventional to write the variables in each term in alphabetical order.

To multiply a polynomial by a monomial, multiply each term of the polynomial by the monomial.

EXAMPLE 2   **(a)** $4(2x + 7) = \boxed{4} \cdot 2x + \boxed{4} \cdot 7$
$$= 8x + 28$$

(b) $3x(6x^2 + 2x + 5) = \boxed{3x} \cdot 6x^2 + \boxed{3x} \cdot 2x + \boxed{3x} \cdot 5$
$$= 18x^3 + 6x^2 + 15x$$

(c) $4x^4(3x^3 + 2x^2 - 4x + 9)$
$$= \boxed{4x^4} \cdot 3x^3 + \boxed{4x^4} \cdot 2x^2 + \boxed{4x^4} \cdot (-4x) + \boxed{4x^4} \cdot 9$$
$$= 12x^7 + 8x^6 - 16x^5 + 36x^4$$

(d) $-10x^2(-4x^2 + 2x - 9)$
$$= \boxed{(-10x^2)} \cdot (-4x^2) + \boxed{(-10x^2)} \cdot (2x) + \boxed{(-10x^2)} \cdot (-9)$$
$$= 40x^4 - 20x^3 + 90x^2$$

(e) $(-7n^3 + 2n^2 - 8)n^2 = -7n^3 \cdot \boxed{n^2} + 2n^2 \cdot \boxed{n^2} - 8 \cdot \boxed{n^2}$
$$= -7n^5 + 2n^4 - 8n^2$$

Multiplying a Polynomial by a Polynomial

• • • • •

Consider the product of two polynomials, $(2x - 4)(3x^2 + 2x - 5)$. Applying the distributive rule, we will multiply each term in one polynomial by each term in the other polynomial and combine similar terms.

$$
\begin{array}{rl}
3x^2 + 2x - 5 & \text{Arrange the problem vertically.} \\
\underline{ 2x - 4} & \\
6x^3 + 4x^2 - 10x & \longleftarrow \text{Product of } 3x^2 + 2x - 5 \text{ and } 2x \\
\underline{ - 12x^2 - 8x + 20} & \longleftarrow \text{Product of } 3x^2 + 2x - 5 \text{ and } -4 \\
6x^3 - 8x^2 - 18x + 20 & \longleftarrow \text{Combine like terms, which have been} \\
& \text{conveniently arranged in vertical columns.}
\end{array}
$$

To multiply a polynomial by a polynomial:

1. Multiply each term in one polynomial by each term in the other polynomial.

2. Arrange like terms under one another in the products.

3. Combine like terms in the products.

E X A M P L E 3   Multiply: $(6x^2 - 4x + 7)(3x + 2)$.

S O L U T I O N

$$
\begin{array}{r}
6x^2 - 4x + 7 \\
3x + 2 \\
\hline
18x^3 - 12x^2 + 21x \\
12x^2 - 8x + 14 \\
\hline
18x^3 + 0x^2 + 13x + 14 \\
= 18x^3 + 13x + 14
\end{array}
$$

Arrange the problem vertically.

$(3x)(6x^2 - 4x + 7)$
$(2)(6x^2 - 4x + 7)$

Combine like terms.

E X A M P L E 4   Find the product $(3x^3 + 4x^2 - 5x + 6)(4x^2 - x + 1)$.

S O L U T I O N

$$
\begin{array}{r}
3x^3 + 4x^2 - 5x + 6 \\
4x^2 - x + 1 \\
\hline
12x^5 + 16x^4 - 20x^3 + 24x^2 \\
- 3x^4 - 4x^3 + 5x^2 - 6x \\
3x^3 + 4x^2 - 5x + 6 \\
\hline
12x^5 + 13x^4 - 21x^3 + 33x^2 - 11x + 6
\end{array}
$$

$(4x^2) \cdot$ (top polynomial)
$(-x) \cdot$ (top polynomial)
$(1) \cdot$ (top polynomial)

Combine like terms.

E X A M P L E 5   Multiply:

$$
\begin{array}{r}
7x^3 + 3x - 2 \\
- 6x^2 - 3 \\
\hline
-42x^5 - 18x^3 + 12x^2 \\
- 21x^3 - 9x + 6 \\
\hline
-42x^5 - 39x^3 + 12x^2 - 9x + 6
\end{array}
$$

$(-6x^2) \cdot$ (top polynomial)
$(-3) \cdot$ (top polynomial)

Combine like terms.

Notice how we left spaces for missing terms so we could add like terms in columns.

3.5   Exercises

▼ Multiply:

1. $(6x)(-5x^2)$

2. $(7x^3)(-5x^4)$

3. $(2y^2)(8y^2)$

4. $(3a^2b)(-4a^3b^4)$

5. $(-3x^2yz)(-2xyz^4)$

6. $(-3x^2y)(-2xy^2)(-7xy)$

7. $2x(3x + 5)$

8. $-4x(-x - 8)$

9. $3x^2(x^2 + 3x - 9)$

10. $-11x^2(-2x^2 + 2x - 1)$

11. $6x(4x^4 + 2x^3 + 3x^2 + 2x - 7)$

12. $7x^5(3x^3 + 4x^2 - 6x)$

13. $-10a^2(6a^3 - 7a^2 - 3a + 9)$

14. $-6a^2(-4a^4 + 2a^2 - 9)$

15. $(x + 5)(x^2 + 2x + 7)$

16. $(y - 2)(y^2 + 2y + 4)$

17. $(6x - 1)(x^2 - 8x - 8)$

18. $(5x + 4)(6x^3 - 2x^2 + 3x - 4)$

19. $(7x^2 + 2x - 4)(3x + 7)$

20. $(4x^3 - 3x^2 + 3x + 8)(6x - 4)$

21. $(x^2 + 2x + 1)(x^2 - 5x + 6)$

22. $(2x^2 - 3x + 4)(x^2 - 4x - 4)$

23. $(x^3 + 2x^2 - 9x - 1)(x^2 + 2x + 3)$

24. $(2x^3 - 3x^2 - 3x + 5)(2x^2 - 3x - 1)$

▼ In Exercises 25–28, multiply the polynomials.

25. $\begin{array}{r} 4x^3 + 3x^2 - 6x - 6 \\ \underline{x^2 + 3} \end{array}$

26. $\begin{array}{r} 6y^3 + 7y^2 - 4 \\ \underline{-3y^2 - 6} \end{array}$

27. $\begin{array}{r} 7a^3 - 7a^2 - a - 5 \\ \underline{3a^2 + a + 5} \end{array}$

28. $\begin{array}{r} 7x^4 + 3x^3 + 2x^2 + 4x + 5 \\ \underline{x^2 + 2x - 3} \end{array}$

29. Explain in your own words how to multiply two monomials.

30. Explain in your own words how to multiply two polynomials.

3.6   Special Products

The Product of Binomials

A product like $(x + 3)(x + 4)$ was calculated previously as follows:

$$
\begin{array}{r}
x + 3 \\
x + 4 \\
\hline
x^2 + 3x \\
4x + 12 \\
\hline
x^2 + 7x + 12
\end{array}
$$

We can obtain the same result horizontally by a method called the **FOIL** method.

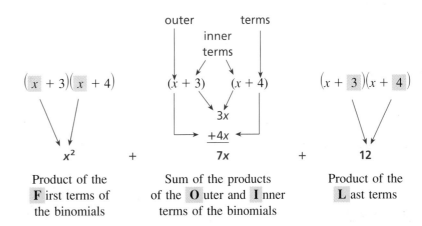

| Product of the **F**irst terms of the binomials | Sum of the products of the **O**uter and **I**nner terms of the binomials | Product of the **L**ast terms |

As you can see, the result is the same. However, you will find that the **FOIL** method, once learned and practiced, is much faster.

To multiply two binomials by the FOIL method:

The first term = the product of the **F**irst terms of the binomials.
The middle term = the sum of the **O**uter and **I**nner products.
The last term = the product of the **L**ast terms of the binomials.

EXAMPLE 1   Multiply the binomials $(x + 2)(x + 5)$.

SOLUTION

$$
\begin{array}{ccccc}
\text{Product of} & \textbf{O}\text{uter} & \textbf{I}\text{nner} & \text{Product of} \\
\textbf{F}\text{irst terms} & \text{product} & \text{product} & \textbf{L}\text{ast terms} \\
\downarrow & \downarrow & \downarrow & \downarrow
\end{array}
$$

$$(x + 2)(x + 5) = x^2 \ + \ \underbrace{5x \ + \ 2x}_{\text{sum}} \ + \ 10$$

$$= x + 7x + 10$$

EXAMPLE 2   Multiply the following binomials:

(a) $(x + 6)(x - 2)$ $= x^2 \boxed{- 2x + 6x} - 12$
$$= x^2 + 4x - 12$$

(b) $(2x + 3)(x - 2)$ $= 2x^2 \boxed{- 4x + 3x} - 6$
$$= 2x^2 - x - 6$$

(c) $(3x - 4)(2x + 1) = 6x^2 \boxed{+ 3x - 8x} - 4$
$$= 6x^2 - 5x - 4$$

(d) $(6y - 5)(2y - 3) = 12y^2 \boxed{- 18y - 10y} + 15$
$$= 12y^2 - 28y + 15$$

(e) $(3x + 2)(3x - 2) = 9x^2 \boxed{- 6x + 6x} - 4$
$$= 9x^2 - 4$$

▮▮

This method should be practiced until the result can be obtained without any intermediate steps.

EXAMPLE 3   Multiply the given binomials:

(a) $(4x - 3)(2x + 3) = 8x^2 + 6x - 9$

(b) $(3z + 5)(3z - 5) = 9z^2 - 25$

(c) $(a + 6)^2 = (a + 6)(a + 6) = a^2 + 12a + 36$

(d) $(2x - 7)(3x - 5) = 6x^2 - 31x + 35$

(e) $(7x + 9)(2x + 7) = 14x^2 + 67x + 63$

▮▮

Squaring a Binomial

• • • • •

The procedure for finding the square of a binomial can be shortened even further. For example, consider $(2x + 3)^2$. We could apply the **FOIL** method, obtaining

$$(2x + 3)^2 = (2x + 3)(2x + 3) = 4x^2 \boxed{+ 6x + 6x} + 9$$
$$= 4x^2 + 12x + 9$$

Noting that the outer product and the inner product are the same, we can state the following rule:

To square a binomial:

1. The square of the first term = the first term of the product.

2. Twice the product of the two terms = the middle term of the product.

3. The square of the last term = the last term of the product.

Learning this rule will be best accomplished by memorizing the following statement:

> The **square of a binomial** equals the sum of the square of the first term, twice the product of the terms, and the square of the last term.

E X A M P L E 4   Square the binomial $(3x + 5)^2$.

SOLUTION

$$(3x + 5)^2 = (3x)^2 \quad + \quad 2(3x \cdot 5) \quad + \quad (5)^2$$

$$= 9x^2 \quad + \quad 30x \quad + \quad 25$$

$\uparrow$ $\uparrow$ $\uparrow$

square of the   twice the product   square of the

first term   of the terms   last term

E X A M P L E 5   Square the following binomials:

(a) $(x + 6)^2 = x^2 + 12x + 36$

(b) $(5x - 1)^2 = 25x^2 - 10x + 1$

(c) $(2x + 7)^2 = 4x^2 + 28x + 49$

(d) $(2y - 3)^2 = 4y^2 - 12y + 9$

(e) $(4b + 5)^2 = 16b^2 + 40b + 25$

STOP   A common error when squaring a binomial is to leave out the middle term:

$$(a + b)^2 \neq a^2 + b^2$$

$$(a + b)^2 = a^2 \;+ 2ab\; + b^2$$

$\uparrow$

This is frequently left out.

$(x + 3)^2 = x^2 + 9$ *is incorrect.*

$$(x + 3)^2 = x^2 + 6x + 9$$

3.6 Exercises

▼ Write the final answers to the following products without writing down any intermediate steps.

1. $(x + 2)(x + 4)$

2. $(y + 3)(y + 2)$

3. $(a - 5)(a - 2)$

4. $(x - 4)(x - 5)$

5. $(x + 3)(x - 5)$

6. $(x + 7)(x - 6)$

7. $(n - 7)(n - 2)$

8. $(m - 3)(m - 7)$

9. $(x - 2)(x + 2)$

10. $(x - 8)(x + 8)$

27. $(5n + 3)(5n - 3)$

28. $(3x + 2)^2$

11. $(3x + 4)(2x - 3)$

12. $(2x - 1)(4x + 1)$

29. $(7n - 1)^2$

30. $(4y - 3)^2$

13. $(3x + 2)(3x - 2)$

14. $(4y + 3)(y - 8)$

31. $(7x + 2)^2$

32. $(2m - 9)^2$

15. $(x + 3)^2$

16. $(x + 4)^2$

33. $(x + y)(x + 3y)$

34. $(2a + b)(3a - b)$

17. $(x - 7)^2$

18. $(x - 4)^2$

35. $(3a + b)(3a - b)$

36. $(7x - 2y)(4x + 3y)$

19. $(y - 9)^2$

20. $(y + 9)^2$

37. $(10m - 3n)(5m + 7n)$

38. $(2x - 3y)^2$

21. $(3y - 4)(4y + 3)$

22. $(2a - 5)(5a + 2)$

39. $(4a - 5b)^2$

40. $(5x + 7y)^2$

23. $(7x - 2)(7x + 2)$

24. $(4x - 7)(4x + 7)$

 41. Explain why $(4x + 3)^2 \neq 16x^2 + 9$.

42. What is the difference between $(x + y)^2$ and $x^2 + y^2$?

25. $(3x - 2)^2$

26. $(h + 4)(7h - 4)$

3.7   Factoring: Common Factors, Grouping, The Difference of Two Squares, The Sum and Difference of Two Cubes

▮ ▮ ▮ ▮ ▮

Common Factors

• • • • •

Previously we multiplied polynomials by monomials by applying the distributive rule. For example:

multiplying
$\longrightarrow$

$$3x(4x + 5) = 12x^2 + 15x$$

$\longleftarrow$
factoring

Our task now is to reverse the process. given $12x^2 + 15x$, we must write it in factored form as $3x(4x + 5)$. The process is called **factoring** and uses the distributive law in the reverse order.

EXAMPLE 1   Factor $6x^2 + 3x^2y + 9x$.

SOLUTION   The largest common factor is $3x$, since it is the greatest monomial that is common to each term. We apply the distributive property to write the polynomial in factored form:

$$6x^2 + 3x^2y + 9x = 3x(2x + xy + 3)$$

EXAMPLE 2   Factor each of the following polynomials:

(a) $14a^2 - 21a^4 = 7a^2(2 - 3a^2)$

(b) $27x^4 - 18x^2y^2 + 36x^3 = 9x^2(3x^2 - 2y^2 + 4x)$

(c) $-28x^5 - 14x^4 - 7x^3 = -7x^3(4x^2 + 2x + 1)$

(d) $36m^2n^3 + 9m^3n^2 - 18m^4n^2 = 9m^2n^2(4n + m - 2m^2)$

(e) $14x^3y - 9a^2y + 6ax^4$ is not factorable because there is no factor common to all of the terms. We call it a **prime polynomial.**

(f) $x(a + b) + y(a + b)$ has a common factor of $(a + b)$ and can be factored as

$$x(a + b) + y(a + b) = (a + b)(x + y)$$

Factoring by Grouping

The polynomial $xa + xb + 2a + 2b$ has no factor common to all terms. However, by grouping terms that have the same common factor, some polynomials can be factored.

$$\underbrace{xa + xb} + \underbrace{2a + 2b} \qquad \text{Group separately.}$$
$$= x(a + b) + 2(a + b) \qquad a + b \text{ is a common factor.}$$
$$= (x + 2)(a + b)$$

EXAMPLE 3   Factor $ab + 2b + 4a + 8$.

SOLUTION
$$\underbrace{ab + 2b} + \underbrace{4a + 8}$$
$$= b(a + 2) + 4(a + 2) \qquad a + 2 \text{ is a common factor.}$$
$$= (b + 4)(a + 2)$$

EXAMPLE 4   Factor $8ax^2 - 4ax - 14x + 7$.

SOLUTION
$$\underbrace{8ax^2 - 4ax} \underbrace{- 14x + 7} \qquad \text{Factor } 4ax \text{ from first group.}$$
$$= 4ax(2x - 1) - 7(2x - 1) \qquad \text{Factor } -7 \text{ from second group.}$$
$$= (4ax - 7)(2x - 1) \qquad 2x - 1 \text{ is a common factor.}$$

Factoring the Difference of Two Squares

• • • • •

When we multiply two binomials of the form $(x - y)(x + y)$, we get an interesting result.

EXAMPLE 5   **(a)** $(x + 2)(x - 2) = x^2 - 4$

(b) $(3x - 2)(3x + 2) = 9x^2 - 4$

(c) $(5y + 3)(5y - 3) = 25y^2 - 9$

(d) $(a + b)(a - b) = a^2 - b^2$ ▮▮

In every case the middle term is 0 and the product consists of *two perfect squares* separated by a negative sign. We call this the **difference of two squares.**

To factor, we reverse the procedure.

Factoring the difference of two squares:

A polynomial of the form $a^2 - b^2$ is factored as $a^2 - b^2 = (a + b)(a - b)$.

EXAMPLE 6   Factor $4x^2 - 9$.

SOLUTION   Since $4x^2 = (2x)^2$ and $9 = 3^2$ and they are separated by a negative sign, we have the *difference of two squares.*

Square root of first term goes here

$$4x^2 - 9 = (2x + 3)(2x - 3)$$

Square root of second term goes here

The sign in one of the factors is $+$ and in the other it's $-$. Since multiplication is commutative, they can be reversed:

$$(2x + 3)(2x - 3) = (2x - 3)(2x + 3) = 4x^2 - 9$$ ▮▮

EXAMPLE 7   Factor the following polynomials:

(a) $16x^2 - 25 = (4x + 5)(4x - 5)$

(b) $4y^2 - 1 = (2y + 1)(2y - 1)$

(c) $h^2 - 9g^2 = (h + 3g)(h - 3g)$

(d) $49x^2 - 16y^2 = (7x - 4y)(7x + 4y)$

(e) $9x^2 - 36y^2 = 9(x^2 - 4y^2)$ Always remove the common factor first; the result is still the difference of two squares.

$$= 9(x - 2y)(x + 2y)$$

(f) $9x^2 + 4$ is prime. ▮▮

The *sum* of two squares is *not* factorable. A common error is to factor $x^2 + y^2$ as $(x + y)(x + y)$. However $(x + y)(x + y) = x^2 + 2xy + y^2$. Repeat: $x^2 + y^2$ is not factorable.

Other Difference-of-Two-Squares Problems

Consider *even* powers of a variable, for example, y^4, x^6, and t^{14}. Since they can be written

$$y^4 = (y^2)^2$$
$$x^6 = (x^3)^2$$
$$t^{14} = (t^7)^2$$

they are all perfect squares. In fact, the square root of any even-powered variable is equal to that variable to one-half the given exponent. This concept is used to factor the difference of two squares, as illustrated in the following example.

EXAMPLE 8   Factor:

(a) $y^4 - 9 = (y^2)^2 - 9 = (y^2 + 3)(y^2 - 3)$

(b) $x^6 - 4 = (x^3)^2 - 4 = (x^3 + 2)(x^3 - 2)$

(c) $t^{14} - 25 = (t^7 + 5)(t^7 - 5)$

(d) $64x^6 - 49y^8 = (8x^3 + 7y^4)(8x^3 - 7y^4)$

(e) $16h^{10} + 9y^2$ is prime. Even though both terms are perfect squares, the *sum* of two squares is prime.

(f) $x^8 - 16 = (x^4 + 4)(x^4 - 4)$
$\qquad\qquad = (x^4 + 4)(x^2 + 2)(x^2 - 2)$ $x^4 - 4$ is factorable as the difference of two squares.   ∎

The Sum and Difference of Two Cubes

Notice what happens when we multiply the following two polynomials:

$$(a + b)(a^2 - ab + b^2) = a^3 - a^2b + ab^2 + a^2b - ab^2 + b^3$$
$$= a^3 + b^3$$

Looking at the problem in reverse gives us a formula for factoring **the sum of two cubes, $a^3 + b^3$.**

To factor the sum of two cubes:
$a^3 + b^3 = (a + b)(a^2 - ab + b^2)$

Similarly, if we multiply

$$(a - b)(a^2 + ab + b^2) = a^3 + a^2b + ab^2 - a^2b - ab^2 - b^3$$
$$= a^3 - b^3$$

and look at the product in reverse, we obtain a formula for the **differences of two cubes, $a^3 - b^3$.**

To factor the difference of two cubes:

$$a^3 - b^3 = (a - b)(a^2 + ab + b^2)$$

EXAMPLE 9   Factor $8x^3 + 27$.

SOLUTION   We have the sum of two cubes, because $8x^3 = (2x)^3$ and $27 = 3^3$. Let a represent $2x$ and b represent 3 in the formula.

$$a^3 + b^3 = (a + b)(a^2 - ab + b^2)$$
$$8x^3 + 27 = (2x)^3 + 3^3 = (2x + 3)[(2x)^2 - (2x)(3) + 3^2]$$
$$= (2x + 3)(4x^2 - 6x + 9) \qquad ▌▌$$

EXAMPLE 10   Factor $64x^3 - 125y^3$.

SOLUTION $64x^3 - 125y^3 = (4x)^3 - (5y)^3 \qquad a = 4x$ and $b = 5y$ in the formula.

$$= (4x - 5y)[(4x)^2 + 4x \cdot 5y + (5y)^2]$$
$$= (4x - 5y)(16x^2 + 20xy + 25y^2) \qquad ▌▌$$

EXAMPLE 11   Factor $10x^7 + 80x^4$.

SOLUTION   Always factor out any common factor first.

$$10x^7 + 80x^4 = 10x^4(x^3 + 8) \qquad\qquad \text{Common factor}$$
$$= 10x^4(x + 2)(x^2 - 2x + 4) \qquad \text{Sum of two cubes} \qquad ▌▌$$

EXAMPLE 12   Factor $x^6 - y^6$.

SOLUTION   We can write $x^6 - y^6$ either as the difference of two squares, $(x^3)^2 - (y^3)^2$, or as the difference of two cubes, $(x^2)^3 - (y^2)^3$. Whenever this occurs, it is always easier to factor as the difference of two squares first:

$$x^6 - y^6 = (x^3)^2 - (y^3)^2$$
$$= (x^3 - y^3)(x^3 + y^3)$$

The first factor is the difference of two cubes and the second is the sum of two cubes, giving us

$$x^6 - y^6 = (x^3 - y^3)(x^3 - y^3)$$
$$= (x - y)(x^2 + xy + y^2)(x + y)(x^2 - xy + y^2) \qquad ▌▌$$

3.7   Exercises

▼ Factor the following polynomials. If not factorable, label the expression *prime.*

1. $36y - 27$

2. $6a^3 + 12a^2$

3. $4a^3 + 16a^2$

4. $24x^2 + 12y^2$

5. $3a^2b + 6x^2y$

6. $14a^2x - 7ay^2$

7. $x^5 - x^4 + x^2$

8. $2a^6 + 7a^4 - 3a$

9. $25x^6 + 15x^4 + 10x^2$

10. $15a^2b - 60a^3b^2$

11. $-12x^3 - 16x^2 - 24x$

12. $-42x^2y^3 + 35x^4y^2 - 21x^5y^3$

13. $54m^2nq - 36m^4n^2q + 18m^2n^2q^2$

14. $-27x^3y^8z^3 - 9xy^4z^{12} - 45xy^6z^9$

15. $48r^3t^2 - 32r^4t^4 + 24r^6$

16. $x(x + 3) + y(x + 3)$

17. $(y + 2)6 - (y + 2)x$

18. $x(x - 1) - 5(x - 1)$

19. $t^2(t + 5) + (t + 5)$

20. $3xy + 6x + 4y + 8$

21. $x^2 - xy - 6x + 6y$

22. $6x^3 + 12x^2 - 5x - 10$

23. $x^5 + x^4 + x + 1$

24. $t^2 - 49$

37. $9x^6y^2 - 49z^{10}$

25. $4p^2 - 81$

26. $16x^2 - 9$

38. $8x^3 + 1$

39. $x^3 + 27y^3$

27. $12x^3 - 27x$

28. $4x^2 + 9$

40. $27x^3 - 8$

29. $9x^2 - 4y^2$

30. $81x^2 - 49y^2$

41. $125x^3 - 27$

31. $y^4 - y^2$

32. $9x^2 + 25y^2$

42. $y^6 - 64$

33. $x^4 - 9$

34. $2x^{10} - 50$

43. $7a^3 + 7b^3$

35. $81x^6 - y^{12}$

44. $9s^6 - 9t^6$

36. $25x^2 - 36x^6y^4$

45. $7s^3 - 56t^3$

3.8   Factoring Trinomials

I I I I I

Trinomials With Leading Coefficient Equal to 1

• • • • •

Previously we multiplied two binomials together and in general we got a trinomial as a result. For example,

$$(x + 4)(x + 3) = x^2 + 7x + 12$$

$$\xrightarrow{\hspace{2cm}}$$

multiplying

Our task now will be to reverse the procedure and write the trinomial as the product of two binomials:

$$x^2 + 7x + 12 = (x + 4)(x + 3)$$

$$\xrightarrow{\hspace{2cm}}$$

factoring

The easiest type of trinomial to factor is one where the coefficient of the x^2 term is 1 (remember: $x^2 = 1 \cdot x^2$). Since $x \cdot x = x^2$, the first term of each factor will always be x. We set up our work, like this:

$$x^2 + 7x + 12 = (x\quad)(x\quad)$$

$$x^2 = x \cdot x$$

The product of the last two terms must equal 12, and the sum of these same last two terms must equal the coefficient of the middle term, 7. So what we are looking for are two integers with product 12 and sum 7. A little thought tells us quickly that $+4$ and $+3$ are the integers.

product is 12, the last term

$$x^2 + 7x + 12 = (x + 4)(x + 3)$$

4x

+3x

7x

sum is 7x, the middle term

EXAMPLE 1   Factor $x^2 + 6x + 8$.

SOLUTION   We set it up like this: $x^2 + 6x + 8 = (x\quad)(x\quad)$. Now we look for two integers with product 8 and sum 6. The integers are $+4$ and $+2$. Our result, then, is $x^2 + 6x + 8 = (x + 4)(x + 2)$. As in any factoring problem, we should always check our answer by multiplying the factors together:

$$(x + 4)(x + 2) = x^2 + 6x + 8$$ ◤   II

> **To factor a trinomial $x^2 + bx + c = (x + p)(x + q)$, find integers p and q such that:**
>
> **1.** $p \cdot q = c$ (the product is c) and
>
> **2.** $p + q = b$ (the sum is b).

Factoring trinomials is basically a trial and error procedure, but the following examples contain some hints on how to minimize the number of trials and errors, so study them carefully. Notice that the order of the binomials in the answer does not matter, since multiplication is commutative.

EXAMPLE 2   Factor $x^2 + 3x - 10$.

SOLUTION   We need two integers with a product of -10 and a sum of $+3$. Since the product is negative, they must have opposite signs. Since the sum is positive, the sign of the larger integer is positive. The only pair of integers that satisfies these conditions is $+5$ and -2.

$$(+5)(-2) = 10 \qquad \text{and} \qquad (+5) + (-2) = +3$$

so our result is

$$x^2 + 3x - 10 = (x + 5)(x - 2)$$

Check $(x + 5)(x + 2) = x^2 + 3x - 10$ ☑   I I

EXAMPLE 3   Factor $x^2 - 8x + 12$.

SOLUTION   We must find integers with a product of $+12$ and a sum of -8. Since the product is positive, their signs must be alike, but since the sum is negative, they must both be negative. Possible pairs of factors are $(-12, -1)$, $(-3, -4)$, and $(-6, -2)$. Of these, only $(-6, -2)$ has a sum equal to -8. Our result is therefore

$$x^2 - 8x + 12 = (x - 6)(x - 2)$$

Check $(x - 6)(x - 2) = x^2 - 8x + 12$ ☑   I I

EXAMPLE 4   Factor $y^2 - 5y + 6$.

SOLUTION   We want two integers that have a product of $+6$ and a sum of -5. Since the product is positive, the signs are alike, but since the sum is negative, we know they are both negative. Possible factors are $(-6, -1)$ and $(-3, -2)$. Since the sum is -5, the correct choice is $(-3, -2)$:

$$y^2 - 5y + 6 = (y - 3)(y - 2)$$

Check $(y - 3)(y - 2) = y^2 - 5y + 6$ ☑   I I

EXAMPLE 5   Factor $x^2 + 9x + 20$.

SOLUTION   We're looking for integers with a product of $+20$ and a sum of $+9$. Since both the product and sum are positive, they will both be positive. Choices of factors are $(+20, +1)$, $(+10, +2)$, and $(+4, +5)$. Since the sum is $+9$, the correct choice is $(+4, +5)$ and the result is

$$x^2 + 9x + 20 = (x + 4)(x + 5)$$

Check $(x + 4)(x + 5) = x^2 + 9x + 20$ ✔   ▌▌

EXAMPLE 6   Factor $y^2 - 9x + 12$.

SOLUTION   We need two integers with product $+12$ and sum -9. The positive product indicates that the signs are alike, but the negative sum tells us that they are both negative. Possible factors of 12 are $(-6, -2)$, $(-4, -3)$, and $(-12, -1)$. None of these factors, when added, have a sum of -9, so we conclude that the given trinomial is not factorable and we say that it is **prime**.   ▌▌

Trinomials with Leading Coefficient Greater than 1

• • • • •

Next we must concern ourselves with factoring trinomials when the coefficient of the squared term is greater than 1.

EXAMPLE 7   Factor $3x^2 + 7x + 2$.

SOLUTION   Since all of the signs in the trinomial are positive, all of the signs in the factors will be positive. The products of the two first terms of our factors must equal $3x^2$, which gives us

$$(3x + \quad)(x + \quad)$$

It remains for us to find the two last terms of the binomials. Since their product must be equal to 2, and they are positive, our only choices are $(2, 1)$ or $(1, 2)$. Trying each of these to see which gives the correct middle term, $+7x$, we have

$$(3x + 2)(x + 1)$$

$+2x$
$+3x$
$+5x$ ← wrong middle term

$$(3x + 1)(x + 2)$$

$+1x$
$+6x$
$+7x$ ← correct middle term

Our result is $3x + 7x + 2 = (3x + 1)(x + 2)$.   ▌▌

> **To factor trinomials of the form $ax^2 + bx + c$, $a > 1$:**
>
> **1.** The product of the first terms must equal ax^2.
>
> **2.** The product of the last terms must equal c.
>
> **3.** The sum of the outer product and inner product must equal bx.

Again, this is a trial and error procedure, but with sufficient practice you will learn to cut down on the number of trials by disregarding certain ones without actually trying them.

EXAMPLE 8   Factor $2x^2 + 11x + 15$.

SOLUTION   Since all of the signs are positive, all of the signs in the factors will be positive.

FACTORS OF 2   **FACTORS OF 15**
2, 1   3, 5
                        1, 15

wrong middle term

Trial 1: $(2x + 3)(x + 5) = 2x^2 + \boxed{13}\,x + 15$ Incorrect
Trial 2:   Let's reverse the order of the factors 3, 5 and try again.

$(2x + 5)(x + 3) = 2x^2 + 11x + 15$ Correct

EXAMPLE 9   Factor $3x^2 - 5x - 12$.

SOLUTION   Since the last term is negative, all pairs of factors of -12 will have different signs.

FACTORS OF 3   **FACTORS OF -12**
3, 1 -2, 6 or 2, -6
                        -3, 4 or 3, -4
                        -1, 12 or 1, -12

Trial 1: $(3x - 2)(x + 6) = 3x^2 + 16x - 12$ Incorrect
Trial 2: $(3x + 6)(x - 2) = 3x^2 - 12$ Incorrect
Trial 3: $(3x - 3)(x + 4) = 3x^2 + 9x - 12$ Incorrect
Trial 4: $(3x + 4)(x - 3) = 3x^2 - 5x - 12$ Correct

Notice that in Trial 2, a common factor of 3 occurred in the first binomial that we tried, $(3x + 6)$. If all common factors have been removed from the given trinomial, then common factors will *never* occur in any of the binomial factors. This means that Trial 2 could have been eliminated immediately, since the trial binomial $3x + 6$ contains a common factor. Similarly, we can see that Trial 3 will not work since the trial binomial $3x - 3$ contains a common factor, 3. This hint can save you a lot of work, so it should be remembered.

EXAMPLE 10   Factor $6x^2 - 7x + 2$.

SOLUTION   Since the last term of the trinomial is positive, the signs of the last terms of the binomial factors must be alike. However, since the middle term is negative, they will both be negative.

FACTORS OF 6   **FACTORS OF 2**
      2, 3 $-2, -1$
      6, 1

Trial 1: $(2x - 2)(3x - 1)$ is immediately incorrect, since $2x - 2$ contains a
            common factor of 2.
Trial 2: $(2x - 1)(3x - 2) = 6x^2 - 7x + 2$ Correct   ✔

EXAMPLE 11   Factor $4x^2 + 9x + 6$.

SOLUTION   Since all signs in the trinomial are positive, all signs in both binomial factors will be positive.

FACTORS OF 4   **FACTORS OF 6**
      2, 2   2, 3
      4, 1   6, 1

Trial 1: $(2x + 2)(2x + 3)$ contains a common factor.   Incorrect
Trial 2: $(2x + 6)(2x + 1)$ contains a common factor.   Incorrect
Trial 3: $(4x + 2)(x + 3)$ contains a common factor.   Incorrect
Trial 4: $(4x + 3)(x + 2) = 4x^2 + 11x + 6$ Incorrect
Trial 5: $(4x + 6)(x + 1)$ contains a common factor.   Incorrect
Trial 6: $(4x + 1)(x + 6) = 4x^2 + 25x + 6$ Incorrect

None of the trials produces a correct result, so, since all of the possible combinations have been tried, we conclude that the trinomial is not factorable, that is, it is *prime*.

Factoring Trinomials with Leading Coefficient Greater than 1— An Alternative Method

• • • • •

Do you remember how to multiply two binomials using the **FOIL** method?

$$\quad\quad\quad\quad\quad\text{F}\quad\quad\text{O}\quad\text{I}\quad\quad\text{L}$$
$$(2x + 3)(x + 4) = 2x^2 + 8x + 3x + 12$$
$$= 2x^2 + 11x + 12$$

Generalizing this process gives us the following:

$$\quad\quad\quad\quad\quad\text{F}\quad\quad\text{O}\quad\quad\text{I}\quad\quad\text{L}$$
$$(ax + b)(cx + d) = acx^2 + adx + bcx + bd$$
$$= acx^2 + (ad + bc)x + bd$$

Reversing this procedure gives a method for factoring:

$$acx^2 + (ad + bc)x + bd = acx^2 + adx + bcx + bd$$
$$= ax(cx + d) + b(cx + d)$$
$$= (ax + b)(cx + d)$$

This probably seems complicated, but working through a few examples will help make things clear.

EXAMPLE 12   Factor $3x^2 + 7x - 6$.

SOLUTION

1. We multiply the leading coefficient, 3, by the constant term, -6:

$$(3)(-6) = -18$$

2. Next we factor -18 so that the sum of the factors is equal to the coefficient of the middle term, 7.

FACTORS OF −18	SUM OF THE FACTORS
2, −9	−7
−2, 9	7
3, −6	−3
−3, 6	3
18, −1	17
−18, 1	−17

The desired pair of factors is $-2, 9$.

3. We write the middle term of the original trinomial, $7x$, as a sum using the factors -2 and 9:

$$3x^2 + 7x - 6 = 3x^2 - 2x + 9x - 6$$

4. Now we separate this expression into two groups and factor out common factors as follows:

$$3x^2 - 2x + 9x - 6 = (3x^2 - 2x) + (9x - 6)$$
$$= x(3x - 2) + 3(3x - 2) \qquad \text{There is a common factor in each group.}$$
$$= (3x - 2)(x + 3) \qquad \text{Factor out the common factor } (3x - 2).$$

We have successfully factored the trinomial:

$$3x^2 + 7x - 6 = (3x - 2)(x + 3) \qquad \blacksquare\blacksquare$$

Generalizing the procedure that we have just illustrated gives us the following rule.

To factor trinomials of the form $ax^2 + bx + c$, $a \neq 1$:

1. Find the factors of the product $a \cdot c$ that have a sum equal to b.

2. Rewrite the middle term of the trinomial being factored as the sum, using the factors found in Step 1.

3. Group the first two terms and the last two terms.

4. Factor out a common factor from each group and form a common binomial factor from that result.

EXAMPLE 13 Factor $4x^2 - 15x - 4$.

1. First we multiply the first and last numbers:

$$(4)(-4) = -16$$

2. Next we factor -16 so that the sum of the factors is -15:

FACTORS OF -16	SUM OF THE FACTORS
2, -8	-6
-2, 8	6
4, -4	0
16, -1	15
1, -16	-15

3. Now we rewrite $4x^2 - 15x - 4$, writing $-15x$ as the sum of $1 \cdot x$ and $-16x$:

$$4x^2 - 15x - 4 = 4x^2 + x - 16x - 4$$

4. We separate this polynomial into two groups and factor out the common factors:

$$
\begin{aligned}
4x^2 + x - 16x - 4 &= (4x^2 + x) + (-16x - 4) \\
&= x(4x + 1) - 4(4x + 1) \qquad \text{Factor each group.} \\
&= (4x + 1)(x - 4) \qquad\qquad 4x + 1 \text{ is a common factor.}
\end{aligned}
$$

We are finished:

$$4x^2 - 15x - 4 = (4x + 1)(x - 4)$$

3.8   Exercises

▮ ▮ ▮ ▮ ▮

▼ Factor the following trinomials. If it is not possible to factor a trinomial, label it *prime*.

1. $x^2 - 4x + 3$ **2.** $x^2 + 7x + 6$

3. $y^2 + 12y + 35$ **4.** $y^2 - y - 6$

5. $y^2 + 11y + 10$ **6.** $y^2 - 3y - 10$

7. $a^2 + 13a - 14$ **8.** $a^2 - 12a + 35$

9. $a^2 - 9a + 14$ **10.** $a^2 + 15a + 12$

11. $a^2 - 10a + 16$ **12.** $b^2 - 8b + 16$

13. $h^2 - 12h + 27$ **14.** $h^2 - 9h + 10$

15. $h^2 + 7h + 10$ **16.** $h^2 - 7h + 10$

17. $t^2 + 13t - 30$ **18.** $t^2 + 11t + 30$

19. $m^2 - 5m - 24$ **20.** $x^2 - 30x - 64$

21. $y^2 + 15y + 56$ **22.** $u^2 + 10u - 75$

23. $3x^2 + x - 4$ **24.** $4x^2 - 3x - 10$

25. $12x^2 + 5x - 2$ **26.** $15x^2 - x - 6$

27. $3y^2 - 8y + 4$ **28.** $2y^2 + y - 6$

29. $4h^2 - 12h + 5$ **30.** $2h^2 + 11h + 5$

31. $5x^2 + 14x + 3$

32. $7y^2 - 4y + 5$ prime

41. $6x^2 - 13x + 6$

42. $8x^2 - 6x - 5$

33. $3y^2 - 7y - 6$

34. $3x^2 - 22x + 7$

43. $25b^2 - 15b - 4$

44. $6m^2 - 5m - 14$

35. $5x^2 - 12x + 7$

36. $4x^2 - 8x - 5$

45. $8a^2 - 22a - 21$

46. $9x^2 + 18x + 8$

37. $2y^2 + 3y + 1$

38. $8y^2 + y - 7$

47. $6h^2 - 11h + 6$

48. $9x^2 - 32x - 16$

39. $7w^2 - 20w - 3$

40. $4x^2 + 15x - 16$ prime

49. $8y^2 - 14y - 9$

50. $12x^2 - 40x - 7$

3.9 Other Factoring Methods

Trinomials Reducible to Quadratic Form

A **quadratic trinomial** is a trinomial of the form $ax^2 + bx + c$, for example, $2x^2 + 5x + 3$. The trinomial $2x^4 + 5x^2 + 3$ is *not* quadratic but it is reducible to quadratic form. If we temporarily make the substitution $y = x^2$, then $y^2 = x^4$ and we have

$$2x^4 + 5x^2 + 3 = 2y^2 + 5y + 3 \qquad \text{This } is \text{ quadratic and factorable.}$$

$$= (2y + 3)(y + 1) \qquad \text{Now substitute } x^2 \text{ back for } y.$$

$$= (2x^2 + 3)(x^2 + 1)$$

Check $\qquad (2x^2 + 3)(x^2 + 1) = 2x^4 + 5x^2 + 3$

EXAMPLE 1   Factor $8h^4 + 2h^2 - 3$.

SOLUTION   We write the given trinomial in quadratic form by making the temporary substitution $y = h^2$ and $y^2 = h^4$:

$$8h^4 + 2h^2 - 3 = 8y^2 + 2y - 3 \qquad \text{It is now quadratic and factorable.}$$

$$= (4y + 3)(2y - 1) \qquad \text{Substitute } h^2 \text{ back for } y.$$

$$= (4h^2 + 3)(2h^2 - 1) \qquad \blacksquare\blacksquare$$

Combining Different Types of Factoring

• • • • •

In many cases a polynomial can be factored more than once. Usually we are instructed to factor a polynomial completely, which means we must look at the factors to see if they can be factored further.

> Always look for a greatest common factor first.

EXAMPLE 2   Factor $3x^2 + 21x + 36$.

SOLUTION   We remove the greatest common factor, 3, first:

$$3x^2 + 21x + 36 = 3(x^2 + 7x + 12)$$

Next we factor the trinomial $x^2 + 7x + 12$:

$$x^2 + 7x + 12 = (x + 3)(x + 4)$$

Putting both of these steps together, we have

$$3x^2 + 21x + 36 = 3(x^2 + 7x + 12) = 3(x + 3)(x + 4)$$

As always, check by multiplying the factors together. $\blacksquare\blacksquare$

EXAMPLE 3   Factor $24x^2 - 6y^2$.

SOLUTION $$24x^2 - 6y^2 = 6(4x^2 - y^2) \qquad \text{The greatest common factor is 6.}$$

$$= 6(2x + y)(2x - y) \qquad \text{Difference of two squares: } 4x^2 - y^2. \qquad \blacksquare\blacksquare$$

EXAMPLE 4   Factor $3a^5 - 48a$.

SOLUTION $3a^5 - 48a = 3a(a^4 - 16)$ The greatest common factor is $3a$.

                 $= 3a(a^2 + 4)(a^2 - 4)$ Difference of two squares: $a^4 - 16$.

                 $= 3a(a^2 + 4)(a + 2)(a - 2)$ Difference of two squares: $a^2 - 4$.   ∎

EXAMPLE 5   Factor $18y^4 + 24y^3 - 24y^2$.

SOLUTION $18y^4 + 24y^3 - 24y^2 = 6y^2(3y^2 + 4y - 4)$ The greatest common factor is $6y^2$.

                 $= 6y^2(3y - 2)(y + 2)$ Factor the trinomial.   ∎

EXAMPLE 6   Factor $-x^2 + 6x + 16$.

SOLUTION   When we are factoring any trinomial, the squared term should always be positive. To accomplish this we will factor out -1 and then factor the resulting trinomial.

                 $-x^2 + 6x + 16 = -1(x^2 - 6x - 16)$ The common factor is -1.

                 $= -1(x - 8)(x + 2)$ Factor the trinomial.

                 $= -(x - 8)(x + 2)$ Usually only the negative sign is written instead of -1.   ∎

EXAMPLE 7   Factor $2x^4 - 2x^2 - 24$.

SOLUTION $2x^4 - 2x^2 - 24 = 2(x^4 - x^2 - 12)$ The common factor is 2.

                 $= 2(y^2 - y - 12)$ Substitute $y = x^2$ and $y^2 = x^4$.

                 $= 2(y + 3)(y - 4)$ Factor the trinomial.

                 $= 2(x^2 + 3)(x^2 - 4)$ Substitute x^2 back for y.

                 $= 2(x^2 + 3)(x - 2)(x + 2)$ Difference of two squares: $x^2 - 4$.   ∎

A knowledge of factoring is essential to your future success in algebra. These skills will be called upon time and time again in the following chapters. The following exercises offer an opportunity to improve these necessary factoring skills.

3.9   Exercises

▮▮▮▮▮

▼ Factor completely. If a problem is not factorable, label it *prime*.

17. $8 - y^3$

1. $2x^2 + 6x + 4$ **2.** $5y^2 - 10y - 15$

18. $2a^6 - 2b^6$

3. $12a^2 - 3b^2$ **4.** $7x^2 - 28y^2$

19. $x^2 + y^2$ **20.** $ab + 5a + b + 5$

5. $8y^2 - 12y - 8$ **6.** $6x^2 - 10x - 24$

21. $ab + 5a - b - 5$ **22.** $-3x^2 - 7x - 2$

7. $3x^2 - 6x - 24$ **8.** $4x^2 - 36$

23. $12a^4 + 4a^2 - 1$ **24.** $2y^2 - y + 7$

9. $6x^2 - 19x + 15$ **10.** $4a^3 - 4ab^2$

25. $x^2(s - t) - y^2(s - t)$

11. $(x + 6)x - (x + 6)y$ **12.** $p^6 - p^4$

13. $x^4 - 4x^2 - 12$ **14.** $x^4 - 5x^2 + 6$

26. $6x^3 + 48y^3$

15. $16t^2 - 44t + 30$ **16.** $h^3 + 27$

27. $2a^2 - a^2b + 2c^2 - bc^2$

28. $x^2y - y + 3x^2 - 3$

29. $x^2y - 4y + 2x^2 - 8$

30. $4d^3 - 108$

31. $6d^2 + 12c^2$

32. $18y^3 - 39y^2 - 15y$

33. $12h^4 - 14h^3 - 6h^2$ 34. $xy^2 + xy - 12x$

35. $x^8 - 81$

36. $(x + 3) - y^2(x + 3)$

37. $5x^{13} - 5$ 38. $3x^2 + 12x + 4$

39. $32x^3 - 50xy^2$ 40. $16a^3 - 36ab^2$

41. $14x^2 + 21x - 35$

42. $48 - 3x^8$

43. $t^4 - 16$ 44. $4c^3 - 100c$

45. $8t^2 - 6t + 15$ 46. $14x^2 - 11x + 12$

47. $9x^2 - 52x - 12$

48. $2x^4 + 250x$

49. $8t^2 - 30t - 27$

50. $8x^4 + 4x^2 - 12$

3.10   Division of Polynomials

Dividing a Polynomial by a Monomial

We will first determine how to divide a polynomial by a monomial. Recall that when we **multiplied** a polynomial by a monomial, we multiplied *each term* in the polynomial by the monomial. Since division is defined in terms of multiplication, it seems reasonable, then, that to divide a polynomial by a monomial, we **divide** each *term* of the polynomial by the monomial.

> **To divide a polynomial by a monomial:**
>
> Divide *each term* of the polynomial by the monomial, using the rule $\dfrac{x^a}{x^b} = x^{a-b}$ to divide the variables.

EXAMPLE 1 Divide the polynomial $6x^3 + 12x^2 - 18x + 9$ by the monomial 3.

SOLUTION
$$\frac{6x^3 + 12x^2 - 18x + 9}{3}$$
$$= \frac{6x^3}{3} + \frac{12x^2}{3} - \frac{18x}{3} + \frac{9}{3} \qquad \text{Divide each term by 3.}$$
$$= 2x^3 + 4x^2 - 6x + 3$$

EXAMPLE 2 $(8x^4 - 24x^3 + 40x - 16) \div (-4x^2)$

SOLUTION
$$\frac{8x^4 - 24x^3 + 40x - 16}{-4x^2}$$
$$= \frac{8x^4}{-4x^2} - \frac{24x^3}{-4x^2} + \frac{40x}{-4x^2} - \frac{16}{-4x^2} \qquad \text{Divide each term by } -4x^2.$$
$$= -2x^2 + 6x - 10x^{-1} + 4x^{-2}$$

$$\left[\frac{40x}{-4x^2} = -10x^{1-2} = -10x^{-1} \quad \text{and} \right.$$
$$\left. \frac{-16}{-4x^2} = \frac{-16x^0}{-4x^2} = 4x^{0-2} = 4x^{-2} \right]$$

$$= -2x^2 + 6x - \frac{10}{x} + \frac{4}{x^2} \qquad \text{Write the answer with positive exponents only.}$$

EXAMPLE 3 $(3x^3 + 7x^2 - 4x + 2) \div 2x$

SOLUTION
$$\frac{3x^3 + 7x^2 - 4x + 2}{2x} = \frac{3x^3}{2x} + \frac{7x^2}{2x} - \frac{4x}{2x} + \frac{2}{2x}$$
$$= \frac{3x^2}{2} + \frac{7x}{2} - 2 + \frac{1}{x}$$

Division of a Polynomial
by a Polynomial

• • • • •

To divide one polynomial by a second polynomial, we go through the following step-by-step procedure that is used for division of whole numbers: estimate, multiply, subtract, bring down the next term. An example is worked out alongside the procedure to help you see exactly what is done in each step.

To divide a polynomial by a polynomial:

Divide: $(-10 + 6x^2 + 11x) \div (2x + 5)$

PROCEDURE

WORKED EXAMPLE

1. Arrange the polynomials in descending order.

$$2x + 5 \overline{)6x^2 + 11x - 10}$$

2. Divide the first term of the dividend by the first term of the divisor.

$$\frac{6x^2}{2x} = 3x$$

$$\begin{array}{r} 3x \\ 2x + 5 \overline{)6x^2 + 11x - 10} \end{array}$$

3. Multiply the entire divisor by the first term of the quotient.

$$3x(2x + 5) = 6x^2 + 15x$$

$$\begin{array}{r} 3x \\ 2x + 5 \overline{)6x^2 + 11x - 10} \\ 6x^2 + 15x \end{array}$$

4. Subtract this product from the dividend. To subtract a polynomial, we add its negative or change the signs and add. Then bring down the next term from the dividend.

$$\begin{array}{r} 3x \\ 2x + 5 \overline{)6x^2 + 11x - 10} \\ \ominus 6x^2 \mp 15x \\ \hline -4x - 10 \end{array}$$

5. Divide the first term of the new dividend by the first term of the divisor.

$$\frac{-4x}{2x} = -2$$

$$\begin{array}{r} 3x - 2 \\ 2x + 5 \overline{)6x^2 + 11x - 10} \\ +6x^2 \mp 15x \\ \hline -4x - 10 \end{array}$$

6. Multiply the entire divisor by the second term of the quotient.

$$-2(2x + 5) = -4x - 10$$

$$\begin{array}{r} 3x - 2 \\ 2x + 5 \overline{)6x^2 + 11x - 10} \\ +6x^2 \mp 15x \\ \hline -4x - 10 \\ -4x - 10 \end{array}$$

7. Subtract this product from the new dividend to obtain the remainder.

$$\begin{array}{r} 3x - 2 \\ 2x + 5 \overline{)6x^2 + 11x - 10} \\ +6x^2 \mp 15x \\ \hline -4x - 10 \\ \oplus 4x \oplus 10 \\ \hline 0 \end{array}$$

EXAMPLE 4   Divide $(-30 - x + 2x^2)$ by $(x - 4)$.

SOLUTION   1. $x - 4\overline{)2x^2 - x - 30}$ Arrange in descending order.

2. $\begin{array}{r} 2x \\ x - 4\overline{)2x^2 - x - 30} \end{array}$ Divide $2x^2$ by x.

3. $\begin{array}{r} 2x \\ x - 4\overline{)2x^2 - x - 30} \\ 2x^2 - 8x \end{array}$ Multiply $x - 4$ by $2x$.

4. $\begin{array}{r} 2x \\ x - 4\overline{)2x^2 - x - 30} \\ 2x^2 - 8x \\ \hline 7x - 30 \end{array}$ Subtract: $-x - (-8x) = -x + 8x = 7x$.
Bring down the -30.

5. $\begin{array}{r} 2x + 7 \\ x - 4\overline{)2x^2 - x - 30} \\ 2x^2 - 8x \\ \hline 7x - 30 \end{array}$ Divide $7x$ by x.

6. $\begin{array}{r} 2x + 7 \\ x - 4\overline{)2x^2 - x - 30} \\ 2x^2 - 8x \\ \hline 7x - 30 \\ 7x - 28 \end{array}$ Multiply $x - 4$ by 7.

7. $\begin{array}{r} 2x + 7 \\ x - 4\overline{)2x^2 - x - 30} \\ 2x^2 - 8x \\ \hline -7x - 30 \\ 7x - 28 \\ \hline -2 \end{array}$ Subtract $-30 - (-28) = -30 + 28 = -2$.

The remainder is -2.

Our answer is $2x + 7$ with a remainder of -2 or $2x + 7 + \dfrac{-2}{x + 4}$. To check, we multiply the quotient by the divisor and add in the remainder; we should get the dividend.

Check
$$\begin{array}{r} 2x + 7 \\ x - 4 \\ \hline 2x^2 + 7x \\ -8x - 28 \\ \hline 2x^2 - x - 28 \\ -2 \\ \hline 2x^2 - x - 30 \end{array}$$

← Quotient
← Divisor

← Add the remainder.
← Dividend

EXAMPLE 5 $(5x + 5x^2 - 3 + x^3) \div (x + 3)$

SOLUTION

$$\begin{array}{r} x^2 + 2x - 1 \\ x + 3\overline{)x^3 + 5x^2 + 5x - 3} \\ x^3 + 3x^2 \\ \hline 2x^2 + 5x \\ 2x^2 + 6x \\ \hline -x - 3 \\ -x - 3 \\ \hline 0 \end{array}$$

Check

$$
\begin{array}{r}
x^2 + 2x - 1 \\
x + 3 \\
\hline
x^3 + 2x^2 - x \\
3x^2 + 6x - 3 \\
\hline
x^3 + 5x^2 + 5x - 3
\end{array}
$$

EXAMPLE 6 $(x^3 + 8) \div (x + 2)$

SOLUTION

$$x + 2 \overline{)x^3 + 0x^2 + 0x + 8}$$

Always leave space for missing powers of the variable by using 0 as the coefficient.

$$
\begin{array}{r}
x^2 - 2x + 4 \\
x + 2 \overline{)x^3 + 0x^2 + 0x + 8} \\
+x^3 + 2x^2 \\
\hline
-2x^2 + 0x \\
-2x^2 - 4x \\
\hline
4x + 8 \\
+4x + 8 \\
\hline
0
\end{array}
$$

Check

$$
\begin{array}{r}
x^2 - 2x + 4 \\
x + 2 \\
\hline
x^3 - 2x^2 + 4x \\
2x^2 - 4x + 8 \\
\hline
x^3 \qquad\quad + 8
\end{array}
$$

3.10   Exercises

▼ Divide:

1. $(4x^3 + 6x^2 - 8x + 16) \div 2$

2. $(-6x^3 + 9x^2 - 12) \div (-3)$

3. $(14x^4 + 7x^3 - 21x + 35) \div 7x$

4. $(36y^4 - 6y^2 + 18y + 24) \div 6y^2$

5. $(8a^4 + 12a^3 - 16a^2 + 24) \div (-4a^4)$

6. $(7c^3 - 21c^2 + 14c - 42) \div 7c$

7. $(x^2 - 13x + 40) \div (x - 5)$

8. $(x^2 - 11x + 28) \div (x - 7)$

9. $(a^2 - 6a + 9) \div (a - 3)$

15. $(6x^2 - 5x - 21) \div (3x + 5)$

10. $(x^2 + 6x + 8) \div (x + 2)$

16. $(10 - 13x + 4x^2) \div (4x - 5)$

11. $(2y^2 + y - 6) \div (y + 2)$

17. $(17x + 1 + 12x^2) \div (5 + 3x)$

12. $(2x^2 - 13x + 20) \div (2x - 5)$

18. $(6 + 37t + 4t^2) \div (4t + 1)$

13. $(6n^2 + 11n + 3) \div (3n + 1)$

19. $(x^3 - 3x^2 + 100) \div (x + 4)$

14: $(3x^2 - 13x - 10) \div (x - 5)$

20. $(x^3 - 8x + 3) \div (x + 3)$

26. $(z^3 - 8) \div (z - 2)$

21. $(6x^2 + 41x) \div (6x - 1)$

27. $(x^4 + x^3 - 9x^2 - 14x - 4) \div (x^2 + 3x + 1)$

22. $(t^3 - 27) \div (t - 3)$

28. $(-7x^2 - 3 + 12x + x^4) \div (x^2 + 3 - 3x)$

23. $(x^3 - x^2 - 5x - 3) \div (x - 3)$

29. The area of the pictured rectangle is $6x^2 + 7x - 20$ and the width is $3x - 4$. Find the length L.

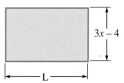
$3x - 4$

L

24. $(-5x^2 - 5 - 13x + 2x^3) \div (2x + 1)$

30. The volume of the box shown here is $6x^3 + 13x^2 - 5x$. The width is x and the height is $3x - 1$. What is its length L?

25. $(y^4 + 2y^2 - 2y^3 - 8) \div (y - 2)$

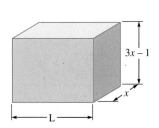
$3x - 1$

x

L

3.11   Synthetic Division

▮ ▮ ▮ ▮ ▮

When dividing polynomials, there are many steps in the method that are duplicated. When the divisor is of the form $x + k$, where k is an integer, we can streamline the process and save ourselves a considerable amount of work and time.

Consider the following division problem:

$$
\begin{array}{r}
2x^2 + x + 2 \\
x + 2 \overline{)\, 2x^3 + 5x^2 + 4x + 9} \\
\underline{2x^3 + 4x^2} \\
x^2 + 4x \\
\underline{x^2 + 2x} \\
2x + 9 \\
\underline{2x + 4} \\
5
\end{array}
$$

We can rewrite this same problem without the variable, showing only the coefficients. If there is a missing term, we write 0 for its coefficient.

$$
\begin{array}{r}
2 + 1 + 2 \\
1 + 2 \overline{)\, 2 + 5 + 4 + 9} \\
\underline{2 + 4} \\
1 + 4 \\
\underline{1 + 2} \\
2 + 9 \\
\underline{2 + 4} \\
5
\end{array}
$$

The numbers in red are repetitions of numbers written above them and can be eliminated:

$$
\begin{array}{r}
2 \quad\;\; 1 \quad\;\; 2 \longrightarrow \text{duplicated in this line} \\
1 + 2 \overline{)\, 2 \quad +5 \quad +4 \quad +9 } \\
\underline{\quad 4 \quad +2 \quad +4} \\
2 \quad\;\; 1 \quad\;\; 2 \quad\;\; 5
\end{array}
$$

The top line can be eliminated, since it is duplicate of the first three numbers in the bottom line. The 1, which is the coefficient of the x in the divisor; can also be eliminated, since we are considering only divisors of the form $x + k$. Remember that when we subtract an integer, we add its opposite. We will therefore use the opposite of the 2 in $x + 2$, namely; -2, and add instead of subtract. In its most compact form, the problem now looks like the following:

$$
\begin{array}{r|rrrr}
-2 & 2 & +5 & +4 & +9 \\
   &   & -4 & -2 & -4 \\
\hline
   & 2 & 1 & 2 & 5
\end{array}
$$

The division, called *synthetic* division, is performed by a series of multiplications and additions. Follow the arrows in the diagram.

$$
\begin{array}{r|rrrr}
-2 & 2 & +5 & +4 & +9 \\
   & \downarrow & -4 & -2 & -4 \\
\hline
   & 2 & 1 & 2 & 5
\end{array}
$$

We bring down the first coefficient, 2, multiply it by the divisor, -2 ($= -4$), and write the result under the next coefficient, 5. Adding gives us 1, which we multiply by -2 ($= -2$) and write under the next coefficient, 4. Adding these gives us 2, so multiply by -2 ($= -4$), put this under the 9, add again, giving us 5, and we are finished. The last term, 5, is the remainder.

EXAMPLE 1   Divide $x^4 + x^3 - 6x^2 + 2x - 2$ by $x + 3$, using synthetic division.

SOLUTION   We first change the sign of the 3 in $x + 3$ to -3. Then we proceed as follows:

$$
\begin{array}{r|rrrrr}
-3 & 1 & 1 & -6 & 2 & -2 \\
   & \downarrow & -3 & 6 & 0 & -6 \\
\hline
   & 1 & -2 & 0 & 2 & -8
\end{array}
$$

The last line gives us the result. The quotient is $x^3 - 2x^2 + 2$ with a remainder of -8. The 0 indicates that there is no x term. Notice that the degree of the quotient is one less than the degree of the dividend and the quotient is written in order of descending powers.   ▐▌

EXAMPLE 2   Divide $x^3 - 19x + 12$ by $x - 4$ using synthetic division.

SOLUTION   We use 0 as the coefficient of the missing x^2 term and change -4 to 4:

$$
\begin{array}{r|rrrr}
4 & 1 & 0 & -19 & 12 \\
  & \downarrow & 4 & 16 & -12 \\
\hline
  & 1 & 4 & -3 & 0
\end{array}
$$

The quotient is $x^2 + 4x - 3$ with no remainder.   ▐▌

EXAMPLE 3   Divide $x^3 + 8$ by $x + 2$ using synthetic division.

SOLUTION

$$
\begin{array}{r|rrrr}
-2 & 1 & 0 & 0 & 8 \\
   & \downarrow & -2 & 4 & -8 \\
\hline
   & 1 & -2 & 4 & 0
\end{array}
$$

$$
\frac{x^3 + 8}{x + 2} = x^2 - 2x + 4
$$

▐▌

3.11   Exercises

▼ Use synthetic division to find the quotient and the remainder.

1. $(x^2 + 8x + 15) \div (x + 3)$

2. $(x^2 + 2x - 8) \div (x - 2)$

3. $(3x^2 + 4x - 15) \div (x + 3)$

4. $(2x^2 - 15x + 18) \div (x - 6)$

5. $(x^3 + 4x^2 + 3x + 1) \div (x + 2)$

6. $(x^3 - 7x^2 + 13x - 1) \div (x - 4)$

7. $(x^3 + 6x^2 + 3x - 6) \div (x + 2)$

8. $(x^3 - 4x^2 - 7x + 2) \div (x - 1)$

9. $(2x^3 - 11x^2 + 13x - 4) \div (x - 4)$

10. $(4x^3 + 2x^2 + x + 3) \div (x + 1)$

The **degree of a polynomial** is the largest degree of any one term in the polynomial.

The degree of $3x^4 - 8x - 5$ is 4.

Descending order: When a polynomial in one variable is written with the highest-degree term first, then the next-highest-degree term; and so on.

$6x^5 + 2x^4 - 4x^2 + 7$ is in descending order.

To add polynomials horizontally:

1. Write the polynomials in descending order;

2. Combine the like terms.

$(3x^3 + 2x^2 - 2x + 5) + (4x^3 - 6x^2 + 3x - 7)$
$= 7x^3 - 4x^2 + x - 2$

To add polynomials vertically:

1. Write the polynomials in descending order with like terms under one another;

2. Combine the like terms.

$$\begin{array}{r} 3x^3 + 2x^2 - 2x + 5 \\ +4x^3 - 6x^2 + 3x - 7 \\ \hline 7x^3 - 4x^2 + x - 2 \end{array}$$

To subtract polynomials:

1. Write each polynomial in descending order;

2. Change the sign of *each* term in the polynomial being subtracted;

3. Combine the like terms.

$(5x^2 + 2x - 1) - (7x^2 + 3x - 5)$
$= 5x^2 + 2x - 1 - 7x^2 - 3x + 5$
$= -2x^2 - x + 4$

To multiply monomials:

1. Multiply the numerical coefficients;

2. Multiply the variables using the rule $x^a \cdot x^b = x^{a+b}$.

$(-7x^4)(4x^6) = (-7)(4)(x^4)(x^6)$
$= -28x^{10}$

To multiply a polynomial by a monomial: Multiply each term of the polynomial by the monomial.

$5x^2(3x^2 + 2x - 1)$
$= 5x^2 \cdot 3x^2 + 5x^2 \cdot 2x + 5x^2 \cdot (-1)$
$= 15x^4 + 10x^3 - 5x^2$

To multiply a polynomial by a polynomial:

1. Multiply each term in one polynomial by each term in the other polynomial;

2. Arrange like terms under one another in the products;

3. Combine the like terms.

$$\begin{array}{r} 7x^2 - 3x - 6 \\ 4x^2 - 7 \\ \hline 28x^4 - 12x^3 - 24x^2 \\ -49x^2 + 21x + 42 \\ \hline 28x^4 - 12x^3 - 73x^2 + 21x + 42 \end{array}$$

To multiply two binomials using the FOIL method:

1. The first term = the product of the **F**irst terms of the binomials.

2. The middle term = the sum of the **O**uter and **I**nner products.

3. The last term = the product of the **L**ast terms of the binomials.

$(3x + 4)(2x - 5) = 6x^2 - 7x - 20$

To square a binomial:

1. The square of the first term = the first term of the product;

2. Twice the product of the two terms = the middle term of the product;

3. The square of the last term = the last term of the product.

Factoring is the process of writing an expression as the product of two expressions, each called **factors**.

A **common factor** is the largest factor that divides each term of the given polynomial

The **difference of two squares:**
$a^2 - b^2 = (a + b)(a - b)$

The **square root of an even-powered variable** is equal to that variable raised to one-half of the given exponent.

The **sum of two cubes:**
$a^3 + b^3 = (a + b)(a^2 - ab + b^2)$

The **difference of two cubes:**
$a^3 - b^3 = (a - b)(a^2 + ab + b^2)$

To factor a trinomial of the form $x^2 + bx + c =$ $(x + p)(x + q)$, find integers p and q such that:

1. $p \cdot q = c$ (their product is c) and

2. $p + q = b$ (their sum is b).

When **factoring a trinomial of the form $ax^2 + bx + c$** where $a > 1$:

1. The product of the first terms must equal ax^2;

2. The product of the last terms must equal c;

3. The sum of the outer products and inner products must equal bx.

Always look for a greatest common factor first.

To divide a polynomial by a monomial; divide *each* term of the polynomial by the monomial, using the rule

$$\frac{x^a}{x^b} = x^{a-b}$$

to divide the variables.

EXAMPLES

$(2x - 3)^2 = 4x^2 - 12x + 9$

$14 = 2 \cdot 7$

$18x^4 + 12x^3 - 12x^2$
$\quad = 6x^2(3x^2 + 2x - 2)$

$36x^2 - 25 = (6x + 5)(6x - 5)$

$x^{10} - 49 = (x^5 + 7)(x^5 - 7)$

$x^3 + 8 = (x + 2)(x^2 - 2x + 4)$

$27x^3 - 1 = (3x - 1)(9x^2 + 3x + 1)$

$x^2 + 3x - 28 = (x - 4)(x + 7)$

$6x^2 - 13x + 5 = (2x - 1)(3x - 5)$

$20x^3 + 5x^2 - 15x$
$\quad = 5x(4x^2 + x - 3)$
$\quad = 5x(4x - 3)(x + 1)$

$(7x^3 + 4x^2 - 8x + 12) \div 2x$
$= \dfrac{7x^3}{2x} + \dfrac{4x^2}{2x} - \dfrac{8x}{2x} + \dfrac{12}{2x}$
$= \dfrac{7}{2}x^2 + 2x - 4 + 6x^{-1}$
$= \dfrac{7}{2}x^2 + 2x - 4 + \dfrac{6}{x}$

EXAMPLES

To divide a polynomial by a polynomial; see the detailed explanation in Section 3.10

$$\begin{array}{r} 3x + 2 \\ x - 2 \overline{\smash{)}3x^2 - 4x - 6} \\ \underline{+3x^2 - 6x} \\ 2x - 6 \\ \underline{+2x - 4} \\ -2 \end{array}$$

remainder ↗

See the detailed explanation for **synthetic division** in Section 3.11

$(3x^2 - 4x - 6) \div (x - 2)$

$$\begin{array}{r|rrr} 2 & 3 & -4 & -6 \\ & \downarrow & 6 & 4 \\ \hline & 3 & 2 & -2 \end{array}$$

The quotient is $3x + 2$ with remainder -2.

A **prime polynomial** is one that is not factorable.

$7x^4y - 8x^2z + 4yz^5$ is prime.
$x^2 - 6x + 2$ is prime.
$5x^2 - 3x + 4$ is prime.
$4x^2 + 9$ is prime.

REVIEW EXERCISES—CHAPTER 3

▮ Simplify, using the rules of exponents.

1. $a^6 \cdot a^5$

2. $n^5 \cdot n^7$

3. $(x^3)^7$

4. $\dfrac{x^8}{x^3}$

5. $(x^5)^4$

6. $h \cdot h^6$

7. $\dfrac{y^7}{y}$

8. $(xy)^6$

9. $\left(\dfrac{x}{y}\right)^8$

10. $(3mn)^2$

11. $x^4 \cdot y^7$

12. $\dfrac{x^{3a}}{x^{2a}}$

▮ Simplify, leaving your answers with positive exponents only.

13. x^{-5}

14. y^{-3}

15. $\dfrac{1}{x^{-4}}$

16. $\dfrac{1}{h^{-1}}$

17. 8^0

18. $4x^0$

19. $\left(\dfrac{m}{n}\right)^{-5}$

20. $\left(\dfrac{a}{b}\right)^{-c}$

21. $x^{-3}y^5$

22. $\dfrac{m^0}{n^{-1}}$

23. $\left(\dfrac{2}{3}\right)^{-2}$

24. $\dfrac{a^{-5}b^4}{c^{-2}}$

25. $(3x^3y^{-2})^3$

26. $(4x^3y^{-4})^{-2}$

31.
$$\begin{array}{r} -3y^3 + 4y^2 \quad\quad - 12 \\ 2y^2 + 8y - 4 \\ 4y^3 \quad\quad + 2y - 4 \\ \hline \end{array}$$

27. $\dfrac{x^3y^{-2}}{xy^4}$

28. $\left(\dfrac{x^{-3}y}{x^2y^{-2}}\right)^{-3}$

▼ Subtract the polynomials.

32. $(-5a^3 + 2a^2 - 6a + 3) - (2a^3 + 6a^2 + 3a + 5)$

▼ Add the polynomials, writing your answer in descending order.

29. $-2x + 5x^3 + 2x^2 - 6$ and $x^3 + x^2 - 6 - 12x$

33. $(12x^3 + 6x + 4) - (4x^2 + 3x)$

30. $-14x - 13x^3 + 7x^2$, $4x - 3x^2 + x^3 - 3$, and $6 + x^3 - 2x^2$

34.
$$\begin{array}{r} -6x^3 + 2x^2 - 4x + 9 \\ -(3x^3 \quad\quad - 6x - 1) \\ \hline \end{array}$$

35. Given the polynomials $3x + 14x^3 - 7x^2 + 3$, $4x^3 - 3x + 2$, and $-6x^3 + 3 - 4x^2 + x$, subtract the second polynomial from the sum of the first and the third.

▼ **Multiply:**

36. $(-3x^4)(-2x^3)$

37. $(-4a^3y)(-2a^3y)(2ay^3)$

38. $-4x(6x - 1)$ **39.** $(6x + 5)(3x - 8)$

40. $(7x + 1)(2x + 5)$ **41.** $(x + 4)(2x^2 + 3x - 4)$

42. $(3x + 4)(4x^3 + 2x^2 - 3x + 5)$

43. $(3x - 2)^2$

$(3x-2)(3x-2)$

▼ Factor each of the following polynomials completely. If not factorable, label the expression **prime.**

44. $4ax^3 - 12a^2x^2$ **45.** $35x^5 - 15x^4 + 10x^2$

46. $y(x + 1) - 7(x + 1)$ **47.** $x^2(x + 2) - (x + 2)$

48. $a^2 - ab - 3a + 3b$ **49.** $a^6 + a^5 + a + 1$

50. $4x^2 - 49$ **51.** $9x^2 + 4y^2$

52. $3x^{12} - 27$ **53.** $y^8 - 16$

54. $5x^3 + 40y^3$ **55.** $64 - x^3$

56. $x^2 + x - 20$ **57.** $y^2 + 7y - 12$

58. $8x^2 - 2x - 15$ **59.** $24x^2 + 46x + 7$

60. $x^4 - 3x^2 - 28$ **61.** $15h^4 + 14h^2 - 8$

62. $y^2(a - b) - x^2(a - b)$

69. $(4y^3 + y + 27) \div (3 + 2y)$

63. $x^6 - 64$

70. $(3x^3 + 17x^2 + 11x - 20) \div (3x + 5)$

64. $a^2b + 5a^2 - b - 5$

▼ Divide:

65. $(-8y^4 - 12y^3 + 28y + 8) \div 4y$

▼ Divide using synthetic division:

71. $(2x^3 - 17x^2 + 33x - 14) \div (x - 6)$

66. $(6x^5 + 22x^4 - 8x^2 + 3x + 9) \div (-3x^3)$

72. $(6x^3 - 21x^2 - 10x - 8) \div (x - 4)$

67. $(x^2 - 6x - 27) \div (x - 9)$

73. $(2x^3 - 10x + 24) \div (x + 3)$

68. $(15x^2 + x - 2) \div (5x + 2)$

74. $(3x^3 + 17x^2 - 50) \div (x + 5)$

CRITICAL THINKING EXERCISES—CHAPTER 3

▸ There is an error in each of the following problems. Can you find it?

1. $21^0 = 0$ $= 1$

2. $x \cdot x = 2x$ $= x^2$

3. $5^2 \cdot 5^3 = 25^5$ $= 5^5$

4. $\dfrac{6^3}{6^3} = 0$ $= 1$

5. $4^{-2} = -16$ $+16$

6. $4^{-2} = -8$ $+8$

7. $6^{-1} = \dfrac{1}{-6}$ $= \dfrac{1}{6^1}$

8. $(y^5)^4 = y^9$ $= y^{20}$

9. $x^a \cdot x^b = x^{ab}$ $= x^{a+b}$

10. $9x^0 = 1$ $= 9$

11. $(4x)^3 = 4x^3$ $= 64x^3$

12. $x^3 \cdot y^5 = (xy)^8$ $= (x^3)(y^5)$

FOIL

13. $(2x - 5)^2 = 4x^2 - 25$ $= 4x^2 - 20x + 25$
   $8x^3$

14. $(2x - 5)^2 = 4x^2 + 25$ $= 4x^2 - 20x + 25$

15. $x^2 - 5x + 6 = (x - 6)(x + 1)$

16. $4x^2 + 9 = (2x + 3)(2x + 3)$ $= $ prime.

17. $-(2x^2 - 3x + 4) = -2x^2 - 3x + 4$

NAME _____   **| | | | |** **CLASS**

ACHIEVEMENT TEST—CHAPTER 3

• • • • • • • ▼

▼ Simplify, leaving answers with positive exponents only.

1. $a^7 \cdot a^8$ 1. _____

2. $\dfrac{x^5}{x^{11}}$ 2. _____

3. $(3x^4)^3$ 3. _____

4. $\left(\dfrac{3}{4}\right)^{-2}$ 4. _____

5. $\dfrac{x^4 y^{-7}}{x^0 y^3}$ 5. _____

6. $\left(\dfrac{2x^3 y^{-2}}{x^4 y}\right)^{-2}$ 6. _____

▼ Add or subtract as indicated.

7. $(-7x^4 - 3x^2 - 2x - 1) + (6x^3 + 3x^2 - 4x + 9)$ 7. _____

8. $(6y^3 - 3y^2 - 3y + 7) - (4y^3 - 3y + 5)$ 8. _____

▼ Multiply:

9. $(-4x^5)(-3x^3)(-2x^3)$ 9. _____

10. $(4x - 3)(2x + 5)$ 10. _____

11. $(6x + 1)^2$ 11. _____

12. $(x + 3)(x^2 + 4x - 5)$

12. _____

▼ Factor completely. If not factorable, write *prime*.

13. $6a^3xy - 12a^2y^2 + 6a^2y$

13. _____

14. $y(x - 5) + 4(x - 5)$

14. _____

15. $y^3 + y^2 + y + 1$

15. _____

16. $36x^2 - 49y^2$

16. _____

17. $9a^2 + 16y^2$

17. _____

18. $8x^3 - 27y^3$

18. _____

19. $x^2 - 13x + 42$

19. _____

20. $15x^2 + 41x + 14$

20. _____

21. $4s^4 + 108st^3$

21. _____

22. $3x^4 - 8x^2 - 3$

22. _____

23. $x^2y - x^2 - 4y + 4$

23. _____

▼ Divide:

24. $(9y^6 + 15y^4 - 3y^3 + 27) \div 3y^2$

24. _____

25. $(2x^2 - 9x - 15) \div (x - 6)$

25. _____

26. $(6x^2 - 5x - 25) \div (3x - 7)$

26. _____

▼ Divide using synthetic division.

27. $(5x^3 - 18x^2 + 10x - 3) \div (x - 3)$

27. _____

28. $(6x^3 - 17x + 14) \div (x + 2)$

28. _____

NAME _____   **I I I I I CLASS** _____

CUMULATIVE REVIEW—CHAPTERS 1, 2, AND 3

• • • • • • • • ▼

1. Write out the set of odd integers between -5 and 5.

2. What is the cardinal number of the set $E = \{e, x, a, m\}$?

▼ In Exercises 3–10, let $E = \{e, x, a, m\}$ and $F = \{a, x, e\}$.

3. Find $E \cap F$.

4. Find $E \cup F$.

▼ Answer true or false:

5. $F \subseteq E$

6. $m \in F$

7. $\{e, x\} \subseteq F$

8. $\varnothing \subseteq E$

9. $F \in E$

10. $a \subseteq E$

▼ In Exercises 11–14, classify each number as rational or irrational.

11. $\sqrt{14}$

12. $\dfrac{3}{17}$

13. $0.31\overline{918}$

14. π

15. Write 126 as a product of prime factors.

▼ Perform the indicated operations.

16. $6 - 8$

17. $(6)(-8)$

18. $6 - (-8)$

1. _____

2. _____

3. _____

4. _____

5. _____

6. _____

7. _____

8. _____

9. _____

10. _____

11. _____

12. _____

13. _____

14. _____

15. _____

16. _____

17. _____

18. _____

19. $\dfrac{0}{0}$ 19. _____

20. $\dfrac{-3}{0}$ 20. _____

21. $\dfrac{0}{16}$ 21. _____

22. $3 - (-7) + 4 - 8$ 22. _____

23. $(-1)(-4)(2)(-2)(-2)(3)$ 23. _____

24. $4 - \{2 - [3 - (2 + 4)]\}$ 24. _____

▼   In Exercises 25–31, solve for x.

25. $3x - 4 = 5(x - 6)$ 25. _____

26. $\frac{1}{6}x + \frac{2}{3} = x - \frac{1}{4}$ 26. _____

27. $2x - 4y - 3z = 0$ 27. _____

28. $3x - 4 \leq x + 12$ 28. _____

29. $|4x - 1| = 7$ 29. _____

30. $|2x - 3| < 5$ 30. _____

31. $|4x + 5| < -2$ 31. _____

32. Three women share an apartment while going to college. All three own cars.   32. _____
Donna has a Toyota that gets 28 miles per gallon (mpg). Diane drives a Ford
Escort that gets 21 mpg. Their roommate, Marcella, drives a beat-up old
station wagon and doesn't want to talk about the terrible mileage it gets. If
the average number of miles per gallon that the three cars get is 19 mph,
how many miles per gallon does "Marcella's Monster" get?

▼   In Exercises 33–38, simplify each expression, using the rules of exponents.
Leave your answers with positive exponents only.

33. $x^5 \cdot x^4$ 33. _____

34. $\dfrac{x^6}{x^{11}}$ 34. _____

35. $(2x^2)^4$ 35. _____

36. $\left(\dfrac{2}{3}\right)^{-3}$ 36. _____

37. $6x^0$

37. _____

38. $\dfrac{8x^4y^3}{2x^6y^{-5}}$

38. _____

39. Subtract: $(6x^4 - 3x^2 + 5x - 1) - (5x^4 + 3x^3 - 3x^2 + 7)$.

39. _____

40. Multiply: $(3x^2 - 2x + 5)(2x - 3)$.

40. _____

41. Square the binomial: $(4x - 5)^2$.

41. _____

▼ In Exercises 42–48, factor completely. It not factorable, write *prime*.

42. $x(y + 2) + 5(y + 2)$

42. _____

43. $5x^2 - 45y^2$

43. _____

44. $8x^3 - 125$

44. _____

45. $5x^2 - 23x - 10$

45. _____

46. $2x^4 - x^2 - 15$

46. _____

47. $10x^2 - 3x + 4$

47. _____

48. $9x^2 + 25$

48. _____

49. Divide: $(8x^2 - 2x - 17) \div (2x - 3)$.

49. _____

50. Divide using synthetic division: $(x^3 - 13x - 12) \div (x - 4)$.

50. _____

Rational Expressions

INTRODUCTION

Previously we learned how to combine polynomials using the operations of addition, subtraction, multiplication, and division. We also learned how to factor polynomials of various kinds.

Now, in this chapter, we will form algebraic fractions from polynomials, then investigate how to simplify them, add, subtract, multiply, and divide them, and solve equations containing them.

CHAPTER 4–NUMBER KNOWLEDGE

Buying a Car—Rebate vs. Lower Financing Rate

Ford offers either a $500 rebate or 2.9% financing. You want to purchase a Ford Mustang at a purchase price of $13,924. What's the best way to pay for it?

Case 1: You take the $500 rebate and finance $13,924 − $500 = $13,424 through a conventional bank car loan that charges an annual percentage rage (APR) of 11.88%. If you finance the car for five years, the total finance charge will be $4,443.76. Your total out-of-pocket cost for the car after five years will be:

$$
\begin{array}{ll}
\$13,924.00 & \\
\underline{-\quad 500.00} & \text{Rebate} \\
13,424.00 & \\
\underline{+\ 4,443.76} & \text{Finance charge} \\
\text{Total:}\quad \$17,867.76 &
\end{array}
$$

Case 2: You take the $500 rebate and pay $13,924 − $500 = $13,424 cash for the car. Using the same five-year period as in Case 1, and assuming that you could have invested the $13,424 in a certificate of deposit at 3.5% interest, compounded quarterly, that would have produced $2,555.12 that you now will not receive.

Your total out-of-pocket cost for the car will be:

$$
\begin{array}{ll}
\$13,924.00 & \\
\underline{-\quad 500.00} & \text{Rebate} \\
13,424.00 & \\
\underline{+\ 2,555.12} & \text{Lost income} \\
\text{Total:}\quad \$15,979.12 &
\end{array}
$$

Case 3: You forego the $500 rebate and choose the 2.9% financing for five years. Your out-of-pocket cost for the car will be

$$
\begin{array}{ll}
\$13,924.00 & \\
\underline{+\ 1,050.68} & \text{Finance charge} \\
\text{Total:}\quad 14,974.68 &
\end{array}
$$

As you can see, the least cost to you will be Case 3.

4.1   Basic Concepts

An **algebraic fraction** or a **rational expression** is an expression of the form

$$
\frac{N}{D} \qquad \text{where } N \text{ and } D \text{ are polynomials}
$$

Your knowledge of arithmetic fractions is an excellent base on which to build the theory of algebraic fractions. You will find that the procedures for simplifying, multiplying, dividing, adding, and subtracting *algebraic* fractions are basically the same as they are for *arithmetic* fractions.

Any value of the variable that makes the denominator of an algebraic fraction equal to zero cannot be allowed, since division by zero is not defined. For example, the algebraic fraction

$$
\frac{2x + 3}{x - 4}
$$

is not defined when $x = 4$ because the value of the denominator would be equal to zero; we call $x = 4$ an **excluded value.** You will frequently see

$$
\frac{2x + 3}{x - 4}, \qquad x \neq 4
$$

and, even if it is not explicitly stated, it is assumed that $x \neq 4$. It should be understood, then, that any value(s) of the variable that makes the denominator equal to zero must be excluded.

> An **excluded value** of an algebraic fraction is a value of the variable that makes the denominator equal to zero.

E X A M P L E 1   Find the excluded values (if any) of each of the following algebraic fractions.

(a) $\dfrac{4x - 1}{x + 7}$

x cannot equal -7 because it would make the denominator zero.

(b) $\dfrac{x - 5}{x^2 + 2x - 3} = \dfrac{x - 5}{(x + 3)(x - 1)}$

x cannot equal either -3 or 1, since either value would make the denominator equal to zero.

(c) $\dfrac{3x + 5}{7}$

No value of *x* will make the denominator zero, so no value of *x* is excluded.

(d) $\dfrac{-9}{x}$

$x \neq 0$ ▌▌

The Three Signs of a Fraction

• • • • •

Every fraction has three signs associated with it: the sign of the numerator, the sign of the denominator, and the sign of the fraction itself.

$$\text{The sign of fraction} \rightarrow +\frac{-4}{+7}$$

sign of numerator

sign of denominator

From the rule of signs for division, we know that

$$\frac{-a}{b} = \frac{a}{-b} = -\frac{a}{b}$$

This leads us to the **rule of signs for fractions:**

> If exactly two of the three signs of a fraction are changed, the value of the fraction remains the same.

E X A M P L E 2

(a) $+\dfrac{+6}{+2} = \begin{cases} +\dfrac{-6}{-2} = +(+3) = 3 & \text{Change the signs of the numerator and the denominator.} \\[2mm] -\dfrac{-6}{+2} = -(-3) = 3 & \text{Change the signs of the fraction and the numerator.} \\[2mm] -\dfrac{+6}{-2} = -(-3) = 3 & \text{Change the signs of the fraction and the denominator.} \end{cases}$

(b) $-\dfrac{3}{-x} = +\dfrac{3}{+x} = \dfrac{3}{x}$ Change the signs of the fraction and the denominator.

(c) $\dfrac{-3}{4-x} = \dfrac{+3}{-(4-x)} = \dfrac{3}{-4+x} = \dfrac{3}{x-4}$ Change the signs of the numerator and the denominator.   ▮▮

This last example illustrates a property that we should look at further:

$$\boxed{(x - y) = -(y - x)}$$

E X A M P L E 3

(a) $\dfrac{3}{a-b} = \dfrac{?}{b-a}$

$\dfrac{3}{a-b} = \dfrac{-3}{-(a-b)} = \dfrac{-3}{b-a}$ Change the signs of the numerator and the denominator.

(b) $-\dfrac{6}{2-x} = +\dfrac{6}{-(2-x)} = \dfrac{6}{x-2}$ Change the signs of the fraction and the denominator.   ▮▮

4.1   Exercises

▮▮▮▮▮

▼ In Exercises 1–12, find the value(s) of the variable (if any) for which the fraction is undefined.

1. $\dfrac{2x-1}{x-3}$

2. $\dfrac{-7}{2x}$

3. $\dfrac{x}{x+3}$

4. $\dfrac{x}{6}$

5. $\dfrac{4-x}{x-4}$

6. $\dfrac{3}{4}$

7. $\dfrac{x-3}{x^2+2x-3}$

8. $\dfrac{x-1}{x^2-x-2}$

9. $\dfrac{-3}{x}$

10. $\dfrac{2}{x^2-x}$

11. $\dfrac{-7}{x^2+3x-10}$

12. $\dfrac{3x}{2x-7}$

▼ In Exercises 13–21, find the missing term using the rule of signs for fractions.

13. $\dfrac{3}{5} = \dfrac{?}{-5}$

14. $\dfrac{-1}{4} = \dfrac{?}{-4}$

15. $-\dfrac{3}{4} = \dfrac{-3}{?}$

16. $\dfrac{x}{-y} = \dfrac{?}{y}$

17. $\dfrac{5}{1-x} = \dfrac{?}{-(1-x)}$

18. $\dfrac{5}{1-x} = \dfrac{?}{x-1}$

19. $\dfrac{2-y}{-3} = \dfrac{?}{3}$

20. $-\dfrac{3}{x-1} = \dfrac{?}{1-x}$

22. Explain in your own words why $-(3-x) = x - 3$.

21. $\dfrac{1-x}{1-y} = \dfrac{?}{y-1}$

4.2   Reducing Fractions to Lowest Terms

In arithmetic we learned that we can divide both the numerator and the denominator of any fraction by the same nonzero number without changing the value of the fraction.

To reduce an arithmetic fraction to its lowest terms, we factor both the numerator and the denominator completely and then divide both the numerator and the denominator by all factors common to both:

$$\frac{6}{9} = \frac{2 \cdot \cancel{3}^{\,1}}{3 \cdot \cancel{3}_{\,1}} = \frac{2}{3}$$

This procedure holds true in algebra, and is illustrated as follows:

$$\frac{ay}{by} = \frac{a \cdot \cancel{y}^{\,1}}{b \cdot \cancel{y}_{\,1}} = \frac{a}{b}$$

> **To reduce a fraction to lowest terms:**
>
> **1.** Factor both the numerator and denominator completely.
>
> **2.** Divide both the numerator and denominator by any common factors.
>
> **3.** Multiply the factors that remain.

EXAMPLE 1   Reduce each fraction to lowest terms. (Assume that no denominator is equal to zero.)

(a) $\dfrac{6a^2b^3}{30a^3b} = \dfrac{\cancel{6}^{\,1} \cdot \cancel{a^2}\, \cdot \cancel{b^3}^{\,b^2}}{\cancel{30}_{\,5} \cdot \cancel{a^3}_{\,a} \cdot \cancel{b}} = \dfrac{b^2}{5a}$

(b) $\dfrac{10x^2yz}{35xy^4z} = \dfrac{\cancel{10}^{\,2} \cdot \cancel{x^2}^{\,x} \cdot \cancel{y} \cdot \cancel{z}}{\cancel{35}_{\,7} \cdot \cancel{x} \cdot \cancel{y^4}_{\,y^3} \cdot \cancel{z}} = \dfrac{2x}{7y^3}$

(c) $\dfrac{x-2}{x^2-4} = \dfrac{\overset{1}{\cancel{(x-2)}}}{\underset{1}{\cancel{(x-2)}}(x+2)} = \dfrac{1}{x+2}$ The number 1 remains in the numerator after being divided by $x-2$.

(d) $\dfrac{2x+6}{x^2-9} = \dfrac{2\cancel{(x+3)}}{(x-3)\cancel{(x+3)}} = \dfrac{2}{x-3}$ The 1's are usually not written when dividing out factors (cancelling).

(e) $\dfrac{a+b}{a}$ Cannot be reduced, since there is no *factor* common to both numerator and denominator.

(f) $\dfrac{a-b}{b-a} = -\dfrac{a-b}{-(b-a)} = -\dfrac{a-b}{a-b} = -1$

         Change the signs of the fraction and the denominator.

(g) $\dfrac{x^2+2x-3}{x^2-2x-8} = \dfrac{(x-1)(x+3)}{(x-4)(x+2)}$ There are no common factors; the fraction is already in lowest terms.

(h) $\dfrac{2x^2-5x-3}{2x^2-x-1} = \dfrac{(x-3)\cancel{(2x+1)}}{\cancel{(2x+1)}(x-1)} = \dfrac{x-3}{x-1}$

(i) $-\dfrac{x^2-2x-3}{6+x-x^2} = -\dfrac{x^2-2x-3}{-(x^2-x-6)}$ Factor -1 out of the denominator and rearrange the terms.

$\qquad = \dfrac{x^2-2x-3}{x^2-x-6}$ Change the signs of the fraction and the denominator.

$\qquad = \dfrac{\cancel{(x-3)}(x+1)}{(x+2)\cancel{(x-3)}}$ Cancel $x-3$ from the numerator and denominator.

$\qquad = \dfrac{x+1}{x+2}$ ▮▮

STOP   Only **factors** of both the numerator and denominator may be divided out (cancelled). For example:

(a) Here is a common error:

                       not a factor

$\dfrac{2+9}{2} = \dfrac{\overset{1}{\cancel{2}}+9}{\underset{1}{\cancel{2}}} = 10$ Incorrect, since 2 is *not* a factor of the numerator.

$\dfrac{2+9}{2} = \dfrac{11}{2} = 5\dfrac{1}{2}$ Correct

(b) $\dfrac{y+\overset{2}{\cancel{4}}}{y+\underset{1}{\cancel{2}}} = \dfrac{y+2}{y+1}$ Incorrect, since 2 is *not* a factor of either the numerator or the denominator

$\dfrac{\overset{1}{\cancel{y}}+4}{\underset{1}{\cancel{y}}+2} = \dfrac{5}{3}$ Incorrect, since y is *not* a factor of either the numerator or the denominator

$\dfrac{y+4}{y+2}$ Cannot be reduced further.

4.2   Exercises

▮ ▮ ▮ ▮ ▮

▼ Reduce the fractions to lowest terms:

1. $\dfrac{12}{18}$

2. $\dfrac{16}{36}$

3. $\dfrac{6x^3y^4}{12x^5y}$

4. $\dfrac{x^7y^3}{x^7y^4}$

5. $\dfrac{3xyz}{-9xy^2z}$

6. $\dfrac{24a^3bc}{6a^4c}$

7. $\dfrac{2xy}{4xy}$

8. $\dfrac{-72x^6y^3}{-6x^2y^2}$

9. $\dfrac{x^2y^4}{a^2b^3}$

10. $\dfrac{x^2-1}{2x+2}$

11. $\dfrac{6}{3x+9}$

12. $\dfrac{x+3}{x-3}$

13. $\dfrac{7}{7x+14}$

14. $\dfrac{x+6}{6}$

15. $\dfrac{3+6}{3}$

16. $\dfrac{x^2-2x}{x^3-3x^2+2x}$

17. $-\dfrac{3x-2}{2-3x}$

18. $\dfrac{x^2-4x-5}{x^2+5x+4}$

19. $\dfrac{3x^2+5x+2}{3x^2-x-2}$

20. $\dfrac{6-5h}{5h-6}$

21. $\dfrac{x^2+3x}{x^3+9x^2+18x}$

22. $\dfrac{4x-4y}{y^2-x^2}$

23. $-\dfrac{2x^2+3x-20}{16-x^2}$

24. $\dfrac{x^2-5x+6}{6+x-x^2}$

25. $\dfrac{t^2-25}{t^2-2t-15}$

26. $\dfrac{(x+y)^2}{x^2+y^2}$

27. $\dfrac{x^3+x^2-12x}{x^3+9x^2+20x}$

28. $\dfrac{3x^2+2xy-y^2}{6x^2-5xy+y^2}$

29. $\dfrac{3a^2-3ab}{b^2-a^2}$

30. $\dfrac{x^2-4y^2}{2x^2-3xy-2y^2}$

31. Which of the following equal -1?

  a. $\dfrac{a+b}{b+a}$

  b. $\dfrac{a-b}{b-a}$

  c. $\dfrac{a+b}{-a-b}$

  d. $\dfrac{a-b}{a+b}$

4.3 Multiplication and Division of Algebraic Fractions

▮ ▮ ▮ ▮ ▮

Multiplying Algebraic Fractions

• • • • •

In arithmetic we multiplied fractions as follows, making sure that all common factors were cancelled before multiplying:

$$\frac{3}{4} \cdot \frac{8}{7} = \frac{3}{\overset{}{\underset{1}{4}}} \cdot \frac{\overset{2}{8}}{7} \qquad \text{Cancel completely.}$$

$$= \frac{6}{7} \qquad \text{Multiply the numerators together and multiply the denominators together.}$$

In algebra we use a similar procedure.

> **To multiply fractions:**
>
> 1. Factor all numerators and denominators completely.
>
> 2. Divide any pair of numerators and denominators by a factor that is common to both.
>
> 3. The answer is the product of the remaining numerators divided by the product of the remaining denominators.

E X A M P L E 1   Multiply:

$$\frac{3x^2}{5y} \cdot \frac{2y}{9} = \frac{3x^2}{5\overset{}{y}} \cdot \frac{2\overset{}{y}}{\underset{3}{9}} \qquad \text{Cancel completely.}$$

$$= \frac{2x^2}{15} \qquad \text{Multiply.}$$

▮▮

E X A M P L E 2   Multiply:

$$\frac{x^2 + 6x + 5}{x^2 - 1} \cdot \frac{x^2 - 4x + 3}{x^2 - 5x + 6}$$

$$= \frac{(x + 5)(x + 1)}{(x - 1)(x + 1)} \cdot \frac{(x - 1)(x - 3)}{(x - 3)(x - 2)} \qquad \begin{array}{l}\text{Factor all numerators}\\ \text{and all denominators}\\ \text{completely.}\end{array}$$

$$= \frac{(x + 5)\cancel{(x + 1)}}{\cancel{(x - 1)}\cancel{(x + 1)}} \cdot \frac{\cancel{(x - 1)}\cancel{(x - 3)}}{\cancel{(x - 3)}(x - 2)} \qquad \text{Cancel.}$$

$$= \frac{x + 5}{x - 2}$$

▮▮

EXAMPLE 3　Multiply:

$$\frac{x}{3x-6} \cdot \frac{5x-10}{x^3} = \frac{x}{3(x-2)} \cdot \frac{5(x-2)}{x^3} \qquad \text{Factor.}$$

$$= \frac{\cancel{x}}{3\cancel{(x-2)}} \cdot \frac{5\cancel{(x-2)}}{\underset{x^2}{\cancel{x^3}}} \qquad \text{Cancel.}$$

$$= \frac{5}{3x^2}$$ ▌▌

EXAMPLE 4　Multiply:

$$\frac{x^2-4}{x^2-x-6} \cdot \frac{x^2+9x+18}{x^2-x-2} \cdot \frac{x^2-4x-5}{x^2+x-30}$$

$$= \frac{(x-2)(x+2)}{(x-3)(x+2)} \cdot \frac{(x+6)(x+3)}{(x-2)(x+1)} \cdot \frac{(x-5)(x+1)}{(x+6)(x-5)} \qquad \text{Factor.}$$

$$= \frac{\cancel{(x-2)}\cancel{(x+2)}}{(x-3)\cancel{(x+2)}} \cdot \frac{\cancel{(x+6)}(x+3)}{\cancel{(x-2)}\cancel{(x+1)}} \cdot \frac{\cancel{(x-5)}\cancel{(x+1)}}{\cancel{(x+6)}\cancel{(x-5)}} \qquad \text{Cancel.}$$

$$= \frac{x+3}{x-3}$$ ▌▌

EXAMPLE 5　Multiply:

$$\frac{x^2-4x-12}{x^2+4x+4} \cdot \frac{x^2-4}{6-x} = \frac{(x-6)(x+2)}{(x+2)(x+2)} \cdot \frac{(x-2)(x+2)}{(-1)(x-6)} \qquad \begin{array}{l}\text{Factor and write}\\ 6-x \text{ as } (-1)(x-6).\end{array}$$

$$= \frac{\cancel{(x-6)}\cancel{(x+2)}}{\cancel{(x+2)}\cancel{(x+2)}} \cdot \frac{(x-2)\cancel{(x+2)}}{(-1)\cancel{(x-6)}} \qquad \begin{array}{l}\text{Cancel but don't}\\ \text{forget } -1 \text{ in the}\\ \text{denominator.}\end{array}$$

$$= \frac{x-2}{-1}$$

$$= -x+2 \quad \text{or} \quad 2-x \qquad \begin{array}{l}\text{Either form is}\\ \text{acceptable.}\end{array}$$ ▌▌

Division of Algebraic Fractions

• • • • •

In arithmetic we divide fractions by inverting the second fraction (the *divisor*) and multiplying. You may have learned an equivalent rule, "multiply by the reciprocal of the second fraction," which accomplishes the same result.

$$\frac{3}{4} \div \frac{5}{8} = \frac{3}{\underset{1}{\cancel{4}}} \cdot \frac{\overset{2}{\cancel{8}}}{5} = \frac{6}{5}$$

The same procedure applies to division of **algebraic** fractions.

To divide fractions:

Invert the second fraction (the divisor) and multiply.

$$\frac{a}{b} \div \frac{c}{d} = \frac{a}{b} \cdot \frac{d}{c}$$

E X A M P L E 6   Divide:

$$\frac{3x + 6}{x^2} \div \frac{x^2 - 4}{x^3} = \frac{3x + 6}{x^2} \cdot \frac{x^3}{x^2 - 4} \qquad \text{Invert the divisor and multiply.}$$

$$= \frac{3(x + 2)}{x^2} \cdot \frac{x^3}{(x - 2)(x + 2)} \qquad \text{Factor.}$$

$$= \frac{3\cancel{(x + 2)}}{\cancel{x^2}} \cdot \frac{\overset{x}{\cancel{x^3}}}{(x - 2)\cancel{(x + 2)}} \qquad \text{Cancel.}$$

$$= \frac{3x}{x - 2} \qquad\qquad\qquad\qquad\qquad ▮▮$$

E X A M P L E 7   Divide:

$$\frac{2x^2 - 5x - 3}{x^2 - 2x - 3} \div 2x^2 - 9x - 5 \qquad 2x^2 - 9x - 5 \text{ can}$$
$$\text{be written as } \frac{2x^2 - 9x - 5}{1}.$$

$$= \frac{2x^2 - 5x - 3}{x^2 - 2x - 3} \cdot \frac{1}{2x^2 - 9x - 5} \qquad \text{Invert and multiply.}$$

$$= \frac{(2x + 1)(x - 3)}{(x - 3)(x + 1)} \cdot \frac{1}{(2x + 1)(x - 5)} \qquad \text{Factor.}$$

$$= \frac{\cancel{(2x + 1)}\cancel{(x - 3)}}{\cancel{(x - 3)}(x + 1)} \cdot \frac{1}{\cancel{(2x + 1)}(x - 5)} \qquad \text{Cancel.}$$

$$= \frac{1}{(x + 1)(x - 5)} \qquad\qquad\qquad ▮▮$$

E X A M P L E   8   Divide:

$$\frac{y^2 - x^2}{6xy - 6y^2} \div \frac{x - y}{3x - 3y}$$

$$= \frac{y^2 - x^2}{6xy - 6y^2} \cdot \frac{3x - 3y}{x - y} \qquad \text{Invert and multiply.}$$

$$= \frac{(y - x)(y + x)}{6y(x - y)} \cdot \frac{3(x - y)}{(x - y)} \qquad \text{Factor.}$$

$$= \frac{(-1)\cancel{(x - y)}(y + x)}{\underset{2}{\cancel{6y}\cancel{(x - y)}}} \cdot \frac{\overset{1}{\cancel{3}\cancel{(x - y)}}}{\cancel{(x - y)}} \qquad \begin{array}{l}\text{Write } y - x \text{ as } (-1)(x - y)\\ \text{in the numerator and cancel.}\end{array}$$

$$= \frac{-(y + x)}{2y} \qquad \text{Don't forget the } -1.$$

$$= \frac{-(x + y)}{2y} \qquad \begin{array}{l}\text{Variables are usually written}\\ \text{in alphabetical order.}\end{array} \qquad \blacksquare\blacksquare$$

4.3   Exercises

▮ ▮ ▮ ▮ ▮

▼ Multiply or divide as indicated.

1. $\dfrac{9}{16} \cdot \dfrac{2}{3}$

2. $\dfrac{32x^3}{24y} \cdot \dfrac{10y^4}{12x^5}$

3. $\dfrac{12x^2y^3}{21x} \div \dfrac{4xy^2}{7x^3}$

4. $\dfrac{6a}{8b} \div \dfrac{3ab}{4a^3b}$

5. $\dfrac{14x^2y}{7xy^3} \div \dfrac{2xy^2}{y^4}$

6. $\dfrac{36x^3y^4z}{18xyz^4} \cdot \dfrac{4x}{15z^2}$

7. $\dfrac{b^2}{a - b} \cdot \dfrac{a^2 - b^2}{3ab + 3b^2}$

8. $\dfrac{y - 4}{y + 4} \div \dfrac{3y^2 - 12y}{y^2 + 7y + 12}$

9. $\dfrac{4x + 32}{2x + 10} \div \dfrac{x^2 + 3x - 40}{x^2 - 25}$

10. $\dfrac{4n - 12}{4n} \cdot \dfrac{2n^2}{16}$

11. $\dfrac{a^3}{a^2b - ab} \div \dfrac{a^2}{ab}$

12. $\dfrac{(x + 1)^2}{x^2 - 64} \cdot \dfrac{x^2 - 6x - 16}{x^2 + 3x + 2}$

13. $\dfrac{x^2 - 9}{x^4 - 5x^3 + 6x^2} \cdot \dfrac{16x^2}{8x + 24}$

20. $\dfrac{4x^2 + 7x + 3}{4x + 3} \div (x^2 - 1)$

14. $\dfrac{x^2 + 2x + 4}{x^2 + 4x - 5} \cdot \dfrac{x^2 + 3x - 10}{x^3 - 8}$

21. $\dfrac{(x - 5)^2}{(x - 5)^3} \div \dfrac{x^2 - x - 6}{x^2 - 3x - 10}$

15. $\dfrac{x^2 - 2x - 63}{x^2 - 4x - 45} \div \dfrac{2x^2 - 5x - 12}{2x^2 + 13x + 15}$

22. $(3 - x) \div \dfrac{x^2 - x - 6}{x^2 - 5x - 14}$

16. $\dfrac{x^2 - 1}{x^2 + 4x + 3} \div \dfrac{x(x - 3) + (x - 3)}{x^2 - 9}$

23. $\dfrac{x^2 + 3x - 10}{x^2 - x - 2} \cdot \dfrac{x^2 - 10x + 9}{x^2 - 4x - 45} \cdot \dfrac{x^2 + 2x + 1}{1 - x^2}$

17. $\dfrac{9x^2 - 1}{6x^2 + 17x + 5} \div (1 - 3x)$

24. $\dfrac{x^3 + x^2 - x - 1}{x^2 - x - 20} \cdot \dfrac{x^2 + 10x + 24}{x^2 - 1} \div \dfrac{x^2 + 7x + 6}{x^2 - 5x}$

18. $\dfrac{x - y}{2x - y} \cdot \dfrac{2x^2 + xy - y^2}{2y^2 - 3xy + x^2}$

25. $\dfrac{6x^2 + x - 1}{x^2 + 5x + 6} \cdot \dfrac{x^3 - 2x^2 - 8x}{4x^2 - 1} \cdot \dfrac{2x^2 - 7x + 3}{3x^3 - 13x^2 + 4x}$

19. $\dfrac{x^2 - 7x + 12}{x^2 + 6x - 7} \cdot \dfrac{x^2 - 2x - 8}{x^2 - x - 6} \cdot \dfrac{x^2 + 4x - 21}{x^2 - 8x + 16}$

26. When dividing fractions, what happens if you invert the first fraction instead of the second? Try an example and see.

4.4 Adding and Subtracting Algebraic Fractions

The same basic procedure is followed for adding or subtracting algebraic fractions as is used for adding or subtracting fractions in arithmetic.

Like fractions are those that have the same denominators. We can add only like fractions.

To add or subtract like fractions:

1. Add or subtract the numerators.

2. Write this sum or difference over the common denominator.

3. Reduce the resulting fraction to lowest terms (when possible).

EXAMPLE 1 Add or subtract the following fractions.

(a) $\dfrac{2}{7} + \dfrac{3}{7} = \dfrac{2+3}{7} = \dfrac{5}{7}$ Add the numerators. Keep the same denominator.

(b) $\dfrac{4}{9} - \dfrac{1}{9} = \dfrac{3}{9} = \dfrac{1}{3}$ Subtract the numerators. Keep the same denominator.

(c) $\dfrac{3}{x^2} + \dfrac{7}{x^2} = \dfrac{10}{x^2}$

(d) $\dfrac{4}{x-1} + \dfrac{3}{x-1} = \dfrac{7}{x-1}$

(e) $\dfrac{6a}{2a-b} - \dfrac{3b}{2a-b} = \dfrac{6a-3b}{2a-b}$

$\qquad = \dfrac{3\cancel{(2a-b)}}{\cancel{(2a-b)}}$ Factor 3 out of the numerator and cancel.

$\qquad = 3$

(f) $\dfrac{4}{x-2} - \dfrac{2x}{x-2} = \dfrac{4-2x}{x-2}$ Subtract the numerators.

$\qquad = \dfrac{2(2-x)}{x-2}$ Factor 2 out of the numerator.

$\qquad = \dfrac{-2\cancel{(x-2)}}{\cancel{(x-2)}}$ Factor −1 out of the numerator and cancel.

$\qquad = -2$

(g) $\dfrac{x + 8}{(x + 2)(x - 2)} \;\boxed{-}\; \dfrac{x + 5}{(x + 2)(x - 2)}$

The entire numerator is being subtracted here.

$= \dfrac{x + 8 \;\boxed{-}\; (x + 5)}{(x + 2)(x - 2)}$

This affects the sign of each term in the numerator being subtracted. This is a common place for an error.

$= \dfrac{x + 8 \;\boxed{-}\; x \;\boxed{-}\; 5}{(x + 2)(x - 2)}$

$= \dfrac{3}{(x + 2)(x - 2)}$

(h) $\dfrac{x^2 - 2x + 1}{(x - 1)(x + 5)} - \dfrac{x^2 - 3x + 2}{(x - 1)(x + 5)}$

$= \dfrac{x^2 - 2x + 1 - (x^2 - 3x + 2)}{(x - 1)(x + 5)}$

Subtract the numerators.

$= \dfrac{x^2 - 2x + 1 - x^2 + 3x - 2}{(x - 1)(x + 5)}$

Change each sign in the numerator being subtracted.

$= \dfrac{x - 1}{(x - 1)(x + 5)}$

Combine like terms.

$= \dfrac{\cancel{(x - 1)}}{\cancel{(x - 1)}(x + 5)}$

Simplify the result.

$= \dfrac{1}{x + 5}$ ▮▮

The Least Common Denominator

• • • • •

Since we can add or subtract fractions only when their denominators are the same, what should we do if we are asked to add or subtract fractions with different denominators (*unlike* fractions)? In this case we find a **least common denominator (LCD)** and rewrite each fraction as an equivalent fraction with the LCD as its denominator.

To find the LCD of a set of fractions:

1. Factor each denominator.

2. Write down each factor the greatest number of times it appears in any one factorization.

3. The LCD is the product of the factors written in Step 2.

EXAMPLE 2 Find the LCD for $\dfrac{1}{24}$ and $\dfrac{5}{36}$.

SOLUTION 1. We factor each denominator:

$$24 = \boxed{2 \cdot 2 \cdot 2} \cdot 3$$
$$36 = 2 \cdot 2 \cdot \boxed{3 \cdot 3}$$

2. In the first factorization, 2 occurs three times, and in the second factorization, 3 occurs twice. This is the *greatest* number of times that they occur in any one denominator.

3. LCD = $\boxed{2 \cdot 2 \cdot 2} \cdot \boxed{3 \cdot 3}$ = 72   ▮▮

EXAMPLE 3 Find the LCD for $\dfrac{2}{9x^2y}$, $\dfrac{7}{18xy}$, and $\dfrac{1}{12xy^2}$.

SOLUTION
$$9x^2y = \boxed{3 \cdot 3} \cdot \boxed{x^2} \cdot y$$
$$18xy = 2 \cdot 3 \cdot 3 \cdot x \cdot y \qquad\qquad \text{Factor each denominator.}$$
$$12xy^2 = \boxed{2 \cdot 2} \cdot 3 \cdot x \cdot \boxed{y^2}$$

Notice that the greatest number of times that a *variable* occurs will always be the highest power of that variable.

$$\text{LCD} = 2 \cdot 2 \cdot 3 \cdot 3 \cdot x^2 \cdot y^2 = 36x^2y^2 \qquad ▮▮$$

EXAMPLE 4 Find the LCD for $\dfrac{x}{x^2 - 1}$ and $\dfrac{5}{x^2 + 3x + 2}$.

SOLUTION
$$x^2 - 1 = \boxed{(x - 1)(x + 1)}$$
$$x^2 + 3x + 2 = \boxed{(x + 2)} \boxed{(x + 1)}$$

Each of the factors $x - 1$, $x + 1$, and $x + 2$ occurs only once in each denominator. Therefore

$$\text{LCD} = (x - 1)(x + 1)(x + 2) \qquad ▮▮$$

EXAMPLE 5 Find the LCD for $\dfrac{a}{a + 2}$, $\dfrac{b}{a^2 - a - 6}$, and $\dfrac{c}{a^2 - 6a + 9}$.

SOLUTION $\boxed{a + 2}$ Cannot be factored further.
$$a^2 - a - 6 = (a + 2) \cdot (a - 3)$$
$$a^2 - 6a + 9 = \boxed{(a - 3) \cdot (a - 3)} \qquad \text{The factor } a - 3 \text{ occurs twice here.}$$
$$\text{LCD} = (a + 2)(a - 3)(a - 3) \qquad ▮▮$$

Changing to Equivalent Fractions

· · · · ·

In order to add or subtract fractions with different denominators, we must first express each fraction as an equivalent fraction using the LCD as the new denominator.

> **To change a fraction to an equivalent fraction with the LCD as the new denominator:**
>
> 1. Divide the LCD by the denominator of the original fraction.
>
> 2. Multiply both the numerator and the denominator of the original fraction by the quotient obtained in Step 1.

EXAMPLE 6 $\dfrac{3}{4} = \dfrac{?}{12}$

SOLUTION   1. We divide 12 by 4, which gives us $\dfrac{12}{4} = 3$.

2. Now we multiply numerator and denominator of the original fraction, $\dfrac{3}{4}$, by 3:

$$\frac{3}{4} \cdot \frac{3}{3} = \frac{9}{12} \qquad \text{so} \qquad \frac{3}{4} = \frac{9}{12}$$

Notice that we multiplied the original fraction by $\dfrac{3}{3}$ or 1, so we can expect our result to be *equivalent* to the original fraction.   ∎∎

EXAMPLE 7 $\dfrac{5}{8x^2y} = \dfrac{?}{48x^4y^4}$

SOLUTION   1. $\dfrac{48x^4y^4}{8x^2y} = 6x^2y^3$ Divide the LCD by the denominator.

2. $\dfrac{5}{8x^2y} \cdot \dfrac{6x^2y^3}{6x^2y^3} = \dfrac{30x^2y^3}{48x^4y^4}$ ∎∎

EXAMPLE 8 $\dfrac{x - 5}{x^2 + 2x - 15} = \dfrac{?}{(x + 5)(x - 3)(x + 1)}$

SOLUTION   1. We factor the original denominator in order to divide:

$$x^2 + 2x - 15 = (x + 5)(x - 3)$$

$$\frac{\text{LCD}}{\text{original denominator}} = \frac{\cancel{(x + 5)}\cancel{(x - 3)}(x + 1)}{\cancel{(x + 5)}\cancel{(x - 3)}} = x + 1$$

2. $\dfrac{x - 5}{x^2 + 2x - 15} = \dfrac{(x - 5)(x + 1)}{(x + 5)(x - 3)(x + 1)}$

$\qquad\qquad\quad = \dfrac{x^2 - 4x - 5}{(x + 5)(x - 3)(x + 1)}$ ∎

EXAMPLE 9 $\dfrac{a + 6}{(a + 2)(a + 2)} = \dfrac{?}{(a + 2)(a + 2)(a - 1)}$

SOLUTION 1. $\dfrac{\cancel{(a + 2)}\cancel{(a + 2)}(a - 1)}{\cancel{(a + 2)}\cancel{(a + 2)}} = a - 1$

2. $\dfrac{a + 6}{(a + 2)(a + 2)} = \dfrac{(a + 6)(a - 1)}{(a + 2)(a + 2)(a - 1)}$

$\qquad\qquad\qquad = \dfrac{a^2 + 5a - 6}{(a + 2)(a + 2)(a - 1)}$ ∎

Adding and Subtracting Unlike Fractions

• • • • •

We have now learned everything that is needed to add or subtract fractions with different denominators.

To add or subtract unlike fractions:

1. Find the LCD for all the given fractions.

2. Change all the given fractions to equivalent fractions with the LCD as their denominators.

3. Add or subtract the like fractions as before.

4. Reduce the result to its lowest terms.

EXAMPLE 10 Add: $\dfrac{2}{x - 1} + \dfrac{5}{x + 2}$.

SOLUTIONS 1. LCD $= (x - 1)(x + 2)$

2. $\dfrac{2}{x - 1} + \dfrac{5}{x + 2} = \dfrac{2(x + 2)}{(x - 1)(x + 2)} + \dfrac{5(x - 1)}{(x + 2)(x - 1)}$ Form equivalent fractions.

3. $\qquad\qquad\quad = \dfrac{2(x + 2) + 5(x - 1)}{(x - 1)(x + 2)}$ Add the numerators.

$\qquad\qquad\quad = \dfrac{2x + 4 + 5x - 5}{(x - 1)(x + 2)}$ Simplify and combine.

$\qquad\qquad\quad = \dfrac{7x - 1}{(x - 1)(x + 2)}$ The result is in lowest terms. ∎

E X A M P L E   1 1   Subtract: $\dfrac{5}{3x^2y} - \dfrac{1}{4xy^2}$.

S O L U T I O N   **1.** LCD $= 12x^2y^2$

2. $\dfrac{12x^2y^2}{3x^2y} = 4y$ and $\dfrac{12x^2y^2}{4xy^2} = 3x$

$\dfrac{5}{3x^2y} - \dfrac{1}{4xy^2} = \dfrac{5 \cdot 4y}{3x^2y \cdot 4y} - \dfrac{1 \cdot 3x}{4xy^2 \cdot 3x}$ Form equivalent fractions.

$= \dfrac{20y}{12x^2y^2} - \dfrac{3x}{12x^2y^2}$ Subtract the numerators.

3. $= \dfrac{20y - 3x}{12x^2y^2}$ The result is in lowest terms. ▌▌

E X A M P L E   1 2   Add: $\dfrac{20}{a^2 - 4} + \dfrac{5}{a + 2}$.

S O L U T I O N   **1.** $a^2 - 4 = (a + 2)(a - 2)$, so the LCD is $(a + 2)(a - 2)$.

2. $\dfrac{20}{(a + 2)(a - 2)} + \dfrac{5}{a + 2} = \dfrac{20}{(a + 2)(a - 2)} + \dfrac{5\,(a - 2)}{(a + 2)\,(a - 2)}$

3. $= \dfrac{20 + 5a - 10}{(a + 2)(a - 2)} = \dfrac{5a + 10}{(a + 2)(a - 2)}$

4. $= \dfrac{5\cancel{(a + 2)}}{\cancel{(a + 2)}(a - 2)} = \dfrac{5}{a - 2}$ Factor the numerator and reduce to lowest terms. ▌▌

STOP   It is beneficial not to multiply the factors in the denominator of your result together, even though it would not be incorrect to do so. This is because we are always looking for ways to reduce our fractions. Keep the denominator (LCD) in factored form, then factor the numerator, if possible, to see if any of the factors in the numerator and denominator can be divided out.

This was illustrated in Steps 3 and 4 in Example 12.

E X A M P L E   1 3   Subtract: $\dfrac{x - 2}{x^2 + 6x + 9} - \dfrac{x + 4}{x^2 - 9}$.

S O L U T I O N   **1.** Since $x^2 + 6x + 9 = (x + 3)(x + 3)$ and $x^2 - 9 = (x - 3)(x + 3)$, the LCD is $(x + 3)(x + 3)(x - 3)$.

2. $\dfrac{x - 2}{(x + 3)(x + 3)} - \dfrac{x + 4}{(x + 3)(x - 3)}$

$= \dfrac{(x - 2)\ (x - 3)}{(x + 3)(x + 3)\ (x - 3)} - \dfrac{(x + 4)\ (x + 3)}{(x + 3)(x - 3)\ (x + 3)}$

$= \dfrac{x^2 - 5x + 6}{(x + 3)(x + 3)(x - 3)} - \dfrac{x^2 + 7x + 12}{(x + 3)(x + 3)(x - 3)}$

The entire numerator is being subtracted, which affect the sign of each term in it.

3. $= \dfrac{x^2 - 5x + 6 - x^2 - 7x - 12}{(x + 3)(x + 3)(x - 3)}$

4. $= \dfrac{-12x - 6}{(x + 3)(x + 3)(x - 3)} = \dfrac{-6(2x - 1)}{(x + 3)(x + 3)(x - 3)}$

In Step 4 we factored the numerator to see if the fraction could be reduced. It is acceptable to leave your answer in either multiplied or factored form.

4.4   Exercises

▼ Add or subtract the fractions as indicated and reduce them to lowest terms.

1. $\dfrac{5}{12} + \dfrac{1}{12}$

2. $\dfrac{a}{6} - \dfrac{4a}{6}$

3. $\dfrac{2}{3x} + \dfrac{4}{3x}$

4. $\dfrac{4}{a + b} - \dfrac{3}{a + b}$

5. $\dfrac{6}{5x} + \dfrac{7}{5x} - \dfrac{3}{5x}$

6. $\dfrac{3y}{y - 4} - \dfrac{12}{y - 4}$

7. $\dfrac{12a}{3a - b} - \dfrac{4b}{3a - b}$

8. $\dfrac{x + 1}{6x^2 + 3} - \dfrac{x + 4}{6x^2 + 3}$

9. $\dfrac{a}{x} + \dfrac{b}{x} + \dfrac{c}{x}$

10. $\dfrac{4}{z - 1} - \dfrac{4z}{z - 1}$

11. $\dfrac{3m}{m - 2n} - \dfrac{2n}{m - 2n} - \dfrac{2m}{m - 2n}$

12. $\dfrac{x + 2}{2x^2 + 4x} + \dfrac{3x - 5}{2x^2 + 4x} - \dfrac{6x - 1}{2x^2 + 4x}$

▼ Find the LCD for each group of fractions.

13. $\dfrac{5}{8}, \dfrac{1}{12}, \dfrac{1}{10}$

14. $\dfrac{1}{x^2 y}, \dfrac{3}{x^3 y}$

15. $\dfrac{1}{6xy^3}, \dfrac{5}{8x^2}$

16. $\dfrac{3}{a}, \dfrac{4}{a + 1}$

17. $\dfrac{6}{x}, \dfrac{-2}{x-4}$

18. $\dfrac{x}{3a^2bc}, \dfrac{y}{12a^3b}, \dfrac{z}{15ab^4c^2}$

19. $\dfrac{5}{x+1}, \dfrac{3x}{x^2+2x+1}$

20. $\dfrac{-4}{x^2-2x}, \dfrac{-7}{x^2-2x+1}$

21. $\dfrac{3x}{7x+1}, \dfrac{4}{7x}, \dfrac{-5}{7x-1}$

22. $\dfrac{2x-1}{x^2-x-12}, \dfrac{x-5}{x^2+6x+9}$

23. $\dfrac{a-1}{a^2-4}, \dfrac{-3}{a^2+4a+4}, \dfrac{a^2+1}{a^2+2a}$

24. $\dfrac{6n}{6n+18}, \dfrac{1}{n}, \dfrac{3n+1}{n^2+6n+9}$

✎ **25.** Under what circumstances is the LCD equal to the product of the denominators of the original fractions?

▼ For each of the following, find a new fraction that is equivalent to the fraction on the left by finding the missing numerator.

26. $\dfrac{1}{3} = \dfrac{?}{12}$ **27.** $\dfrac{2}{5} = \dfrac{?}{15}$

28. $\dfrac{3}{4x^3y} = \dfrac{?}{16x^5y^2}$ **29.** $\dfrac{5}{6xy^2} = \dfrac{?}{30x^3y^5}$

30. $\dfrac{2}{5ab} = \dfrac{?}{15a^2bc}$ **31.** $\dfrac{7}{4ab^2} = \dfrac{?}{28ab^3c^2}$

32. $\dfrac{3}{x+2} = \dfrac{?}{(x+2)(x-5)}$

33. $\dfrac{4}{x+3} = \dfrac{?}{(x+3)(x-4)}$

34. $\dfrac{x-3}{x-4} = \dfrac{?}{(x-4)(x+1)}$

35. $\dfrac{x-2}{x-7} = \dfrac{?}{(x-7)(x+2)}$

36. $\dfrac{2}{x+3} = \dfrac{?}{(x+1)(x+2)(x+3)}$

37. $\dfrac{4}{x-2} = \dfrac{?}{(x-4)(x-2)(x+1)}$

47. $\dfrac{a}{bc} + \dfrac{b}{ac} + \dfrac{2c}{ab}$

48. $4x - \dfrac{2}{x}$

38. $\dfrac{x+4}{(x-2)(x+3)} = \dfrac{?}{(x-2)(x+3)(x-1)}$

49. $\dfrac{3}{x} + \dfrac{x}{x+3}$

50. $\dfrac{y}{y-6} - \dfrac{6}{y}$

39. $\dfrac{x-4}{(x+1)(x+5)} = \dfrac{?}{(x+1)(x+5)(x-4)}$

51. $\dfrac{x+y}{y} + \dfrac{y}{x-y}$

52. $\dfrac{2}{x} + \dfrac{1}{x+1}$

40. $\dfrac{x-1}{(x+1)(x+1)} = \dfrac{?}{(x+1)(x+2)(x+1)}$

53. $\dfrac{2a}{5a+10} - \dfrac{3a}{15a+30}$

54. $\dfrac{1}{x+2} + \dfrac{5}{x^2-x-6}$

▼ Add or subtract as indicated. Reduce the answer to lowest terms.

41. $\dfrac{1}{3} + \dfrac{2}{9} + \dfrac{5}{12}$

42. $\dfrac{3}{x} + \dfrac{4}{x^2}$

55. $\dfrac{3}{x+2} - \dfrac{5}{x-5}$

56. $\dfrac{7}{x-5} - \dfrac{3}{5-x}$

43. $\dfrac{2}{x^2 y} - \dfrac{1}{xy^2}$

44. $\dfrac{7}{6x^2 y^2 z} - \dfrac{1}{xy^3 z}$

57. $\dfrac{2y}{5y+10} + \dfrac{4}{y^2-3y-10}$

45. $\dfrac{5}{6y^2} + \dfrac{9}{4y}$

46. $\dfrac{2}{3x} + \dfrac{4}{2x} - \dfrac{5}{9x^2}$

58. $\dfrac{5}{x^2+6x+9} - \dfrac{4}{x^2-9}$

59. $\dfrac{6}{x-1} - \dfrac{4}{x-2} + \dfrac{4}{x^2 - 3x + 2}$

60. $\dfrac{b}{a^2 - ab} - \dfrac{a}{ab - b^2}$

61. $\dfrac{x+2}{3-x} + \dfrac{x-1}{x^2 - 9}$

62. $\dfrac{3x-1}{2x^2 + x - 3} - \dfrac{x+1}{x-1}$

63. $\dfrac{3}{x+1} + \dfrac{2}{x^2 - 1} - \dfrac{1}{x^2 + x - 2}$

64. $\dfrac{x-3}{x+5} - \dfrac{x-5}{x+3}$

65. $\dfrac{2x+3}{2x-6} + \dfrac{x-3}{x^2 - 9}$

66. $\dfrac{4}{x^2 + 2x - 3} + \dfrac{2x}{x^2 + 5x + 6}$

▼ More than one operation may occur in a problem. Use the correct order of operations and the meaning of parentheses in the following:

67. $\dfrac{2}{a} + \dfrac{2}{b} \cdot \dfrac{3b}{a}$ **68.** $\left(\dfrac{2}{a} + \dfrac{2}{b}\right) \cdot \dfrac{3b}{a+b}$

69. $\dfrac{3}{x^2 - 2x - 8} \div \dfrac{1}{x-4} - \dfrac{1}{x+2}$

70. $\dfrac{3}{x^2 - 2x - 8} \div \left(\dfrac{1}{x-4} - \dfrac{1}{x+2}\right)$

71. $\dfrac{2x}{x^2 - 4} + \dfrac{2x-6}{x^2 - 4x + 3} \cdot \dfrac{x^2 - 5x + 4}{x^2 - 2x - 8} - \dfrac{x-3}{x^2 + 3x - 10} \div \dfrac{x-3}{3x+15}$

72. Does $\dfrac{1}{a} + \dfrac{1}{b} = \dfrac{1}{a+b}$? Explain why or why not.

4.5   Complex Fractions

▌ ▌ ▌ ▌ ▌

A fraction with a numerator or denominator (or both) that contains a fraction is called a **complex fraction.** The following are all examples of complex fractions:

$$\frac{\dfrac{2}{3} - \dfrac{1}{2}}{\dfrac{1}{8}} \qquad \frac{x - y}{1 - \dfrac{x}{y}} \qquad \frac{\dfrac{a}{b} + \dfrac{b}{a}}{\dfrac{b}{a} - \dfrac{a}{b}}$$

A fraction such that neither the numerator nor the denominator contains a fraction is called a **simple fraction.** Examples of simple fractions are the following:

$$\frac{3}{4} \qquad \frac{x - 5}{x} \qquad \frac{-17}{x^2 + 2x - 1} \qquad \frac{x - 2}{x^2 + 6x + 5}$$

To simplify a complex fraction, we must change it to a simple fraction and then reduce it if possible. There are two methods for accomplishing this task, both of which we will investigate. We will find that in some cases, Method I will be easier to use, while in other problems, Method II is superior.

To simplify complex fractions:

METHOD I

1. Find the LCD of all of the fractions in the problem.

2. Multiply both the numerator and the denominator by this LCD.

3. Simplify the result.

METHOD II

1. Add or subtract, wherever necessary, to obtain a single fraction in both the numerator and the denominator.

2. Divide the simplified fraction in the numerator by the simplified fraction in the denominator.

E X A M P L E   1   Simplify the following complex fraction:

$$\frac{\dfrac{2}{3} - \dfrac{1}{4}}{1 + \dfrac{1}{6}}$$

S O L U T I O N   Method I

1. The LCD of all the fractions is 12.

2. Multiply the entire numerator and the entire denominator by 12, the LCD.

$$\frac{12 \cdot \left(\dfrac{2}{3} - \dfrac{1}{4}\right)}{12 \cdot \left(1 + \dfrac{1}{6}\right)} = \frac{\overset{4}{\cancel{12}} \cdot \dfrac{2}{\cancel{3}} - \overset{3}{\cancel{12}} \cdot \dfrac{1}{\cancel{4}}}{12 \cdot 1 + \overset{2}{\cancel{12}} \cdot \dfrac{1}{\cancel{6}}} = \frac{8 - 3}{12 + 2} = \frac{5}{14}$$

Method II

1. Obtain a single fraction in both the numerator and the denominator.

- Numerator: $\dfrac{2}{3} - \dfrac{1}{4} = \dfrac{8}{12} - \dfrac{3}{12} = \dfrac{5}{12}$

- Denominator: $1 + \dfrac{1}{6} = 1\dfrac{1}{6} = \dfrac{7}{6}$

2. Divide the simplified numerator by the simplified denominator.

$$\frac{\dfrac{2}{3} - \dfrac{1}{4}}{1 + \dfrac{1}{6}} = \frac{\dfrac{5}{12}}{\dfrac{7}{6}} = \frac{5}{12} \div \frac{7}{6} = \frac{5}{\underset{2}{\cancel{12}}} \cdot \frac{\overset{1}{\cancel{6}}}{7} = \frac{5}{14}$$ ▮ ▮

E X A M P L E   2   Simplify:

$$\frac{\dfrac{3}{x} - \dfrac{4}{x^2}}{\dfrac{3}{x} + 2}$$

S O L U T I O N   ·Method I

The LCD is x^2.

$$\frac{x^2 \cdot \left(\dfrac{3}{x} - \dfrac{4}{x^2}\right)}{x^2 \cdot \left(\dfrac{3}{x} + 2\right)} = \frac{\overset{x}{\cancel{x^2}} \cdot \dfrac{3}{\cancel{x}} - \cancel{x^2} \cdot \dfrac{4}{\cancel{x^2}}}{\overset{x}{\cancel{x^2}} \cdot \dfrac{3}{\cancel{x}} + x^2 \cdot 2} = \frac{3x - 4}{3x + 2x^2} \quad \text{or} \quad \frac{3x - 4}{x(2x + 3)}$$

Method II

$$\frac{\dfrac{3}{x} - \dfrac{4}{x^2}}{\dfrac{3}{x} + 2} = \frac{\dfrac{3 \cdot x}{x \cdot x} - \dfrac{4}{x^2}}{\dfrac{3}{x} + \dfrac{2 \cdot x}{1 \cdot x}} = \frac{\dfrac{3x - 4}{x^2}}{\dfrac{3 + 2x}{x}} = \frac{3x - 4}{x^2} \div \frac{3 + 2x}{x}$$

$$= \frac{3x - 4}{\underset{x}{\cancel{x^2}}} \cdot \frac{\cancel{x}}{3 + 2x} = \frac{3x - 4}{x(2x + 3)}$$ ▮ ▮

E X A M P L E   3   Simplify the following complex fraction:

$$\frac{\dfrac{1}{x^2} - \dfrac{1}{y^2}}{\dfrac{1}{x^2 y^2}}$$

Method I

The LCD is $x^2 y^2$.

$$\frac{x^2 y^2 \cdot \left(\dfrac{1}{x^2} - \dfrac{1}{y^2}\right)}{x^2 y^2 \cdot \dfrac{1}{x^2 y^2}} = \frac{x^2 y^2 \cdot \dfrac{1}{x^2} - \dfrac{1}{y^2} \cdot x^2 y^2}{x^2 y^2 \cdot \dfrac{1}{x^2 y^2}} = \frac{y^2 - x^2}{1} = y^2 - x^2$$

Method II

$$\frac{\dfrac{1}{x^2} - \dfrac{1}{y^2}}{\dfrac{1}{x^2 y^2}} = \frac{\dfrac{y^2}{x^2 y^2} - \dfrac{x^2}{x^2 y^2}}{\dfrac{1}{x^2 y^2}} = \frac{\dfrac{y^2 - x^2}{x^2 y^2}}{\dfrac{1}{x^2 y^2}}$$

$$= \frac{y^2 - x^2}{x^2 y^2} \cdot \frac{x^2 y^2}{1} = y^2 - x^2$$

4.5   Exercises

▼ Simplify the complex fractions.

1. $\dfrac{\dfrac{a}{b}}{\dfrac{a^3}{b^2}}$

2. $\dfrac{\dfrac{1}{x}}{\dfrac{1}{y}}$

3. $\dfrac{\dfrac{a}{b}}{a}$

4. $\dfrac{\dfrac{3}{4} + \dfrac{5}{8}}{\dfrac{1}{2} + \dfrac{3}{8}}$

5. $\dfrac{1 - \dfrac{3}{5}}{4 - \dfrac{2}{15}}$

6. $\dfrac{\dfrac{1}{8} + 2}{5 - \dfrac{3}{4}}$

7. $\dfrac{\dfrac{2x^4 y}{x^3 y}}{\dfrac{4x}{xy^2}}$

8. $\dfrac{\dfrac{12a^2}{3b}}{\dfrac{4a^3}{5b^2}}$

9. $\dfrac{\dfrac{1}{a}+\dfrac{1}{b}}{a}$

10. $\dfrac{\dfrac{1}{x}+\dfrac{1}{y}}{\dfrac{1}{x}-\dfrac{1}{y}}$

21. $\dfrac{\dfrac{2}{x}-\dfrac{1}{x^2-x}}{\dfrac{3}{x-1}}$

11. $\dfrac{\dfrac{x}{y}-2}{2-\dfrac{y}{x}}$

12. $\dfrac{\dfrac{x}{y}-\dfrac{y}{x}}{x+y}$

22. $\dfrac{\dfrac{x}{x+1}-\dfrac{2}{x}}{\dfrac{1}{x}+\dfrac{2}{x+1}}$

13. $\dfrac{\dfrac{1}{x+3}-\dfrac{1}{x}}{\dfrac{2}{x+3}}$

14. $\dfrac{\dfrac{x}{x+y}}{\dfrac{y}{x-y}}$

23. $\dfrac{\dfrac{1}{x^2-x-2}+\dfrac{1}{x+1}}{\dfrac{1}{x-2}+\dfrac{1}{x^2-x-2}}$

15. $\dfrac{\dfrac{1}{x+y}-\dfrac{1}{x-y}}{\dfrac{1}{x^2-y^2}}$

16. $\dfrac{\dfrac{1}{3a}-\dfrac{1}{4b}}{\dfrac{1}{2a}+\dfrac{1}{3b}}$

24. $\dfrac{a}{1+\dfrac{a-1}{a+1}}$

17. $\dfrac{\dfrac{1}{a}+\dfrac{2}{a^2}}{\dfrac{3}{a^2}+4}$

18. $\dfrac{\dfrac{x}{y}+3}{5-\dfrac{x^2}{y^2}}$

25. $\dfrac{x+1}{1+\dfrac{1}{x}}$

26. $\dfrac{1-a+\dfrac{2}{a^2}}{\dfrac{4}{a}-a}$

19. $\dfrac{\dfrac{x}{y}+\dfrac{y}{x}}{\dfrac{1}{xy}}$

20. $\dfrac{\dfrac{1}{4x}+\dfrac{2}{3y}}{\dfrac{3}{4y}-\dfrac{1}{3x}}$

27. $\dfrac{\dfrac{3}{x^2 - 3x + 2} - \dfrac{1}{x - 1}}{\dfrac{1}{x - 1} + 1}$

30. $\dfrac{3 - \dfrac{2}{x + 1}}{\dfrac{3}{x + 1} - 4}$

28. $\dfrac{\dfrac{x + 2}{x - 2} - \dfrac{4}{x + 2}}{1 + \dfrac{1}{x + 2}}$

31. Which of the two methods for simplifying complex fractions do you find easiest to use in most cases?

29. $\dfrac{\dfrac{x}{x + 1} - \dfrac{x^2}{x^2 - 1}}{1 + \dfrac{1}{x + 1}}$

4.6   Fractional Equations

In an earlier section we found that we could clear the fractions from an equation by multiplying both sides of the equation by the LCD of all the fractions in the equation. Now we are going to solve more complicated equations by using essentially the same procedure.

To solve an equation containing fractions:

1. Clear the fractions by multiplying each term on both sides of the equation by the LCD of all the fractions.

2. Solve the resulting equation.

3. Check the solution in the original equation.

E X A M P L E   1   Solve for x: $\dfrac{x + 2}{2} + \dfrac{x + 3}{4} = 4.$

S O L U T I O N   **1.** LCD = 4 by inspection.

2. Multiply both sides of the equation by 4, the LCD.

$$\frac{x + 2}{2} + \frac{x + 3}{4} = 4$$

$$\overset{2}{\cancel{4}} \cdot \left(\frac{x + 2}{\cancel{2}} \right) + \overset{1}{\cancel{4}} \cdot \left(\frac{x + 3}{\cancel{4}} \right) = \boxed{4} \cdot 4$$

$$2(x + 2) + 1(x + 3) = 16$$

$$2x + 4 + x + 3 = 16$$

$$3x + 7 = 16$$

$$3x = 9$$

$$x = 3$$

3. Check $\dfrac{3 + 2}{2} + \dfrac{3 + 3}{4} \overset{?}{=} 4$ Substitute 3 for x.

$$\frac{5}{2} + \frac{6}{4} \overset{?}{=} 4$$

$$\frac{10}{4} + \frac{6}{4} = \frac{16}{4} = 4 \quad \boxed{✓}$$

An equation that contains a variable in the denominator must be checked for excluded values. There can be *apparent solutions* that yield zero in a denominator even though no error was made in the procedure. These are called **extraneous solutions** and they do not satisfy the equation. This is illustrated in the next example.

E X A M P L E   2   Solve for x: $\dfrac{1}{x + 1} = \dfrac{-2}{x^2 - 1}$.

S O L U T I O N   **1.** LCD $= (x - 1)(x + 1)$

2. $\dfrac{1}{x + 1} = \dfrac{-2}{(x - 1)(x + 1)}$

$$\cancel{(x - 1)(x + 1)}\;\frac{1}{\cancel{(x + 1)}} = \cancel{(x - 1)(x + 1)}\;\frac{(-2)}{\cancel{(x - 1)(x + 1)}} \qquad \text{Multiply both sides by the LCD.}$$

$$x - 1 = -2$$

$$x = -1$$

3. Check   We can see by inspection that $x = -1$ is an excluded value, for if we check $x = -1$ in the original equation, we see that it cannot be a solution:

$$\frac{1}{-1 + 1} \overset{?}{=} \frac{-2}{(-1)^2 - 1} \qquad \text{or} \qquad \frac{1}{\boxed{0}} \overset{?}{=} \frac{-2}{\boxed{0}}$$

$$\uparrow \qquad\quad \uparrow$$

Zero as a denominator is not defined.

Since $x = -1$ is an excluded value, it is an extraneous solution. Therefore, the original equation has **no solution.**   ▪ ▪

STOP   The apparent solution to any equation with a variable in the denominator *must* be checked in the original equation to be certain that it does not make any denominator equal to zero.

EXAMPLE 3   Solve for y: $\dfrac{3}{y + 5} + \dfrac{2}{y - 2} = \dfrac{4}{y^2 + 3y - 10}$.

SOLUTION   1. Factoring $y^2 + 3y - 10$ gives us $(y + 5)(y - 2)$, so the LCD is $(y + 5)(y - 2)$.

2.
$$\frac{3}{y + 5} + \frac{2}{y - 2} = \frac{4}{(y + 5)(y - 2)}$$

$$\frac{\cancel{(y + 5)}(y - 2) \cdot 3}{\cancel{(y + 5)}} + \frac{(y + 5)\cancel{(y - 2)} \cdot 2}{\cancel{(y - 2)}} = \frac{\cancel{(y + 5)}\cancel{(y - 2)} \cdot 4}{\cancel{(y + 5)}\cancel{(y - 2)}}$$ Multiply by the LCD.

$$3y - 6 + 2y + 10 = 4$$ Collect like terms.

$$5y + 4 = 4$$

$$5y = 0$$

$$y = 0$$ Solve for y.

3. Check $\dfrac{3}{0 + 5} + \dfrac{2}{0 - 2} \overset{?}{=} \dfrac{4}{0^2 + 3 \cdot 0 - 10}$ Substitute 0 for y.

$$\frac{3}{5} + \frac{2}{-2} \overset{?}{=} \frac{4}{-10}$$

$$\frac{6}{10} - \frac{10}{10} \overset{?}{=} -\frac{4}{10}$$

$$-\frac{4}{10} = -\frac{4}{10} \quad ✔$$

EXAMPLE 4   Solve for x: $\dfrac{3}{x - 2} - \dfrac{4}{x + 1} = \dfrac{5}{x^2 - x - 2}$.

SOLUTION   1. Factoring $x^2 - x - 2$ yields $(x - 2)(x + 1)$, so the LCD is $(x - 2)(x + 1)$.

2.
$$\frac{3}{x - 2} - \frac{4}{x + 1} = \frac{5}{(x - 2)(x + 1)}$$

$$\frac{\cancel{(x - 2)}(x + 1) \cdot 3}{\cancel{(x - 2)}} - \frac{(x - 2)\cancel{(x + 1)} \cdot 4}{\cancel{(x + 1)}} = \frac{\cancel{(x - 2)}\cancel{(x + 1)} \cdot 5}{\cancel{(x - 2)}\cancel{(x + 1)}}$$ Multiply by the LCD.

Be careful with this negative sign.

$$3x + 3 \; - \; (4x - 8) = 5$$

The signs of both terms have been changed.

$$3x + 3 \; - \; 4x \; + \; 8 = 5$$

$$-x + 11 = 5$$

$$-x = -6$$ Solve for x.

$$x = 6$$

3. Check $\dfrac{3}{6-2} - \dfrac{4}{6+1} \stackrel{?}{=} \dfrac{5}{6^2 - 6 - 2}$ Substitute 6 for x.

$$\dfrac{3}{4} - \dfrac{4}{7} \stackrel{?}{=} \dfrac{5}{28}$$

$$\dfrac{21}{28} - \dfrac{16}{28} \stackrel{?}{=} \dfrac{5}{28}$$

$$\dfrac{5}{28} = \dfrac{5}{28} \quad ✔$$

4.6   Exercises

▼ Solve and check.

1. $\dfrac{y}{5} + \dfrac{y}{2} = 14$

2. $\dfrac{5}{6}x + \dfrac{1}{3}x = \dfrac{1}{2}$

3. $\dfrac{2}{3}a - \dfrac{2}{5} = \dfrac{2}{5}a$

4. $\dfrac{1}{x} + \dfrac{1}{2x} = 3$

5. $\dfrac{1}{b} = \dfrac{5}{2b}$

6. $\dfrac{x}{5} + \dfrac{x}{3} = 8$

7. $\dfrac{x+1}{5} - \dfrac{x-1}{2} = 1$

8. $\dfrac{a}{5} + \dfrac{a+2}{4} = \dfrac{1}{4}$

9. $\dfrac{x+2}{4} + \dfrac{x+6}{2} = 3$

10. $\dfrac{6a-3}{9} - \dfrac{2a+6}{4} = \dfrac{1}{3}$

11. $\dfrac{2}{x} = \dfrac{3}{x} - 1$

12. $\dfrac{2y-1}{3} - \dfrac{3y}{4} = \dfrac{5}{6}$

13. $\dfrac{n-3}{3n+2} = \dfrac{1}{4}$

14. $\dfrac{x-6}{x-9} = \dfrac{3}{x-9}$

15. $\dfrac{2}{x+2} + \dfrac{3}{x-2} = \dfrac{12}{x^2-4}$

16. $\dfrac{2y}{4y-4} = \dfrac{7}{8}$

17. $\dfrac{x}{x-3} = \dfrac{3}{x-3} + 4$

18. $\dfrac{4}{x-2} - \dfrac{7}{2x-3} = \dfrac{4}{2x^2-7x+6}$

25. $\dfrac{x-1}{x+3} + \dfrac{8}{x^2+x-6} = \dfrac{x+1}{x-2}$

19. $\dfrac{x+2}{x^2+3x} = \dfrac{3}{x+3} + \dfrac{1}{x}$

26. $\dfrac{2}{3x+6} + \dfrac{5}{4x+8} = \dfrac{1}{3}$

20. $\dfrac{x}{x+1} = 2 - \dfrac{1}{x+1}$

27. $\dfrac{x-1}{x-2} - \dfrac{2x+1}{x-3} = \dfrac{3-x^2}{x^2-5x+6}$

21. $\dfrac{1}{x+5} + \dfrac{1}{x-5} = \dfrac{1}{x^2-25}$

28. $\dfrac{4x-3}{2x-1} = \dfrac{6x}{3x+1}$

22. $\dfrac{4}{y-3} + \dfrac{2y}{y^2-9} = \dfrac{1}{y+3}$

29. The LCD is used in different ways to (**1**) add or subtract algebraic fractions; (**2**) simplify complex fractions; and (**3**) solve equations containing algebraic fractions. In which of these cases does the LCD remain at the end of the problem and in which cases is the LCD "used up" in solving the problem?

23. $\dfrac{x}{x+4} + \dfrac{4}{4-x} = \dfrac{x^2+16}{x^2-16}$

30. When adding algebraic fractions, why can't we just multiply through by the LCD to get rid of the fractions?

24. $\dfrac{a-1}{a} = 1 + \dfrac{1}{a-1}$

4.7   Applications Involving Algebraic Fractions

▮ ▮ ▮ ▮ ▮

Many real-life problems result in equations that contain algebraic fractions.

E X A M P L E   1   The sum of the reciprocals of two consecutive even integers is equal to 7 divided by 4 times the first integer. Find the integers.

S O L U T I O N

- Let x = the first even integer.

- Then $x + 2$ = the second even integer.

- The *reciprocal* of the *first* integer $= \dfrac{1}{x}$.

- The *reciprocal* of the *second* integer $= \dfrac{1}{x + 2}$.

- The *sum* of the reciprocals $= \dfrac{1}{x} + \dfrac{1}{x + 2}$.

The complete equation is

$$\frac{1}{x} + \frac{1}{x + 2} = \frac{7}{4x}$$

Now we solve for x using the methods of this sections.

1. By inspection, the LCD is $4x(x + 2)$.

2.
$$4x(x + 2) \cdot \frac{1}{x} + 4x(x + 2) \cdot \frac{1}{x+2} = 4x(x + 2) \cdot \frac{7}{4x}$$

$$4x + 8 + 4x = 7x + 14$$

$$8x + 8 = 7x + 14$$

$$x = 6$$

Therefore the integers are 6 and 8.

3. **Check**
$$\frac{1}{6} + \frac{1}{8} \stackrel{?}{=} \frac{7}{4 \cdot 6}$$

$$\frac{4}{25} + \frac{3}{24} \stackrel{?}{=} \frac{7}{24}$$

$$\frac{7}{24} = \frac{7}{24} \quad ✔$$

▮ ▮

E X A M P L E   2   A swimming pool takes 45 hours to fill using a garden hose. Using a larger-diameter hose, the same pool can be filled in only 30 hours. How long will it take to fill the pool if both hoses are used?

SOLUTION $45 =$ the number of hours for the garden hose to fill the pool

$\dfrac{1}{45} =$ the part of the pool filled by the garden hose in 1 hour

$30 =$ the number of hours for the larger hose to fill the pool

$\dfrac{1}{30} =$ the part of the pool filled by the larger hose in 1 hour

Let $t =$ the number of hours required to fill the pool if both hoses are being used.

Then

$\dfrac{1}{t} =$ the part of the pool filled by both hoses in 1 hour

part filled by garden hose in 1 hour	$+$	part filled by larger hose in 1 hour	$=$	part filled by both hoses in 1 hour
$\dfrac{1}{45}$	$+$	$\dfrac{1}{30}$	$=$	$\dfrac{1}{t}$

$$90t \cdot \dfrac{1}{45} + 90t \cdot \dfrac{1}{30} = 90t \cdot \dfrac{1}{t} \qquad \text{Multiply by } 90t; \text{ the LCD.}$$

$$2t + 3t = 90$$

$$5t = 90$$

$$t = 18$$

The pool can be filled in 18 hours using both hoses.   ▮▮

Problems involving more than one person doing a particular job are done in the same way as Example 2.

EXAMPLE 3   It takes Ralph 2 hours to cut up a tree using a chain saw. If his friend Charley, using a hand saw, helps him, it takes them only $1\frac{1}{2}$ hours. How long would it take Charley if he cut up the tree by himself?

SOLUTION $2 =$ the number of hours it takes Ralph to cut up the tree

$\dfrac{1}{2} =$ the part Ralph can cut in 1 hour

Let $t =$ the number of hours it takes Charley to cut up the tree.

Then

$\dfrac{1}{t} =$ the part Charley can cut in 1 hour

$1\dfrac{1}{2} = \dfrac{3}{2} =$ the number of hours it takes Ralph and Charley together

$\dfrac{1}{1\frac{1}{2}} = \dfrac{2}{3} =$ the part Ralph and Charley can cut in 1 hour

part cut by Ralph in 1 hour	+	part cut by Charley in 1 hour	=	part cut by both in 1 hour
$\dfrac{1}{2}$	+	$\dfrac{1}{t}$	=	$\dfrac{2}{3}$

$$6t \cdot \frac{1}{2} + 6t \cdot \frac{1}{t} = 6t \cdot \frac{2}{3} \qquad \text{Multiply by } 6t, \text{ the LCD.}$$

$$3t + 6 = 4t$$

$$6 = t$$

It would take Charley 6 hours to cut up the tree with a hand saw.   ▪ ▪

Many application problems involve some form of the basic relationship between **distance, rate,** and **time:**

$$d = r \cdot t$$

EXAMPLE 4   A passenger train travels 20 mph faster than a car. if the train travels 180 miles in the same time that the car travels 120 miles, what is the speed of each?

SOLUTION   Let r = the speed (rate) of the car.
Then:

$$r + 20 = \text{the speed (rate) of the train}$$

$$120 = \text{the distance the car travels}$$

$$180 = \text{the distance the train travels}$$

The statement of the problem tells us that the traveling time is the same for both the car and the train; this will give us our equation. Since $d = r \cdot t$, we have $t = d/r$.

$$\frac{120}{r} = \text{the time the car travels}$$

$$\frac{180}{r + 20} = \text{the time the train travels}$$

$$\frac{180}{r + 20} = \frac{120}{r} \qquad \text{The times are the same.}$$

$$r(r + 20) \cdot \frac{180}{r + 20} = \frac{120}{r} \cdot r(r + 20) \qquad \text{Multiply by } r(r + 20), \text{ the LCD.}$$

$$180r = 120r + 2400$$

$$60r = 2400$$

$$r = 40$$

So the speed of the car is 40 mph and the speed of the train, $t + 20$, is 60 mph.

Some students find that making a chart like the following helps them in problems of this type.

	distance	rate	time = $\dfrac{\text{distance}}{\text{rate}}$
train	180	$r + 20$	$\dfrac{180}{r + 20}$
car	120	r	$\dfrac{120}{r}$

EXAMPLE 5 A boat can travel 4 miles upstream (against the current) in the same time that it can travel 6 miles of downstream (with the current). If the current of the river is flowing at the rate of 3 mph, what is the rate of the boat in still water?

SOLUTION If we let r = rate of the boat in still water, then

- $r - 3$ = rate of the boat upstream   (Why?)

- $r + 3$ = rate of the boat downstream   (Why?)

	distance	rate	time = $\dfrac{\text{distance}}{\text{rate}}$
upstream	4	$r - 3$	$\dfrac{4}{r - 3}$
downstream	6	$r + 3$	$\dfrac{6}{r + 3}$

The time is the same upstream as it is downstream, giving us the equation:

$$\text{time (upstream)} = \text{time (downstream)}$$

or $\qquad \dfrac{d}{r}(\text{upstream}) = \dfrac{d}{r}(\text{downstream})$

$$\frac{4}{r - 3} = \frac{6}{r + 3}$$

$$(r - 3)(r + 3) \cdot \frac{4}{r - 3} = \frac{6}{r + 3} \cdot (r - 3)(r + 3) \qquad \text{LCD} = (r - 3)(r + 3)$$

$$4r + 12 = 6r - 18$$

$$-2r = -30$$

$$r = 15$$

The rate of the boat in still water is 15 mph.

4.7   Exercises

▮▮▮▮▮

▼ Solve each of the following problems.

1. One number is 8 more than a second number. If the larger number is divided by the smaller number, the result is $\frac{7}{3}$. Find the numbers.

2. One number is equal to four times another number. The sum of the reciprocals of the two numbers is $\frac{10}{3}$. Find the two numbers.

3. Find the integer that, when added to both the numerator and denominator of $\frac{7}{9}$, yields a fraction equivalent to $\frac{6}{7}$.

4. One more than two times a certain number, divided by 9 more than the number, is equal to 1. Find the number.

5. One number is 2 less than a second number. If the second number is divided by twice the first number, the result is $\frac{5}{6}$. Find the numbers.

6. Pipe A takes 21 minutes to fill a tank, while pipe B takes 28 minutes to fill the same tank. How long will it take to fill the tank if both pipes are being used at the same time?

7. Three pipes feed into a vat. Pipe A can fill the vat in 2 hours, pipe B in 3 hours, and pipe C in 4 hours. How long will it take to fill the vat if all three pipes are used?

8. A faucet can fill a sink in 10 minutes. It takes 12 minutes to drain the sink. How long does it take to fill the sink if the drain is open?

9. Fred can mow his lawn in 4 hours by himself. His son, Freddie, can do the same job in 2 hours. How long does it take them to mow the lawn if they work together?

10. It takes Ron and Don 6 hours to paint a porch working together. If Ron can do the job alone in 8 hours, how long would it take Don to do the job alone?

11. The formula $\dfrac{1}{s_1} + \dfrac{1}{s_2} = \dfrac{1}{f}$ gives the focal length, f, of a simple camera lens, where s_1 is the object distance and s_2 is the image distance. Find the focal length of the lens if the object distance is 10 in. and the image distance is 15 in.

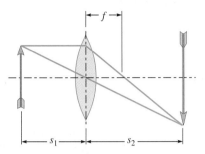

14. A man paddled his canoe upstream a distance of 5 miles and then turned around and in the same time paddled 10 miles downstream. If the man normally paddles the canoe at a rate of 3 mph in still water, what is the rate of the current of the stream?

15. A cyclist can travel 30 km in the same time that it takes a runner to run 12 km. Find the rate of each person if the rate of the cyclist is 8 km/hr faster than the runner.

12. The formula $t = \dfrac{176{,}000}{7v} + \dfrac{a}{v}$ gives the time t, in hours, for a satellite to circle the earth if its altitude above the earth's surface is a miles and it has a velocity of v miles per hour. Find the time t a satellite takes to circle the earth if $a = 700$ miles and $v = 1600$ mph.

16. An airplane flies 220 miles into the wind in the same time that it can fly 280 miles with the wind. If the speed of the wind is 30 mph, what is the *air speed* of the plane (the speed of the plane with no wind velocity)?

17. A small plane averaged 100 mph on a flight from New York to Philadelphia. The average speed on the return trip was 120 mph. If it took $2\frac{1}{2}$ hours for the round trip, how far is it from New York to Philadelphia?

13. The formula $\dfrac{1}{R} = \dfrac{1}{R_1} + \dfrac{1}{R_2}$ determines the resistance R that is equivalent to resistances R_1 and R_2 (measured in ohms) connected in parallel. Find R if $R_1 = 20$ ohms and $R_2 = 30$ ohms.

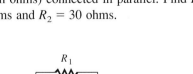

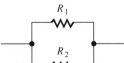

18. A cold water faucet fills a bathtub in 12 minutes, the hot water faucet fills the bathtub in 10 minutes, and if you remove the stopper, the full tub will empty in 6 minutes. How long will it take to fill the tub if both faucets are on and the stopper is removed?

SUMMARY—CHAPTER 4

•••••••• ▼

EXAMPLE

An **excluded value** of a variable is one that makes the denominator of a fraction equal to zero.

The variable x may not equal 4 in $\dfrac{3}{x-4}$, since then the denominator would equal zero.

The rule of signs for fractions: If exactly two of the three signs of a fraction are changed, the value of the fraction remains the same.

$$-\frac{2}{3-x} = \frac{+}{-} \frac{2}{(3-x)}$$

$$= \frac{2}{x-3}$$

To reduce a fraction to lowest terms:

1. Factor both the numerator and the denominator completely;

2. Divide both the numerator and the denominator by a factor that is common to both;

3. Multiply the factors that remain.

$$\frac{3x+6}{x^2-4} = \frac{3(x+2)}{(x-2)(x+2)}$$

$$= \frac{3}{x-2}$$

To multiply fractions:

1. Factor all numerators and denominators completely;

2. Divide any pair of numerators and denominators by a factor that is common to both;

3. The answer is the product of the remaining numerators divided by the product of the remaining denominators

$$\frac{x^2-3x-4}{4x+8} \cdot \frac{2x-2}{x^2-1}$$

$$= \frac{(x-4)(x+1)}{\overset{}{4} \cdot (x+2)} \cdot \frac{\overset{1}{2(x-1)}}{(x-1)(x+1)}$$

$$= \frac{x-4}{2(x+2)}$$

To divide fractions, invert the second fraction (divisor) and multiply.

$$\frac{x^2+x-6}{x^2+2x-3} \div \frac{x^2+2x-8}{x^2-2x+1}$$

$$= \frac{x^2+x-6}{x^2+2x-3} \cdot \frac{x^2-2x+1}{x^2+2x-8}$$

$$= \frac{(x+3)(x-2)}{(x-1)(x+3)} \cdot \frac{(x-1)(x-1)}{(x+4)(x-2)}$$

$$= \frac{x-1}{x+4}$$

To add or subtract *like* fractions:

1. Add or subtract the numerators;

2. Write their sum or difference over the common denominator;

3. Reduce the resulting fraction to lowest terms.

$$\frac{3}{x-2} + \frac{x+1}{x-2} = \frac{x+4}{x-2}$$

To find the LCD (*least common denominator*) of a set of fractions:

1. Factor each denominator;

2. Write each factor the greatest number of times that it appears in any one factorization.

3. The LCD is the product of the factors written in Step 2.

To change a fraction to an equivalent fraction with the LCD as the new denominator:

1. Divide the LCD by the denominator of the original fraction;

2. Multiply both the numerator and the denominator of the original fraction by the quotient obtained in Step 1.

To add or subtract *unlike* fractions:

1. Find the LCD for all the given fractions.

2. Change all the given fractions to equivalent fractions with the LCD as their denominators.

3. Add or subtract the like fractions as before.

4. Reduce the result to lowest terms.

To simplify complex fractions:

Method I

1. Find the LCD of all of the fractions in the problem;

2. Multiply both the numerator and the denominator by this LCD;

3. Simplify the result.

EXAMPLE

Find the LCD for the fractions

$$\frac{2}{x^2 - 9} \quad \text{and} \quad \frac{6}{x^2 - x - 12}$$

$$x^2 - 9 = (x - 3)(x + 3)$$

$$x^2 - x - 12 = (x + 3)(x - 4)$$

$$\text{LCD} = (x - 3)(x + 3)(x - 4)$$

$$\frac{x - 2}{(x + 3)(x + 4)} = \frac{?}{(x + 3)(x + 4)(x + 2)}$$

$$\frac{\cancel{(x + 3)}\cancel{(x + 4)}(x + 2)}{\cancel{(x + 3)}\cancel{(x + 4)}} = x + 2$$

$$\frac{(x - 2)\ (x + 2)}{(x + 3)(x + 4)\ (x + 2)}$$

$$= \frac{x^2 - 4}{(x + 3)(x + 4)(x + 2)}$$

Add: $\dfrac{3}{x - 4} + \dfrac{6}{x - 2}$.

The LCD is $(x - 4)(x - 2)$.

$$\frac{3}{x - 4} + \frac{6}{x - 2}$$

$$= \frac{3\ (x - 2)}{(x - 4)\ (x - 2)} + \frac{6\ (x - 4)}{(x - 2)\ (x - 4)}$$

$$= \frac{3x - 6}{(x - 4)(x - 2)} + \frac{6x - 24}{(x - 4)(x - 2)}$$

$$= \frac{9x - 30}{(x - 4)(x - 2)}$$

$$= \frac{3(3x - 10)}{(x - 4)(x - 2)}$$

Simplify $\dfrac{\dfrac{2}{x} + 3}{4 - \dfrac{5}{x}}$ by Method I.

The LCD is x.

$$\frac{\cancel{x} \cdot \dfrac{2}{\cancel{x}} + x \cdot 3}{x \cdot 4 - \cancel{x} \cdot \dfrac{5}{\cancel{x}}}$$

$$= \frac{2 + 3x}{4x - 5}$$

Method II

EXAMPLE

Simplify $\dfrac{\dfrac{3}{2x}}{1 + \dfrac{1}{x}}$ by Method II.

1. Add or subtract, wherever necessary, to obtain a single fraction in both the numerator and denominator.

$$\dfrac{\dfrac{3}{2x}}{\dfrac{x}{x} + \dfrac{1}{x}} = \dfrac{\dfrac{3}{2x}}{\dfrac{x+1}{x}}$$

2. Divide the simplified fraction in the numerator by the simplified fraction in the denominator.

$$= \dfrac{3}{2\cancel{x}} \cdot \dfrac{\cancel{x}}{x+1}$$

$$= \dfrac{3}{2(x+1)}$$

To solve an equation containing fractions:

Solve for x: $\quad \dfrac{5}{x-1} = \dfrac{4}{x} + \dfrac{2}{3x}.$

The LCD is $3x(x-1)$.

1. Clear of fractions by multiplying each term on both sides of the equation by the LCD of all the fractions;

$$(3x(\cancel{x-1})) \dfrac{5}{(\cancel{x-1})}$$

2. Solve the resulting equation;

$$= 3\cancel{x}(x-1) \dfrac{4}{\cancel{x}} + \cancel{3}\cancel{x}(x-1) \dfrac{2}{\cancel{3x}}$$

$$15x = 12x - 12 + 2x - 2$$

$$15x = 14x - 14$$

$$x = -14$$

3. Check the solution in the original equation.

Check

$$\dfrac{5}{-14-1} \overset{?}{=} \dfrac{4}{-14} + \dfrac{2}{3(-14)}$$

$$-\dfrac{5}{15} \overset{?}{=} -\dfrac{2}{7} - \dfrac{1}{21}$$

$$-\dfrac{1}{3} \overset{?}{=} -\dfrac{6}{21} - \dfrac{1}{21}$$

$$-\dfrac{1}{3} = -\dfrac{7}{21} \quad ✔$$

REVIEW EXERCISES—CHAPTER 4

▼ Find the value(s) of the variable (if any) for which the fraction is undefined.

1. $\dfrac{3}{x-1}$

2. $\dfrac{x-4}{3}$

3. $\dfrac{-16}{x^2-4}$

▼ Find the missing term using the rule of signs for fractions.

4. $\dfrac{5}{8} = \dfrac{?}{-8}$

5. $\dfrac{y+4}{-2} = \dfrac{?}{2}$

6. $-\dfrac{x-5}{3} = \dfrac{5-x}{?}$

▼ Reduce to lowest terms.

7. $\dfrac{12}{32}$

8. $\dfrac{-24x^3yz}{-6xy^4z}$

9. $\dfrac{3x+1}{3x-1}$

10. $\dfrac{3-4x}{4x-3}$

11. $\dfrac{2x^2+5x-3}{2x^2-7x+3}$

▼ Multiply or divide as indicated.

12. $\dfrac{5}{15} \cdot \dfrac{9}{16}$

13. $\dfrac{12x^2y}{3x^4} \div \dfrac{6xy^4}{9xy}$

14. $\dfrac{3x-12}{x} \cdot \dfrac{8x^2}{x-4}$

15. $\dfrac{2x^2+9x+4}{3x^4} \cdot \dfrac{x^3-5x^2}{2x+1}$

16. $\dfrac{x^2-5x+6}{x^2-7x+10} \div \dfrac{x^2-3x-4}{x^2-4x-5}$

17. $\dfrac{2+x-x^2}{3x^2-2x-8} \div \dfrac{x^2-4x-5}{3x^2-11x-20}$

▼ Add or subtract as indicated and reduce to lowest terms.

18. $\dfrac{5}{3x^2} + \dfrac{9}{2x^3}$

19. $\dfrac{x}{x-4} - \dfrac{2}{x+1}$

20. $\dfrac{3x}{x^2-x-2} - \dfrac{2}{x-2}$

21. $\dfrac{x+1}{x-2} + \dfrac{3}{x+3} - \dfrac{4x-3}{x^2+x-6}$

▼ Combine using the correct order of operations.

22. $\dfrac{2x + 6}{x^2 - 2x - 3} + \dfrac{x - 2}{x^2 + x} \cdot \dfrac{x^2 + 3x}{x^2 + x - 6}$

▼ Simplify the complex fractions.

23. $\dfrac{\dfrac{2}{3} + \dfrac{3}{4}}{\dfrac{5}{6} - \dfrac{1}{2}}$

24. $\dfrac{\dfrac{a}{b} - 3}{3 - \dfrac{b}{a}}$

25. $\dfrac{\dfrac{3}{x - y} + \dfrac{2}{x + y}}{\dfrac{1}{x^2 y^2}}$

▼ Solve for **x:**

26. $\dfrac{x}{9} + \dfrac{x}{12} = \dfrac{3}{4}$

27. $\dfrac{1}{4x} - \dfrac{1}{3x} = \dfrac{1}{x^2}$

28. $\dfrac{2}{y + 4} + \dfrac{5}{y - 2} = \dfrac{3y}{y^2 + 2y - 8}$

29. $\dfrac{2x}{4 - x} + \dfrac{8}{x - 4} = 1$

30. $\dfrac{2x + 1}{x + 3} - \dfrac{x - 1}{x - 4} = \dfrac{x^2 + 8}{x^2 - x - 12}$

31. The numerator of a fraction is 4 more than the denominator. When this fraction is added to $\frac{2}{3}$, the result is 3. Find the fraction.

32. A husband and wife working together can do their spring cleaning in 2 days. The wife can do the job alone in 3 days. How long would it take the husband to do the job if he were to work alone?

CRITICAL THINKING EXERCISES—CHAPTER 4

▼ There is an error in each of the following. Can you find it?

1. x may not equal 4 in the fraction $\dfrac{x-4}{9}$.

2. $\dfrac{y+4}{4} = \dfrac{\overset{1}{y+\cancel{4}}}{\underset{1}{\cancel{4}}} = y+1$

3. $\dfrac{x+3}{-2} = \dfrac{-x+3}{2}$

4. $\dfrac{6x-5}{5-6x} = 1$

5. $\dfrac{2}{3y} + \dfrac{4}{2y} = \dfrac{6}{5y}$

6. $\dfrac{a^2+b^2}{(a+b)^2} = 1$

7. $\dfrac{3x-2}{x+5} - \dfrac{x+7}{x+5} = \dfrac{3x-2-x+7}{x+5} = \dfrac{2x+5}{x+5}$

NAME _____   **I I I I I CLASS**

ACHIEVEMENT TEST—CHAPTER 4

▼ For what value of x is the fraction not defined?

1. $\dfrac{2x}{x + 1}$

2. $\dfrac{x - 2}{x^2 + 6x + 5}$

▼ Find the missing term using the rule of signs for fractions.

3. $-\dfrac{2}{3} = \dfrac{2}{?}$

4. $\dfrac{2}{2 - x} = \dfrac{?}{x - 2}$

▼ Reduce to lowest terms.

5. $\dfrac{x^2 - 1}{2x - 2}$

6. $\dfrac{x^2 - 3x - 4}{x^2 + 3x - 10}$

7. $\dfrac{x^2 - 3x + 9}{x^3 + 27}$

▼ Perform the indicated operations. Reduce to lowest terms.

8. $\dfrac{x^2}{x - y} \cdot \dfrac{x^2 - y^2}{5x^3y + x^2}$

1. _____

2. _____

3. _____

4. _____

5. _____

6. _____

7. _____

8. _____

9. $\dfrac{x^2 - 9}{x^2 + 2x - 3} \div \dfrac{x^2 + x - 12}{x^2 + 3x - 4}$

9. _____

10. $\dfrac{7x}{3x - y} - \dfrac{x + 2y}{3x - y}$

10. _____

11. $3x + \dfrac{4}{x}$

11. _____

12. $\dfrac{3}{a - 2} + \dfrac{a}{2 - a}$

12. _____

13. $\dfrac{6}{x^2 + 2x + 1} - \dfrac{4}{x^2 - 1}$

13. _____

14. $\dfrac{x - 2}{x + 1} - \dfrac{x^2 + 6x + 8}{x^2 - 3x - 4} \div \dfrac{x^2 + 8x + 12}{x^2 + 2x - 24}$

14. _____

▼ Simplify the complex fractions.

15. $\dfrac{\dfrac{1}{x - 2} - \dfrac{3}{x}}{\dfrac{4}{x - 2}}$

15. _____

16. $\dfrac{\dfrac{a^2}{a^2-1}+\dfrac{a}{a-1}}{\dfrac{3}{a+1}-\dfrac{1}{a-1}}$

16. _____

▼ Solve and check.

17. $\dfrac{2}{3}x-\dfrac{1}{4}x=\dfrac{5}{12}$

17. _____

18. $\dfrac{3}{x-4}+\dfrac{3}{x+2}=\dfrac{6}{x^2-2x-8}$

18. _____

19. $\dfrac{a}{a-2}+\dfrac{3}{2-a}=\dfrac{a^2-1}{a^2-4}$

19. _____

20. $\dfrac{2x-3}{x-1}+\dfrac{1}{x-1}=3$

20. _____

21. One number is 6 greater than a second number. If the smaller number is divided by the larger number, the result is $\frac{6}{5}$. Find the numbers.

21. _____

22. Pipe A fills a tank in 40 minutes, and pipe B fills the same tank in 60 minutes. How long will it take for both pipes together to fill the tank?

22. _____

Roots, Radicals, Rational Exponents, and Complex Numbers

INTRODUCTION

In the previous chapter we studied *rational* expressions. In this chapter we will investigate *roots* and *radicals*, which often involve *irrational* numbers. We will also extend our knowledge of number systems by introducing a new set of numbers called *complex numbers*.

CHAPTER 5—NUMBER KNOWLEDGE

Fractals

Fractals were first introduced and discussed by Benoit Mandelbroit in 1975. Applications for this new branch of mathematics are being found every day. Scientists have found that clouds follow fractal patterns, which helps to explain the origin of weather conditions. Investigations of the paths that a computer takes when it searches its memory and the way in which our neurons fire when we go searching through our own memories are both related to fractals. High-definition television (HDTV) uses fractals to put the HDTV signal into already existing broadcast signals.

What exactly are fractals? Perhaps the most well known fractal is called the "snowflake fractal," illustrated as follows: Start with a line segment that has been divided into equal thirds

Construct an equilateral triangle on the middle segment and remove the middle third of the original line.

We repeat the process until we have reached the level of complexity that is desired. A finished six-pointed snowflake fractal shown here is aptly named.

Now, for each line segment, we repeat this process.

Problem: Make your own fractal by beginning with a square, each side of which has been divided into equal thirds, and follow the step-by-step process just described, except draw squares instead of equilateral triangles.

5.1   Roots

Square Roots

> b is a **square root** of a if $b^2 = a$.

- 4 is a square root of 16 because $4^2 = 16$.
- -4 is also a square root of 16 because $(-4)^2 = 16$.

In fact, every *positive* number must have two square roots, equal in magnitude but with opposite signs.

Zero has only one square root and that is zero.

Negative numbers do not have square roots that are real numbers because if we square any real number (except zero) we get a *positive* number.

EXAMPLE 1   What are the square roots of 25?

SOLUTION   The square roots of 25 are 5 and -5 because $5^2 = 25$ and $(-5)^2 = 25$.   ❚❚

The symbol $\sqrt{}$ is called a **radical sign** and is used to name the *positive square root* of a number. This positive square root is called the **principal square root.** The number under the radical is called the **radicand.**

EXAMPLE 2   **(a)** $\overset{\text{radical sign}}{\sqrt{\underset{\text{radicand}}{9}}} = 3 \; \leftarrow$ principal square root (always positive)

(b) $-\sqrt{9} = -3$

(c) $\sqrt{49} = 7$

(d) $-\sqrt{49} = -7$

(e) $\pm\sqrt{49} = \pm 7$ Here $\pm$ means "plus or minus," so the result is $+7$ *or* -7.

(f) $\sqrt{-9}$ is not a real number.

(g) $\sqrt{0} = 0$ ❚❚

Square roots of perfect squares such as $\sqrt{a^2}$ present an interesting situation.

- If $a = 4$ we have $\sqrt{4^2} = \sqrt{16} = 4$.
- If $a = -4$ we have $\sqrt{(-4)^2} = \sqrt{16} = 4$.

As you can see, the result is positive in both cases.

If we write $\sqrt{a^2} = a$, we will be incorrect whenever a is negative. To make sure that the result is always positive, we write

$$\sqrt{a^2} = |a| \quad \text{(the absolute value of } a\text{)}$$

EXAMPLE 3   **(a)** $\sqrt{3^2} = |3| = 3$

(b) $\sqrt{(-3)^2} = |-3| = 3$

(c) $\sqrt{x^2} = |x|$ Remember, we don't know whether x represents a positive or a negative number.

(d) $\sqrt{(3x)^2} = |3x|$ or $3|x|$

(e) $\sqrt{x^2 - 2x + 1} = \sqrt{(x-1)^2} = |x - 1|$ ❚❚

Higher-Order Roots

• • • • •

The *n*th root of a number
b is the ***n*th root** of *a* if $b^n = a$.

The notation we use for the *n*th root of *a* is as follows:

index   root

$$\sqrt[n]{a} = b$$

radicand

Absence of the index number implies that we mean square root.

EXAMPLE 4   **(a)** 4 is the third root, called the **cube root** of 64 because $4^3 = 64$. We write $\sqrt[3]{64} = 4$.

(b) -4 is the cube root of -64 because $(-4)^3 = -64$. We write $\sqrt[3]{-64} = -4$.

(c) Both 2 and -2 are fourth roots of 16 because $2^4 = 16$ and $(-2)^4 = 16$. The positive, or *principal*, fourth root of 16 is written $\sqrt[4]{16} = 2$, and the negative fourth root of 16 is written $-\sqrt[4]{16} = -2$.

(d) $\sqrt[4]{-16}$ is not a real number because there is no real number that can be raised to the fourth power to produce -16.   ▮▮

EXAMPLE 5   **(a)** $\sqrt[5]{32} = 2$

(b) $\sqrt[5]{-32} = -2$

(c) $\sqrt[4]{81} = 3$

(d) $\sqrt[4]{-81}$ is not a real number.

(e) $\sqrt[3]{0} = 0$

(f) $\pm\sqrt{25} = \pm 5$ ▮▮

Following is a list of roots that you should know by memory. You probably know many of them already.

Square Roots		**Cube Roots**	**Other Roots**
$\sqrt{0} = 0$	$\sqrt{49} = 7$	$\sqrt[3]{0} = 0$	$\sqrt[n]{0} = 0$
$\sqrt{1} = 1$	$\sqrt{64} = 8$	$\sqrt[3]{1} = 1$	$\sqrt[n]{1} = 1$
$\sqrt{4} = 2$	$\sqrt{81} = 9$	$\sqrt[3]{8} = 2$	$\sqrt[4]{16} = 2$
$\sqrt{9} = 3$	$\sqrt{100} = 10$	$\sqrt[3]{27} = 3$	$\sqrt[5]{32} = 2$
$\sqrt{16} = 4$	$\sqrt{121} = 11$	$\sqrt[3]{64} = 4$	$\sqrt[4]{81} = 3$
$\sqrt{25} = 5$	$\sqrt{144} = 12$	$\sqrt[3]{125} = 5$	
$\sqrt{36} = 6$	$\sqrt{225} = 15$		

5.1   Exercises

▼ Find the square roots of the following numbers.

1. 100　　　　**2.** 36　　　　**3.** 64

4. 81　　　　**5.** 121　　　　**6.** 1

7. 144　　　　**8.** 225

▼ Evaluate the following roots. If not possible, say so.

9. $\sqrt{4}$　　　　**10.** $\sqrt{16}$

11. $\sqrt{1}$　　　　**12.** $\sqrt{36}$

13. $\sqrt{49}$　　　　**14.** $-\sqrt{49}$

15. $\sqrt{-49}$　　　　**16.** $\pm\sqrt{49}$

17. $-\sqrt{81}$　　　　**18.** $\sqrt{-144}$

19. $-\sqrt{169}$　　　　**20.** $\pm\sqrt{400}$

▼ Simplify.

21. $\sqrt{a^2}$　　　　**22.** $\sqrt{h^2}$

23. $\sqrt{9x^2}$　　　　**24.** $\sqrt{(-4)^2}$

25. $\sqrt{(-x)^2}$　　　　**26.** $\sqrt{0}$

27. $\sqrt{(x-2)^2}$　　　　**28.** $\sqrt{x^2-4x+4}$

29. $-\sqrt{x^2-4x+4}$

▼ Find the following roots. If not possible, say so.

30. $\sqrt[3]{27}$　　　　**31.** $\sqrt[3]{64}$

32. $\sqrt[3]{-27}$　　　　**33.** $\sqrt[6]{1}$

34. $\sqrt[6]{-1}$　　　　**35.** $\sqrt[3]{-8}$

36. $\sqrt[4]{81}$　　　　**37.** $\sqrt[6]{64}$

38. $\sqrt[3]{343}$　　　　**39.** $\pm\sqrt[4]{16}$

40. $\sqrt[5]{-243}$　　　　**41.** $\sqrt[4]{-256}$

42. Explain why $\sqrt{-25}$ is not a real number.

43. Why will a radical with an odd index and a real-number radicand always be a real number?

44. Find several values of x such that $\sqrt{(x-1)^2} \neq x - 1$. Can you make a general statement about these values for x?

5.2   Multiplying and Factoring Radicals

▮ ▮ ▮ ▮ ▮

Compare the following statements

1. $\sqrt{4} \cdot \sqrt{9} = 2 \cdot 3 = 6$

2. $\sqrt{4 \cdot 9} = \sqrt{36} = 6$

This suggests that $\sqrt{4} \cdot \sqrt{9} = \sqrt{4 \cdot 9}$ and, indeed, this is true in general.

$$\sqrt[n]{a} \cdot \sqrt[n]{b} = \sqrt[n]{a \cdot b}; \qquad a \ge 0, b \ge 0$$

EXAMPLE 1   Multiply the following:

(a) $\sqrt{2} \cdot \sqrt{3} = \sqrt{2 \cdot 3} = \sqrt{6}$

(b) $\sqrt{2} \cdot \sqrt{32} = \sqrt{2 \cdot 32} = \sqrt{64} = 8$

(c) $\sqrt{7} \cdot \sqrt{7} = \sqrt{7 \cdot 7} = \sqrt{49} = 7$

(d) $\sqrt[3]{2} \cdot \sqrt[3]{4} = \sqrt[3]{2 \cdot 4} = \sqrt[3]{8} = 2$

(e) $\sqrt[3]{6} \cdot \sqrt[3]{2} = \sqrt[3]{12}$ ▮▮

STOP   Notice that $\sqrt{-4} \cdot \sqrt{-9} \ne 6$ even though $\sqrt{36} = 6$, because $\sqrt{-4}$ and $\sqrt{-9}$ are not real numbers. This situation will be addressed later in this chapter.

Our rule can be used for both multiplication and factoring of radicals. When factoring, it is desirable to factor a product so that one of the factors is a perfect square. In this way the radical can be simplified.

EXAMPLE 2   Simplify $\sqrt{28}$ by factoring.

SOLUTION $\sqrt{28} = \sqrt{\boxed{4} \cdot 7}$ 4 is a perfect square.

$= \sqrt{4} \cdot \sqrt{7}$ $\sqrt{a \cdot b} = \sqrt{a} \cdot \sqrt{b}$

$= 2\sqrt{7}$ $\sqrt{4} = 2$

We could also have factored $\sqrt{28}$ as $\sqrt{2 \cdot 14}$, but neither 2 nor 14 is a perfect square, so nothing would have been gained.   ▮▮

EXAMPLE 3   Simplify $\sqrt{75}$.

SOLUTION $\sqrt{75} = \sqrt{25} \cdot \sqrt{3}$ 25 is a perfect square

$= 5\sqrt{3}$ ▮▮

EXAMPLE 4   Simplify $\sqrt[3]{24}$.

SOLUTION   We write 24 as the product of two numbers, one of which is a perfect *cube:*

$$\sqrt[3]{24} = \sqrt[3]{8} \cdot \sqrt[3]{3} \qquad \text{8 is a perfect cube.}$$
$$= 2\sqrt[3]{3}$$

EXAMPLE 5   Simplify $\sqrt{48}$.

SOLUTION
$$\sqrt{48} = \sqrt{16} \cdot \sqrt{3} \qquad \text{16 is a perfect square.}$$
$$= 4\sqrt{3}$$

You may have noticed that $\sqrt{48}$ could also have been factored

$$\sqrt{48} = \sqrt{4} \cdot \sqrt{12} \qquad \text{4 is a perfect square.}$$
$$= 2\sqrt{12}$$
$$= 2\sqrt{4} \cdot \sqrt{3} \qquad \text{4 is a perfect square.}$$
$$= 2 \cdot 2 \cdot \sqrt{3}$$
$$= 4\sqrt{3}$$

which is the same answer as before. However, it is always easier to factor out the *largest* perfect square immediately.

When the radicand involves variables, we must restrict them to positive values, since even roots of negative numbers are not real numbers. **For the remainder of this chapter, we will assume that all variables represent positive numbers, and** therefore we will not need to use absolute values.

EXAMPLE 6   Simplify the following:
   (a) $\sqrt{x^3} = \sqrt{x^2} \cdot \sqrt{x} = x\sqrt{x} \qquad x^2$ is a perfect square.
   (b) $\sqrt{x^5} = \sqrt{x^4} \cdot \sqrt{x} = x^2\sqrt{x} \qquad x^4 = (x^2)^2$ is a perfect square.
   (c) $\sqrt[3]{x^5} = \sqrt[3]{x^3} \cdot \sqrt[3]{x^2} = x\sqrt[3]{x^2}$
   (d) $\sqrt{x^8} = x^4$

Since $x^8 = (x^4)^2$, it is a perfect square already. In fact, every even power of a variable is a perfect square, and its square root is the variable raised to half that power.

EXAMPLE 7   Simplify $\sqrt{12x^3}$.

SOLUTION
$$\sqrt{12x^3} = \sqrt{4 \cdot 3 \cdot x^2 \cdot x} \qquad \text{4 and } x^2 \text{ are perfect squares.}$$
$$= \sqrt{4x^2} \cdot \sqrt{3x} \qquad \text{Separate the perfect squares.}$$
$$= 2x\sqrt{3x}$$

EXAMPLE 8   Simplify $\sqrt{50x^5y^6}$.

SOLUTION

$$\sqrt{50x^5y^6} = \sqrt{\boxed{25}\cdot 2 \cdot \boxed{x^4} \cdot x \cdot \boxed{y^6}}$$ 25, x^4, and y^6 are perfect squares.

$$= \sqrt{25x^4y^6} \cdot \sqrt{2x}$$ Separate the perfect squares.

$$= 5x^2y^3\sqrt{2x}$$ ▮▮

EXAMPLE 9   Multiply and simplify $\sqrt{2x} \cdot \sqrt{14xy}$.

SOLUTION

$$\sqrt{2x} \cdot \sqrt{14xy} = \sqrt{28x^2y}$$ Multiply.

$$= \sqrt{\boxed{4}\cdot 7 \cdot \boxed{x^2} \cdot y}$$ Factor and find the perfect squares.

$$= \sqrt{4x^2} \cdot \sqrt{7y}$$ Separate the perfect squares.

$$= 2x\sqrt{7y}$$ ▮▮

EXAMPLE 10   Simplify $\sqrt[3]{54x^6y^4}$.

SOLUTION

$$\sqrt[3]{54x^6y^4} = \sqrt[3]{\boxed{27}\cdot 2 \cdot \boxed{x^6} \cdot \boxed{y^3} \cdot y}$$ 27, x^6, and y^3 are perfect cubes.

$$= \sqrt[3]{27x^6y^3} \cdot \sqrt[3]{2y}$$ Separate the perfect cubes.

$$= 3x^2y\sqrt[3]{2y}$$ ▮▮

EXAMPLE 11   Multiply and simplify $\sqrt{2a^3b} \cdot \sqrt{12a^3}$.

SOLUTION

$$\sqrt{2a^3b} \cdot \sqrt{12a^3} = \sqrt{24a^6b}$$ Multiply.

$$= \sqrt{\boxed{4}\cdot 6 \cdot \boxed{a^6} \cdot b}$$ Factor and find the perfect squares.

$$= 2a^3\sqrt{6b}$$ ▮▮

5.2 Exercises

▮▮▮▮▮

▼ Simplify. Assume that all variables represent positive numbers.

1. $\sqrt{3} \cdot \sqrt{2}$

2. $\sqrt{5} \cdot \sqrt{7}$

3. $\sqrt{6} \cdot \sqrt{6}$

4. $\sqrt{18}$

5. $\sqrt[3]{40}$

6. $\sqrt{20}$

7. $\sqrt{72}$

8. $\sqrt{50}$

9. $\sqrt[4]{32}$

10. $\sqrt{32}$

11. $\sqrt{y^4}$

12. $\sqrt[3]{x^6}$

13. $\sqrt{16x^6}$

14. $\sqrt{a^9b^7}$

15. $\sqrt[3]{54z^5}$

16. $\sqrt{t^{30}}$

17. $\sqrt{t^{25}}$

18. $\sqrt{28x^7y^4}$

19. $\sqrt{40a^3b^2c}$

20. $\sqrt[4]{32x^5y^6}$

21. $\sqrt[3]{24a^4b^3}$

▼ Multiply and simplify. Assume that all variables
represent positive numbers.

22. $\sqrt{10} \cdot \sqrt{5}$

23. $\sqrt{6} \cdot \sqrt{8}$

24. $\sqrt{5} \cdot \sqrt{15}$

25. $\sqrt{2} \cdot \sqrt{14}$

26. $\sqrt{6} \cdot \sqrt{3}$

27. $\sqrt{x^5} \cdot \sqrt{x^3}$

28. $\sqrt[3]{9} \cdot \sqrt[3]{6}$

29. $\sqrt[3]{4} \cdot \sqrt[3]{4}$

30. $\sqrt[3]{x^5} \cdot \sqrt[3]{x^2}$

31. $\sqrt{a^3} \cdot \sqrt{a}$

32. $\sqrt{3x} \cdot \sqrt{4x}$

33. $\sqrt{6xy} \cdot \sqrt{8x}$

34. $\sqrt[3]{3x^3y} \cdot \sqrt[3]{8xy^5}$

35. $\sqrt{5x^5} \cdot \sqrt{10x^{10}}$

36. $\sqrt[3]{6x^2y} \cdot \sqrt[3]{18xy^2}$

37. $\sqrt{8x^3} \cdot \sqrt{2x} \cdot \sqrt{3x^5}$

38. $\sqrt{10ab} \cdot \sqrt{20a^3b} \cdot \sqrt{50a^4b^4}$

39. $\sqrt{3xy} \cdot \sqrt{2x^3} \cdot \sqrt{6y^5}$

40. $\sqrt[3]{6xy} \cdot \sqrt[3]{4x^2y^2} \cdot \sqrt[3]{2xy^2}$

5.3   Adding and Subtracting Radicals

▌ ▌ ▌ ▌ ▌

Like or **similar radicals** are radicals that have the same *radicand* (the number under
the radical) and the same *index* (the root).

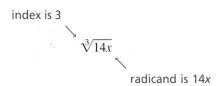

index is 3

$\sqrt[3]{14x}$

radicand is 14x

Adding and subtracting like radicals is similar to adding and subtracting like terms
of polynomials (for example, $5x + 2x = 7x$). We apply the distributive rule and
add or subtract the coefficients.

E X A M P L E   1   Combine: $5\sqrt{3} + 2\sqrt{3}$.

S O L U T I O N $5\sqrt{3} + 2\sqrt{3} = 7\sqrt{3}$ Add the coefficients using the
distributive rule $(5 + 2 = 7)$.   ▌▌

EXAMPLE 2   Combine: $2\sqrt[3]{5} - 8\sqrt[3]{5} + 3\sqrt[3]{5}$.

SOLUTION $2\sqrt[3]{5} - 8\sqrt[3]{5} + 3\sqrt[3]{5} = -3\sqrt[3]{5}$ Combine the coefficients.   ▮▮

In order to add or subtract, the radicals must be *like*. If they are not, we apply the methods of the last section and attempt to make them *like*.

EXAMPLE 3   Combine: $\sqrt{12} + \sqrt{27}$.

SOLUTION $\sqrt{12} + \sqrt{27} = \sqrt{4 \cdot 3} + \sqrt{9 \cdot 3}$ Simplify each radical.

$= 2\sqrt{3} + 3\sqrt{3}$ Combine the like radicals.

$= 5\sqrt{3}$ ▮▮

EXAMPLE 4   Combine: $\sqrt{50} + \sqrt{32} - \sqrt{18}$.

SOLUTION $\sqrt{50} + \sqrt{32} - \sqrt{18} = \sqrt{25 \cdot 2} + \sqrt{16 \cdot 2} - \sqrt{9 \cdot 2}$ Simplify each radical.

$= 5\sqrt{2} + 4\sqrt{2} - 3\sqrt{2}$ Combine.

$= 6\sqrt{2}$ ▮▮

EXAMPLE 5   Combine: $3\sqrt{12} - 2\sqrt{75}$.

SOLUTION $3\sqrt{12} - 2\sqrt{75} = 3\sqrt{4 \cdot 3} - 2\sqrt{25 \cdot 3}$ Simplify each radical

$= 3 \cdot 2\sqrt{3} - 2 \cdot 5\sqrt{3}$

$= 6\sqrt{3} - 10\sqrt{3}$ Combine.

$= -4\sqrt{3}$ ▮▮

EXAMPLE 6   Combine: $\sqrt{72} - \sqrt{48}$.

SOLUTION $\sqrt{72} - \sqrt{48} = \sqrt{36 \cdot 2} - \sqrt{16 \cdot 3}$

$= 6\sqrt{2} - 4\sqrt{3}$

Since the radicals are not *like* radicals, they cannot be combined.   ▮▮

EXAMPLE 7   Combine: $\sqrt[3]{4} + \sqrt{2}$.

SOLUTION   These radicals are not *like* since they have different indices. They cannot be combined.   ▮▮

STOP

A common error is to assume that $\sqrt{x} + \sqrt{y} = \sqrt{x + y}$. Let's try $x = 9$ and $y = 16$. Then we would have

$$\sqrt{9} + \sqrt{16} = \sqrt{9 + 16}$$

$$3 + 4 = \sqrt{25}$$

But $7 \neq 5$!

The sum of the square roots of two (or more) numbers is *not* equal to the square root of the sum.

$$\sqrt{x} + \sqrt{y} \neq \sqrt{x+y}$$

5.3   Exercises

▼ Combine where possible.

1. $3\sqrt{5} + 2\sqrt{5}$

2. $5\sqrt[3]{2} - 7\sqrt[3]{2}$

3. $3\sqrt[3]{x} + 7\sqrt[3]{x}$

4. $6\sqrt{3} + 3\sqrt{x} - 8\sqrt{x}$

5. $2\sqrt{10} - 3\sqrt{3} + 4\sqrt{5}$

6. $5\sqrt{6} + 2\sqrt{3} - 4\sqrt{6} + \sqrt{3}$

7. $\sqrt{18} - \sqrt{8}$

8. $\sqrt{50} - \sqrt{8}$

9. $\sqrt[3]{16} - \sqrt[3]{54}$

10. $2\sqrt{12} + 3\sqrt{48}$

11. $\sqrt{32} - 3\sqrt{27} + 2\sqrt{8}$

12. $2\sqrt{20} + 2\sqrt{80}$

13. $2\sqrt{44} - \sqrt{99}$

14. $\sqrt{98} - 2\sqrt{50} + 3\sqrt{8}$

15. $\sqrt[4]{32x} + \sqrt[3]{16x}$

16. $\sqrt{45} - 2\sqrt{20}$

17. $3\sqrt{8} + 2\sqrt{18} - 4\sqrt{72}$

18. $2\sqrt[3]{24ab} + \sqrt[3]{81ab}$

19. $2\sqrt{40} + 4\sqrt{20} + 4\sqrt{90} - 3\sqrt{125}$

20. $\sqrt{48x} + \sqrt{27x}$

21. $6\sqrt{45ab} - 3\sqrt{5ab} - 2\sqrt{20ab}$

22. $3\sqrt{28} + 4\sqrt{63} - \sqrt{7a}$

23. $2\sqrt{8a} + 2\sqrt{32b} - 4\sqrt{18c}$

5.4   Division of Radicals; Rationalizing Denominators

▮ ▮ ▮ ▮ ▮

Division of Radicals

• • • • •

Compare the following statements:

1. $\sqrt{\dfrac{36}{9}} = \sqrt{4} = 2$

2. $\dfrac{\sqrt{36}}{\sqrt{9}} = \dfrac{6}{3} = 2$

Therefore $\sqrt{\dfrac{36}{9}} = \dfrac{\sqrt{36}}{\sqrt{9}}$, and this is true in general.

$$\sqrt[n]{\frac{a}{b}} = \frac{\sqrt[n]{a}}{\sqrt[n]{b}}; \qquad a \geq 0,\ b > 0$$

EXAMPLE 1 $\dfrac{\sqrt{24}}{\sqrt{6}} = \sqrt{\dfrac{24}{6}} = \sqrt{4} = 2$ ▮▮

EXAMPLE 2 $\dfrac{\sqrt{26x^5}}{\sqrt{13x^3}} = \sqrt{\dfrac{26x^5}{13x^3}} = \sqrt{2x^2} = x\sqrt{2}$ ▮▮

EXAMPLE 3 $\sqrt{\dfrac{36}{25}} = \dfrac{\sqrt{36}}{\sqrt{25}} = \dfrac{6}{5}$ ▮▮

EXAMPLE 4 $\dfrac{\sqrt[3]{56}}{\sqrt[3]{7}} = \sqrt[3]{\dfrac{56}{7}} = \sqrt[3]{8} = 2$ ▮▮

Rationalizing Denominators

• • • • •

If the denominator of a fraction is not a *rational* number, we attempt to make it *rational* by applying the following procedure.

> **To rationalize the denominator of a fraction**
>
> Multiply both the numerator and the denominator of the fraction by a quantity that makes the radicand of the denominator a perfect *n*th power (square, cube, etc.).

EXAMPLE 5 Rationalize the denominator of $\dfrac{\sqrt{5}}{\sqrt{2}}$.

SOLUTION $\dfrac{\sqrt{5}}{\sqrt{2}} = \dfrac{\sqrt{5}}{\sqrt{2}} \cdot \dfrac{\sqrt{2}}{\sqrt{2}}$ Multiply the numerator and the denominator by $\sqrt{2}$.

$= \dfrac{\sqrt{5 \cdot 2}}{\sqrt{2 \cdot 2}}$

$= \dfrac{\sqrt{10}}{\sqrt{4}}$ 4 is a perfect square, $\sqrt{2} \cdot \sqrt{2} = 2$

$= \dfrac{\sqrt{10}}{2}$ The denominator is now a rational number. ▐▐

EXAMPLE 6 Rationalize the denominator of $\sqrt{\dfrac{2}{3}}$.

SOLUTION $\sqrt{\dfrac{2}{3}} = \dfrac{\sqrt{2}}{\sqrt{3}}$ Apply the division rule.

$= \dfrac{\sqrt{2}}{\sqrt{3}} \cdot \dfrac{\sqrt{3}}{\sqrt{3}}$ Rationalize the denominator, $\sqrt{3} \cdot \sqrt{3} = 3$.

$= \dfrac{\sqrt{6}}{3}$ The denominator is now a rational number. ▐▐

EXAMPLE 7 Rationalize the denominator of $\dfrac{3\sqrt{2}}{\sqrt{3}}$.

SOLUTION $\dfrac{3\sqrt{2}}{\sqrt{3}} = \dfrac{3\sqrt{2}}{\sqrt{3}} \cdot \dfrac{\sqrt{3}}{\sqrt{3}}$

$= \dfrac{\cancel{3}\sqrt{6}}{\cancel{3}}$

$= \sqrt{6}$ ▐▐

EXAMPLE 8 Rationalize the denominator of $\sqrt{\dfrac{5}{12x}}$.

SOLUTION $\sqrt{\dfrac{5}{12x}} = \dfrac{\sqrt{5}}{\sqrt{12x}}$ Apply the division rule.

$= \dfrac{\sqrt{5}}{\sqrt{12x}} \cdot \dfrac{\sqrt{3x}}{\sqrt{3x}}$ Rationalize the denominator. We need not multiply by $\sqrt{12x}$. Always multiply by the *smallest* quantity that will make a perfect square.

$= \dfrac{\sqrt{15x}}{\sqrt{36x^2}}$ $36x^2$ is a perfect square.

$= \dfrac{\sqrt{15x}}{6x}$ The denominator is a rational number. ▐▐

EXAMPLE 9   Rationalize the denominator of $\dfrac{\sqrt[3]{3}}{\sqrt[3]{5}}$.

SOLUTION   We must multiply both numerator and denominator of the fraction by a quantity that will make the radicand of the denominator a perfect *cube*.

$$\frac{\sqrt[3]{3}}{\sqrt[3]{5}} = \frac{\sqrt[3]{3}}{\sqrt[3]{5}} \cdot \boxed{\frac{\sqrt[3]{25}}{\sqrt[3]{25}}}$$

$$= \frac{\sqrt[3]{75}}{\sqrt[3]{125}} \qquad \text{125 is a perfect cube.}$$

$$= \frac{\sqrt[3]{75}}{5} \qquad \text{The denominator is a rational number.} \qquad \blacksquare\blacksquare$$

EXAMPLE 10   Rationalize the denominator of $\sqrt[3]{\dfrac{3}{4x}}$.

SOLUTION $\sqrt[3]{\dfrac{3}{4x}} = \dfrac{\sqrt[3]{3}}{\sqrt[3]{4x}} \qquad$ Apply the division rule.

$$= \frac{\sqrt[3]{3}}{\sqrt[3]{4x}} \cdot \boxed{\frac{\sqrt[3]{2x^2}}{\sqrt[3]{2x^2}}}$$

$$= \frac{\sqrt[3]{6x^2}}{\sqrt[3]{8x^3}} \qquad 8x^3 \text{ is a perfect cube.}$$

$$= \frac{\sqrt[3]{6x^2}}{2x} \qquad\qquad\qquad \blacksquare\blacksquare$$

5.4   Exercises

▮▮▮▮▮

▼ Divide and simplify: Assume that all variables represent positive numbers.

1. $\dfrac{\sqrt{45}}{\sqrt{15}}$

2. $\dfrac{\sqrt{27}}{\sqrt{3}}$

3. $\dfrac{\sqrt{20}}{\sqrt{5}}$

4. $\dfrac{\sqrt{75}}{\sqrt{15}}$

5. $\dfrac{\sqrt[3]{18}}{\sqrt[3]{3}}$

6. $\dfrac{\sqrt{100}}{\sqrt{25}}$

7. $\dfrac{\sqrt[3]{48}}{\sqrt[3]{6}}$

8. $\dfrac{\sqrt{64}}{\sqrt{16}}$

9. $\sqrt{\dfrac{49}{4}}$

10. $\sqrt{\dfrac{25}{16}}$

11. $\sqrt[3]{\dfrac{8}{27}}$

12. $\sqrt{\dfrac{49}{81}}$

13. $\dfrac{\sqrt{12x}}{\sqrt{3x}}$

14. $\dfrac{\sqrt{20x^3}}{\sqrt{5x}}$

15. $\dfrac{\sqrt{75x^3y}}{\sqrt{15xy}}$

16. $\dfrac{\sqrt{60x^5}}{\sqrt{15x^3}}$

17. $\dfrac{\sqrt{18x^5y^3}}{\sqrt{3xy^3}}$

18. $\dfrac{\sqrt{15a^7b^3}}{\sqrt{3a^3b}}$

31. $\dfrac{7}{\sqrt{y}}$

32. $\dfrac{3}{\sqrt{x}}$

▼ Rationalize the denominator: Assume that all variables represent positive numbers.

33. $\sqrt{\dfrac{5}{x}}$

34. $\sqrt{\dfrac{4}{y}}$

19. $\dfrac{\sqrt{3}}{\sqrt{2}}$

20. $\dfrac{\sqrt{5}}{\sqrt{3}}$

35. $\dfrac{2\sqrt{5}}{\sqrt{6}}$

36. $\dfrac{12\sqrt{2}}{\sqrt{3}}$

21. $\dfrac{\sqrt{5}}{\sqrt{2}}$

22. $\dfrac{\sqrt{8}}{\sqrt{3}}$

37. $\dfrac{5\sqrt{5}}{\sqrt{2}}$

38. $\dfrac{4\sqrt{3}}{\sqrt{6}}$

23. $\dfrac{10}{\sqrt{5}}$

24. $\dfrac{12}{\sqrt{6}}$

39. $\dfrac{6}{\sqrt{2x}}$

40. $\dfrac{3}{\sqrt{3x}}$

25. $\dfrac{2}{\sqrt[3]{4}}$

26. $\dfrac{3}{\sqrt[3]{9}}$

41. $\dfrac{\sqrt{5}}{\sqrt{12}}$

42. $\dfrac{\sqrt{3}}{\sqrt{8}}$

27. $\sqrt{\dfrac{5}{7}}$

28. $\sqrt{\dfrac{3}{5}}$

43. $\dfrac{\sqrt{2}}{\sqrt{27x}}$

44. $\dfrac{\sqrt{5}}{\sqrt{18y}}$

29. $\sqrt{\dfrac{1}{7}}$

30. $\sqrt{\dfrac{5}{6}}$

45. $\dfrac{\sqrt[3]{2}}{\sqrt[3]{3}}$

46. $\dfrac{\sqrt[3]{2}}{\sqrt[3]{9}}$

47. $\dfrac{\sqrt[3]{5}}{\sqrt[3]{4}}$

48. $\dfrac{\sqrt[3]{3}}{\sqrt[3]{2}}$

53. $\dfrac{\sqrt[4]{3}}{\sqrt[4]{8}}$

54. $\dfrac{\sqrt[5]{3}}{\sqrt[5]{16}}$

49. $\dfrac{\sqrt[3]{3x}}{\sqrt[3]{2x^2}}$

50. $\dfrac{\sqrt[3]{5x}}{\sqrt[3]{32x}}$

55. Without using a calculator or tables, determine which is larger, $\dfrac{1}{\sqrt{2}+3}$ or $3 + \sqrt{2}$.

51. $\sqrt[3]{\dfrac{5}{9x}}$

52. $\sqrt[3]{\dfrac{7}{16x^2}}$

5.5   Other Radical Expressions

▮ ▮ ▮ ▮ ▮

The following examples illustrate still other expressions containing radicals.

Products Containing Radicals

• • • • •

EXAMPLE 1   Multiply: $\sqrt{2}(3\sqrt{2} + \sqrt{7})$.

SOLUTION
$$\sqrt{2}(3\sqrt{2} + \sqrt{7}) = \sqrt{2} \cdot 3\sqrt{2} + \sqrt{2} \cdot \sqrt{7} \qquad \text{Apply the distributive property.}$$
$$= 3\sqrt{2 \cdot 2} + \sqrt{2 \cdot 7}$$
$$= 3 \cdot 2 + \sqrt{14}$$
$$= 6 + \sqrt{14} \qquad\qquad ▮▮$$

EXAMPLE 2   Multiply: $\sqrt{5}(3\sqrt{5} - 2)$.

SOLUTION
$$\sqrt{5}(3\sqrt{5} - 2) = 3\sqrt{5} \cdot \sqrt{5} - 2\sqrt{5} \qquad \text{Apply the distributive property.}$$
$$= 3 \cdot 5 - 2\sqrt{5}$$
$$= 15 - 2\sqrt{5} \qquad\qquad ▮▮$$

Pairs of binomials containing radicals are multiplied using the **FOIL** method in the same way as binomials containing variables.

EXAMPLE 3   Multiply: $(\sqrt{2} - 3)(\sqrt{2} + 4)$.

SOLUTION
$$(\sqrt{2} - 3)(\sqrt{2} + 4) = \underbrace{\sqrt{2} \cdot \sqrt{2}}_{\text{First}} + \underbrace{4 \cdot \sqrt{2}}_{\text{Outer}} - \underbrace{3 \cdot \sqrt{2}}_{\text{Inner}} - \underbrace{3 \cdot 4}_{\text{Last}}$$
$$= 2 + 4\sqrt{2} - 3\sqrt{2} - 12$$
$$= \sqrt{2} - 10 \qquad\qquad ▮▮$$

EXAMPLE 4 Multiply: $\left(2\sqrt{3}-5\right)\left(4\sqrt{3}+2\right)$.

SOLUTION $\left(2\sqrt{3}-5\right)\left(4\sqrt{3}+2\right) = \underbrace{2\sqrt{3}\cdot4\sqrt{3}}_{F} + \underbrace{2\cdot2\sqrt{3}}_{O} - \underbrace{5\cdot4\sqrt{3}}_{I} - \underbrace{(5)(2)}_{L}$

$$= \quad 24 \quad + \quad 4\sqrt{3} \quad - \quad 20\sqrt{3} \quad - \quad 10$$

$$= 14 - 16\sqrt{3} \qquad\qquad ▮▮$$

EXAMPLE 5 Multiply: $\left(\sqrt{3}-4\right)\left(\sqrt{3}+4\right)$.

SOLUTION $\left(\sqrt{3}-4\right)\left(\sqrt{3}+4\right) = \sqrt{3}\cdot\sqrt{3} + \underbrace{4\sqrt{3}-4\sqrt{3}}_{} - 16$

Middle term is zero.

$$= 3 - 16$$

$$= -13 \qquad\qquad ▮▮$$

The binomials in Example 5 are alike except that the signs of the second terms are different. Such binomials are called **conjugates** of each other. Whenever conjugates are multiplied together, the middle term of the result drops out. This property is quite useful in rationalizing a binomial denominator containing square roots.

EXAMPLE 6 **(a)** The conjugate of $\sqrt{3}-4$ is $\sqrt{3}+4$.

(b) The conjugate of $2\sqrt{3}+\sqrt{5}$ is $2\sqrt{3}-\sqrt{5}$.

(c) The conjugate of $4-\sqrt{x}$ is $4+\sqrt{x}$. ▮▮

To rationalize a binomial denominator containing radicals:
Multiply the numerator and denominator by the **conjugate** of the denominator.

EXAMPLE 7 Rationalize the denominator of $\dfrac{4}{\sqrt{3}-2}$.

SOLUTION $\dfrac{4}{\sqrt{3}-2} = \dfrac{4}{\left(\sqrt{3}-2\right)}\cdot\dfrac{\left(\sqrt{3}+2\right)}{\left(\sqrt{3}+2\right)}$

Multiply the numerator and denominator by the conjugate of the denominator.

$$= \frac{4\sqrt{3}+4\cdot2}{\sqrt{3}\cdot\sqrt{3}+2\cdot\sqrt{3}-2\cdot\sqrt{3}-4}$$

The two middle terms combine to make 0.

$$= \frac{4\sqrt{3}+8}{3-4}$$

$$= \frac{4\sqrt{3}+8}{-1}$$

$$= -4\sqrt{3}-8 \qquad\qquad ▮▮$$

EXAMPLE 8   Rationalize the denominator of $\dfrac{\sqrt{5} - 2}{\sqrt{5} + 2}$.

SOLUTION

$$\frac{\sqrt{5} - 2}{\sqrt{5} + 2} = \frac{(\sqrt{5} - 2)}{(\sqrt{5} + 2)} \cdot \frac{(\sqrt{5} - 2)}{(\sqrt{5} - 2)} \qquad \text{Multiply by the conjugate of the denominator.}$$

$$= \frac{\sqrt{5} \cdot \sqrt{5} - 2\sqrt{5} - 2\sqrt{5} + (-2)(-2)}{\sqrt{5} \cdot \sqrt{5} - 2\sqrt{5} + 2\sqrt{5} + (+2)(-2)}$$

$$= \frac{5 - 4\sqrt{5} + 4}{5 - 4}$$

$$= \frac{9 - 4\sqrt{5}}{1}$$

$$= 9 - 4\sqrt{5} \qquad\qquad\qquad\qquad\qquad\qquad ▮▮$$

EXAMPLE 9   Rationalize the denominator of $\dfrac{6}{\sqrt{5} - 2}$.

SOLUTION

$$\frac{6}{\sqrt{5} - \sqrt{2}} = \frac{6}{(\sqrt{5} - \sqrt{2})} \cdot \frac{(\sqrt{5} + \sqrt{2})}{(\sqrt{5} + \sqrt{2})} \qquad \text{Multiply by the conjugate of the denominator.}$$

$$= \frac{6(\sqrt{5} + \sqrt{2})}{5 - 2} \qquad\qquad\qquad \text{Simplify the denominator.}$$

$$= \frac{\overset{2}{\cancel{6}}(\sqrt{5} + \sqrt{2})}{\underset{1}{\cancel{3}}} \qquad\qquad \text{Do not multiply in the numerator until you observe whether the fraction can be reduced.}$$

$$= 2\sqrt{5} + 2\sqrt{2} \qquad\qquad\qquad\qquad\qquad ▮▮$$

5.5   Exercises

▮▮▮▮▮

▼ Multiply and simplify. Assume that all variables represent positive numbers.

1. $\sqrt{3}(4\sqrt{3} + 2)$ **2.** $\sqrt{5}(3 - \sqrt{5})$

3. $2\sqrt{7}(\sqrt{3} - \sqrt{7})$ **4.** $\sqrt{10}(\sqrt{5} - \sqrt{2})$

5. $3\sqrt{2}(\sqrt{2} - \sqrt{6})$ **6.** $\sqrt{x}(\sqrt{x} - 2)$

7. $\sqrt{y}(3 + 2\sqrt{y})$ **8.** $3\sqrt{5}(\sqrt{15} + 2\sqrt{5})$

9. $4\sqrt{2}(2\sqrt{2} + \sqrt{6})$ **10.** $2\sqrt{3x}(\sqrt{3x} - \sqrt{x})$ **21.** $(\sqrt{3} + 5)^2$ **22.** $(2\sqrt{2} - 3)^2$

11. $(\sqrt{2} + 3)(\sqrt{2} - 1)$ **12.** $(\sqrt{5} + 3)(\sqrt{5} - 2)$ **23.** $(2 + \sqrt{3})^2$ **24.** $(3\sqrt{3} - \sqrt{5})^2$

13. $(\sqrt{3} - \sqrt{2})(\sqrt{3} - 2\sqrt{2})$

▼ Write the conjugate of each of the following binomials.

25. $3 - \sqrt{7}$ **26.** $3\sqrt{2} + 5$ **27.** $\sqrt{5} + 3\sqrt{6}$

14. $(2\sqrt{7} - 3)(3\sqrt{7} + 3)$

28. $\sqrt{7} + 2\sqrt{2}$ **29.** $x - \sqrt{5}$ **30.** $3\sqrt{x} + y$

15. $(\sqrt{6} + 2)(\sqrt{6} + 5)$

▼ Rationalize the denominators and simplify.

16. $(\sqrt{5} - 3)(\sqrt{5} + 3)$ **31.** $\dfrac{2}{\sqrt{2} + 1}$ **32.** $\dfrac{6}{\sqrt{2} + 2}$

17. $(\sqrt{5} - 2\sqrt{7})(\sqrt{5} + 2\sqrt{7})$ **33.** $\dfrac{6}{\sqrt{3} + \sqrt{5}}$ **34.** $\dfrac{8}{\sqrt{5} - 2}$

18. $(3 - \sqrt{6})(3 + \sqrt{6})$ **35.** $\dfrac{-2}{1 - \sqrt{3}}$ **36.** $\dfrac{3}{3 + \sqrt{3}}$

19. $(\sqrt{3} + \sqrt{2})(\sqrt{3} - \sqrt{6})$ **37.** $\dfrac{6}{4 - 2\sqrt{3}}$ **38.** $\dfrac{-5}{\sqrt{5} + 2\sqrt{2}}$

20. $(\sqrt{5} - 7)(\sqrt{2} + 3)$

39. $\dfrac{\sqrt{3}-1}{\sqrt{3}+1}$ **40.** $\dfrac{3-\sqrt{5}}{2+\sqrt{5}}$ **43.** $\dfrac{2-3\sqrt{5}}{4+\sqrt{5}}$ **44.** $\dfrac{\sqrt{5}-1}{\sqrt{5}-2}$

41. $\dfrac{3+2\sqrt{2}}{3-\sqrt{2}}$ **42.** $\dfrac{\sqrt{3}+5}{\sqrt{3}+3}$ **45.** $\dfrac{3\sqrt{2}}{\sqrt{5}-\sqrt{2}}$

5.6   Rational Exponents

▮▮▮▮▮

How can we define expressions such as $9^{1/2}$, $8^{2/3}$, and $a^{3/4}$? If we assume that all previous rules of exponents hold for rational exponents, we have

$$a^{1/2} \cdot a^{1/2} = a^{1/2\,+\,1/2} = a^1 = a$$

But we know that $\sqrt{a} \cdot \sqrt{a} = a$. It seems reasonable, then, that we should define $a^{1/2}$ to be $\sqrt{a}$.

Similarly, $a^{1/3} \cdot a^{1/3} \cdot a^{1/3} = a^{1/3\,+\,1/3\,+\,1/3} = a$, and we also know that $\sqrt[3]{a} \cdot \sqrt[3]{a} \cdot \sqrt[3]{a} = a$, which leads us to define $a^{1/3}$ as $\sqrt[3]{a}$.

This is true in general, giving us the following definition:

$$a^{1/n} = \sqrt[n]{a}, \qquad \text{where } n \text{ is a positive integer and } a \geq 0 \text{ if } n \text{ is even.}$$

EXAMPLE 1   Evaluate the following:

(a) $9^{1/2} = \sqrt{9} = 3$

(b) $8^{1/3} = \sqrt[3]{8} = 2$

(c) $(-9)^{1/2} = \sqrt{-9}$, which is not a real number.

(d) $(-8)^{1/3} = \sqrt[3]{-8} = -2$ ▮▮

Now let's consider how we should handle $8^{2/3}$. Again if all previous rules of exponents hold true, then we have

$$8^{2/3} = (8^{1/3})^2 = (\sqrt[3]{8})^2 = 2^2 = 4$$

or

$$8^{2/3} = (8^2)^{1/3} = \sqrt[3]{8^2} = \sqrt[3]{64} = 4$$

This leads to the following definition.

$$a^{m/n} = (\sqrt[n]{a})^m \quad \text{or} \quad a^{m/n} = \sqrt[n]{a^m}, \quad n \neq 0, \text{ and } a \geq 0 \text{ if } n \text{ is even.}$$

Here is the commonly used terminology:

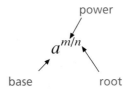

E X A M P L E 2   Evaluate the following:

$$16^{3/4} = \left(\sqrt[4]{16}\right)^3 = 2^3 = 8$$

or

$$16^{3/4} = \sqrt[4]{16^3} = \sqrt[4]{4096} = 8$$

You can see that the first method is much simpler than the second. In general it is easier first to take the root and then raise to the power. ∎

All previous rules for exponents hold for rational exponents and they are summarized as follows:

If *m* and *n* are rational numbers, $a \geq 0$ and $b \geq 0$, then:

1. $a^m \cdot a^n = a^{m+n}$

2. $\dfrac{a^m}{a^n} = a^{m-n};\qquad a \neq 0$

3. $(a^m)^n = a^{mn}$

4. $(ab)^n = a^n b^n$

5. $\left(\dfrac{a}{b}\right)^n = \dfrac{a^n}{b^n},\qquad b \neq 0$

6. $a^{-n} = \dfrac{1}{a^n},\qquad a \neq 0$

7. $\left(\dfrac{a}{b}\right)^{-n} = \left(\dfrac{b}{a}\right)^n,\qquad a, b \neq 0$

E X A M P L E 3   Simplify the following.

(a) $25^{3/2} = \left(\sqrt{25}\right)^3 = 5^3 = 125$

(b) $27^{-1/3} = \dfrac{1}{27^{1/3}} = \dfrac{1}{\sqrt[3]{27}} = \dfrac{1}{3}$

(c) $\left(\dfrac{9}{4}\right)^{-1/2} = \left(\dfrac{4}{9}\right)^{1/2} = \dfrac{\sqrt{4}}{\sqrt{9}} = \dfrac{2}{3}$

(d) $-36^{1/2} = -\sqrt{36} = -6$

(e) $(-36)^{1/2} = \sqrt{-36}$ is not a real number.

(f) $\dfrac{x^{5/3}}{x^{2/3}} = x^{5/3-2/3} = x^{3/3} = x$

(g) $(x^{2/3}y^{1/4})^3 = x^{(2/3)(3)} \cdot y^{(1/4)(3)} = x^2 y^{3/4}$

(h) $x^{1/2} \cdot x^{3/4} = x^{1/2+3/4} = x^{5/4}$

(i) $(x^{3/4})^{2/3} = x^{(3/4)(2/3)} = x^{1/2}$ ▮▮

5.6   Exercises

▮▮▮▮▮

▼ Write in exponential form:

1. $\sqrt{5}$ **2.** $\sqrt[3]{5^2}$ **3.** $\sqrt[4]{y^5}$

4. $\sqrt[3]{a^4}$ **5.** $\left(\sqrt[3]{a}\right)^4$ **6.** $\sqrt[3]{a^2 b}$

▼ Write using radicals and simplify as much as possible.

7. $6^{1/2}$ **8.** $7^{1/3}$ **9.** $5^{2/3}$

10. $x^{2/3}$ **11.** $x^{-2/3}$ **12.** $(x^2 y)^{1/3}$

13. $16^{1/4}$ **14.** $16^{1/3}$

▼ Simplify as much as possible.

15. $9^{3/2}$ **16.** $(-16)^{1/2}$

17. $-16^{1/2}$ **18.** $(-27)^{1/3}$

19. $-27^{1/3}$ **20.** $-27^{-1/3}$

21. $\left(\dfrac{9}{16}\right)^{-1/2}$ **22.** $36^{3/2}$

23. $81^{-3/4}$ **24.** $4^{-3/2}$

25. $\left(\dfrac{16}{49}\right)^{-1/2}$ **26.** $\left(\dfrac{27}{8}\right)^{-1/3}$

27. $16^{1/4} + 32^{1/5}$ **28.** $8^{1/3} + 25^{1/2}$

29. $27^{-1/3} + 49^{-1/2}$ **30.** $36^{-1/2} + 8^{-2/3}$

▼ Assuming that all variables are positive, simplify using the properties of exponents.

31. $(x^{2/3})^{3/2}$ **32.** $x^{2/3} \cdot x^{1/3}$

33. $x^{3/5} \cdot x^{2/5}$ **34.** $x^{1/4} \cdot x^{1/2}$

35. $x^{2/3} \cdot x^{3/4}$

36. $\dfrac{x^{4/3}}{x^{1/3}}$

41. $(x^{3/4} \cdot y^{1/2})^{2/3}$

42. $(x^{3/5} \cdot y^{2/3})^{5/2}$

37. $\dfrac{x^{3/4}}{x^{1/2}}$

38. $\dfrac{x^{3/5}}{x^{1/3}}$

43. $\dfrac{(x^{1/4})^{4/3}}{(x^{1/9})^{3/2}}$

44. $\dfrac{(x^{5/4})^{2/3}}{(x^{4/3})^{1/4}}$

39. $(x^{1/4} \cdot y^{1/3})^2$

40. $(x^2 \cdot y^3)^{1/2}$

45. $\dfrac{(x^{2/3})^{1/2}}{(x^{3/4})^{4/9}}$

5.7   Complex Numbers

In the past we conveniently avoided the problem of the square root of a negative number. For example, $\sqrt{-4}$ is not a real number because there is no real number that we can square that will give us -4.

There are many problems in engineering and science that use square roots of negative numbers. René Descartes (1596–1650), the French philosopher and mathematician, called them "imaginary." A century later, the Swiss mathematician Leonhard Euler (1707–1783) used the letter i to represent $\sqrt{-1}$, the imaginary unit.

Definition: $i = \sqrt{-1}$ and $i^2 = -1.$

What about expressions like $\sqrt{-4}$, $\sqrt{-9}$, and $\sqrt{-5}$? We rewrite them as follows:

$$\sqrt{-4} = \sqrt{(4)(-1)} = \sqrt{4} \cdot \sqrt{-1} = 2 \cdot \sqrt{-1} = 2i$$
$$\sqrt{-9} = \sqrt{(9)(-1)} = \sqrt{9} \cdot \sqrt{-1} = 3 \cdot \sqrt{-1} = 3i$$
$$\sqrt{-5} = \sqrt{(5)(-1)} = \sqrt{5} \cdot \sqrt{-1} = \sqrt{5} \cdot i \quad \text{or} \quad i\sqrt{5}$$

The numbers $2i$, $3i$, and $i\sqrt{5}$ are called **imaginary numbers.**

An **imaginary number** is a number of the form bi, where b is any real number.

EXAMPLE 1 Write the following numbers in imaginary form.

(a) $\sqrt{-25} = 5i$

(b) $\sqrt{-81} = 9i$

(c) $\sqrt{-7} = i\sqrt{7}$

(d) $\sqrt{-12} = \sqrt{4(-3)} = 2\sqrt{-3} = 2i\sqrt{3}$ ▮▮

Assuming that previously learned rules of exponents hold when the base is i, we have the following:

$$i = \sqrt{-1}$$
$$i^2 = -1$$
$$i^3 = i^2 \cdot i = -1 \cdot i = -i$$
$$i^4 = i^2 \cdot i^2 = (-1)(-1) = 1$$
$$i^5 = i^4 \cdot i = 1 \cdot i = i$$
$$i^6 = i^4 \cdot i^2 = 1(-1) = -1$$
$$i^7 = -i$$

.

.

.

As you can see, the cycle $i, -1, -i, 1$ repeats as we evaluate higher powers of i.

EXAMPLE 2 Simplify:

(a) $i^8 = (i^4)^2 = 1^2 = 1$

(b) $i^9 = (i^4)^2 \cdot i = 1 \cdot i = i$

(c) $i^{30} = (i^4)^7 \cdot i^2 = 1(-1) = -1$ ▮▮

Imaginary numbers are members of a larger set of numbers called **complex numbers.**

> A **complex number** is an expression of the form $a + bi$, where a and b are real numbers and $i = \sqrt{-1}$. Here a is called the **real part** and b is called the **imaginary part.**

EXAMPLE 3 $6 - 3i$ is a complex number, where 6 is the real part and -3 is the imaginary part.   ▮▮

All *real* numbers are also complex numbers. For example, the real number 14 can be written with the imaginary part equal to zero: $14 + 0i$.

Addition and Subtraction of Complex Numbers

• • • • •

To add (or subtract) complex numbers:

Add (or subtract) the real parts and add (or subtract) the imaginary parts.

EXAMPLE 4   Add or subtract as indicated.

(a) $(6 + 5i) + (3 + 2i) = 9 + 7i$

(b) $(3 - 4i) + (-2 - i) = 1 - 5i$

(c) $(4 - 2i) - (3 - 3i) = 4 - 2i - 3 + 3i = 1 + i$

(d) $(6 - 3i) - (4 - 3i) = 6 - 3i - 4 + 3i = 2 + 0i = 2$

(e) $(4 + 2i) + (7 - 2i) - (5 - i) = 6 + i$ ▐▐

5.7   Exercises

▐ ▐ ▐ ▐ ▐

▼ Write the following radicals in imaginary form and simplify as much as possible.

1. $\sqrt{-9}$ 2. $\sqrt{-64}$

3. $-\sqrt{-25}$ 4. $-\sqrt{-36}$

5. $\sqrt{-20}$ 6. $\sqrt{-32}$

7. $\sqrt{-75}$ 8. $\sqrt{-72}$

9. $-\sqrt{-27}$ 10. $\sqrt{-288}$

11. $-\sqrt{-48}$ 12. $-\sqrt{-80}$

▼ Write the following as i, -1, $-i$, or 1.

13. i^2 14. i^3 15. i^5

16. i^{10} 17. i^{15} 18. i^{12}

19. i^{50} 20. i^{40} 21. i^{33}

▼ Combine the following complex numbers.

22. $(2 + 5i) + (4 + 2i)$ 23. $(6 + 3i) + (4 - 5i)$

24. $(7 + 3i) + (-6 - 2i)$ 25. $(-4 + i) + (7 - 3i)$

26. $(6 - 2i) - (4 + 2i)$ **27.** $(7 - 3i) - (-6 + 2i)$ **36.** $(3 - 2i) + (5 - 4i) - (6 + 2i)$

28. $(-4 - i) - (6 - 3i)$ **29.** $(-5 - i) - (7 - 5i)$ **37.** $(5 - 2i) + (8 + 8i) - (6 - 3i)$

30. $(8 - 3i) + (4 + 3i)$ **31.** $(6 - 5i) - (7 - 5i)$ **38.** $(4 + 7i) - (6 + 6i) + (2 - 3i)$

32. $(6 + 2i) - (6 - 2i)$ **33.** $(14 - 3i) - (14 - 3i)$ **39.** $(4 + 7i) - [(6 + 6i) + (2 - 3i)]$

34. $(4 + 2i) + (6 - 3i) + (2 - i)$ **40.** $(2 + 8i) - (4 + 2i) - (6 - 9i)$

35. $(7 + 2i) + (4 - 5i) + (6 - 2i)$ **41.** $2 + 8i - [(4 + 2i) - (6 - 9i)]$

5.8   Multiplication and Division of Complex Numbers

▮ ▮ ▮ ▮ ▮

Multiplication of Complex Numbers

• • • • •

Complex numbers are multiplied together by using the **FOIL** method that we used for multiplying binomials.

EXAMPLE 1   Multiply: $(3 + 5i)(4 - 2i)$.

SOLUTION

$$\overset{\text{F} \quad\quad \text{O} \quad\quad\quad \text{I} \quad\quad \text{L}}{(3 + 5i)(4 - 2i) = 3 \cdot 4 + 3(-2i) + 4 \cdot 5i + 5i(-2i)}$$

$$= 12 - 6i + 20i - 10i^2$$

Now we substitute -1 for i^2 and combine similar terms:

$$12 - 6i + 20i - 10i^2 = 12 - 6i + 20i - 10(-1)$$

$$= 12 - 6i + 20i + 10$$

$$= 22 + 14i$$

Notice that the product of the two complex numbers $3 + 5i$ and $4 - 2i$ is again a complex number, $22 + 14i$.   ▮▮

EXAMPLE 2   Multiply: $3i(4 - 5i)$.

SOLUTION $3i(4 - 5i) = 12i - 15i^2$

$= 12i - 15 \cdot (-1)$ Substitute $i^2 = -1$.

$= 12i + 15$

$= 15 + 12i$ In $a + bi$ form   ▮▮

EXAMPLE 3   Multiply: $(3 + 2i)(3 - 2i)$.

SOLUTION $(3 + 2i)(3 - 2i) = 9 - 6i + 6i - 4i^2$ The middle two terms combine to 0.

$= 9 - 4(-1)$ $i^2 = -1$

$= 9 + 4$

$= 13$ ▮▮

In this last example, you should notice that the middle term is 0, and we obtained a real number, 13, for our final result. The complex numbers $3 + 2i$ and $3 - 2i$ are called *conjugates* of each other.

> The complex numbers $a + bi$ and $a - bi$ are called **complex conjugates** of each other. The product of two complex conjugates is always a real number.

Division of Complex Numbers

• • • • •

Complex numbers are divided using a method similar to that used to rationalize binomial denominators containing square roots.

To divide complex numbers:

Write the quotient as a fraction, then multiply both numerator and denominator by the **conjugate** of the denominator.

EXAMPLE 4   Divide: $\dfrac{3 + 4i}{1 - 3i}$.

SOLUTION   Multiply numerator and denominator by the *conjugate* of the denominator, $1 + 3i$:

$$\frac{3 + 2i}{1 - 3i} = \frac{(3 + 2i)(1 + 3i)}{(1 - 3i)(1 + 3i)}$$

$$= \frac{3 + 11i + 6i^2}{1 - 9i^2} \qquad \text{Substitute } i^2 = -1.$$

$$= \frac{3 + 11i + 6(-1)}{1 - 9(-1)} \qquad \text{Simplify.}$$

$$= \frac{-3 + 11i}{10} \qquad \text{Write the result in complex-number form.}$$

$$= -\frac{3}{10} + \frac{11}{10}i \qquad \text{This is again a complex number.} \qquad ▮▮$$

EXAMPLE 5   Divide: $\dfrac{3 - 2i}{3i}$.

SOLUTION   The conjugate of $0 + 3i$ is $0 - 3i$ or $-3i$. However, since $3i^2$ is the real number -3, we can simplify our work by multiplying numerator and denominator by i.

$$\frac{3 - 2i}{3i} = \frac{(3 - 2i) \cdot i}{3i \cdot i}$$

$$= \frac{3i - 2i^2}{3i^2} \qquad \text{Simplify.}$$

$$= \frac{3i + 2}{-3}$$

$$= \frac{3i}{-3} + \frac{2}{-3} \qquad \text{Rewrite as a sum of terms.}$$

$$= -i - \frac{2}{3} \qquad \text{Simplify where possible.}$$

$$= -\frac{2}{3} - i \qquad \text{In } a + bi \text{ form} \qquad ▮▮$$

5.8   Exercises

▮▮▮▮▮

▼ Multiply.

1. $6(2 + 5i)$

2. $-3(4 - i)$

3. $2i(3 + 2i)$

4. $5i(7 - 2i)$

5. $(2 + 3i)(5 + 2i)$

6. $(5 - 3i)(2 + 2i)$

7. $(4 - i)(2 + i)$ **8.** $(7 + 2i)(4 - 3i)$

9. $(5 - 2i)(6 - 2i)$ **10.** $(-2 - 3i)(2 - i)$

11. $(-5 - i)(4 - 3i)$ **12.** $(-6 + 5i)(-5 - 2i)$

13. $(4 + 3i)^2$ **14.** $(3 - 2i)^2$

15. $(5 - i)^2$ **16.** $(4 + 3i)(4 - 3i)$

17. $(3 - 5i)(3 + 5i)$

▼ Write the conjugate of each of the following
complex numbers.

18. $2 + 5i$ **19.** $3 - 4i$ **20.** $-6 - 5i$

21. $-4 + i$ **22.** $-7i$ **23.** $5i$

▼ Divide and write the quotient in complex-number
form.

24. $\dfrac{3 + 5i}{2i}$ **25.** $\dfrac{4 - 3i}{i}$

26. $\dfrac{4 + 5i}{-3i}$ **27.** $\dfrac{6 - i}{-4i}$

28. $\dfrac{4}{2 + i}$ **29.** $\dfrac{6}{3 + i}$

30. $\dfrac{-2}{1 + 2i}$ **31.** $\dfrac{-3}{1 - 2i}$

32. $\dfrac{2 + i}{3 + i}$ **33.** $\dfrac{4 + 2i}{2 + i}$

34. $\dfrac{5 + 4i}{2 + 3i}$ **35.** $\dfrac{6 - i}{4 - i}$

36. $\dfrac{3 - 3i}{2 + 5i}$ **37.** $\dfrac{4 + 7i}{4 - 3i}$

38. Write the reciprocal of $2 + i$ as a complex number.

39. Write the reciprocal of $-3 - i$ as a complex number.

SUMMARY—CHAPTER 5

• • • • • • • • ▼

EXAMPLES

A number b is a **square root** of a if $b^2 = a$. $\sqrt{\ \ }$ is called a **radical sign**.

4 and -4 are square roots of 16, since $4^2 = 16$ and $(-4)^2 = 16$.

$\sqrt{16} = 4$

The number under a radical sign is called the **radicand**. The symbol $\sqrt{\ \ }$ indicates a **positive** square root called the **principal square root**:

$$\sqrt{a^2} = |a|$$

A number b is the **nth root** of a if $b^n = a$.

$\sqrt[5]{32} = 2$ because $2^5 = 32$

$\sqrt[n]{a} \cdot \sqrt[n]{b} = \sqrt[n]{a \cdot b}, \quad a \ge 0, b \ge 0$

a. $\sqrt{3} \cdot \sqrt{12} = \sqrt{3 \cdot 12} = \sqrt{36} = 6$

b. $\sqrt{32} = \sqrt{16 \cdot 2}$
$= \sqrt{16} \cdot \sqrt{2} = 4\sqrt{2}$

Like or **similar radicals** have the same radicand and the same index.

To add (or subtract) like radicals, add (or subtract) their coefficients and keep the same radicand.

a. $3\sqrt{7} + 5\sqrt{7} = 8\sqrt{7}$

b. $\sqrt{18} - \sqrt{32} = \sqrt{9 \cdot 2} - \sqrt{16 \cdot 2}$
$= 3\sqrt{2} - 4\sqrt{2}$
$= -\sqrt{2}$

$\sqrt[n]{\dfrac{a}{b}} = \dfrac{\sqrt[n]{a}}{\sqrt[n]{b}}, \quad a \ge 0, b > 0$

a. $\sqrt{\dfrac{49}{36}} = \dfrac{\sqrt{49}}{\sqrt{36}} = \dfrac{7}{6}$

b. $\dfrac{\sqrt{75}}{\sqrt{3}} = \sqrt{\dfrac{75}{3}} = \sqrt{25} = 5$

To rationalize the denominator of a fraction that contains a radical, multiply the numerator and the denominator of the fraction by the smallest quantity that makes the denominator a perfect nth power (square, cube, etc.).

Rationalize the denominator in $\sqrt{\dfrac{3}{8}}$:

$$\sqrt{\dfrac{3}{8}} = \dfrac{\sqrt{3}}{\sqrt{8}} \cdot \dfrac{\sqrt{2}}{\sqrt{2}}$$

$$= \dfrac{\sqrt{6}}{\sqrt{16}} = \dfrac{\sqrt{6}}{4}$$

The **conjugate** of a binomial containing radicals is the same binomial with the sign of the second term changed.

The conjugate of $\sqrt{3} - 2$ is $\sqrt{3} + 2$.

To rationalize a binomial denominator containing radicals, multiply the numerator and denominator by the conjugate of the denominator.

Rationalize the denominator of $\dfrac{4}{\sqrt{3} - 2}$:

$$\dfrac{4}{\sqrt{3} - 2} = \dfrac{4}{(\sqrt{3} - 2)} \cdot \dfrac{(\sqrt{3} + 2)}{(\sqrt{3} + 2)}$$

$$= \dfrac{4(\sqrt{3} + 2)}{3 - 4} = \dfrac{4\sqrt{3} + 8}{-1}$$

$$= -4\sqrt{3} - 8$$

$a^{1/n} = \sqrt[n]{a}$, where n is a positive integer and $a \ge 0$ if n is even.

$27^{1/3} = \sqrt[3]{27} = 3$

$a^{m/n} = \left(\sqrt[n]{a}\right)^m = \sqrt[n]{a^m}$, for $n \neq 0$ and $a \geq 0$ if n is even.

The **imaginary unit** is $i = \sqrt{-1}$.

An **imaginary number** is a number of the form bi, where b is a real number.

A **complex number** is a number of the form $a + bi$, where a and b are real numbers. Here a is the **real part** and b is the **imaginary part**.

To add (or subtract) complex numbers add (or subtract) the real parts and add (or subtract) the imaginary parts.

Complex numbers are **multiplied** using the **FOIL** method.

The two complex numbers $a + bi$ and $a - bi$ are **complex conjugates** of each other.

To divide complex numbers, write the problem as a fraction, then multiply numerator and denominator by the conjugate of the denominator.

EXAMPLES

$27^{2/3} = \left(\sqrt[3]{27}\right)^2 = 3^2 = 9$

$3i$ and $-7i$ are imaginary numbers.

$6 + 3i$ and $-4 + 5i$ are complex numbers.

$(3 + 4i) + (5 - 2i) = 8 + 2i$
$(6 - 2i) - (3 + 4i) = 3 - 6i$

$$(2 + 3i)(5 - i) = 10 + 13i - 3i^2$$
$$= 10 + 13i - 3(-1)$$
$$= 13 + 13i$$

$4 - 3i$ and $4 + 3i$ are complex conjugates of each other.

$$\frac{4}{3 - i} = \frac{4(3 + i)}{(3 - i)(3 + i)}$$

$$= \frac{12 + 4i}{9 - i^2} = \frac{12 + 4i}{10}$$

$$= \frac{6}{5} + \frac{2}{5}i$$

REVIEW EXERCISES—CHAPTER 5

• • • • • • • ▼

▼ Find the square roots of the following numbers.

1. 36   2. 121

▼ Evaluate the following roots if possible.

3. $\sqrt{81}$ 4. $-\sqrt{81}$ 5. $\sqrt{-81}$

6. $\sqrt{t^2}$ 7. $\sqrt[3]{64}$ 8. $\sqrt[3]{-64}$

▼ Simplify each quantity as much as possible.

9. $\sqrt{24}$ 10. $\sqrt[3]{40}$ 11. $\sqrt{12x^5}$

▼ Multiply and simplify:

12. $\sqrt{6} \cdot \sqrt{8}$ 13. $\sqrt[3]{x^4} \cdot \sqrt[3]{x^2}$

14. $2\sqrt{3}(\sqrt{5} + 2\sqrt{3})$ 15. $3\sqrt{2}(\sqrt{2} - 3\sqrt{6})$

▼ Combine where possible:

16. $7\sqrt{3} - 2\sqrt{3}$ 17. $3\sqrt{2} + 4\sqrt{3}$

18. $2\sqrt{28} + 3\sqrt{63}$ 19. $2\sqrt[3]{3} + 3\sqrt{2}$

▼ Divide and simplify:

20. $\dfrac{\sqrt{28}}{\sqrt{7}}$ 21. $\dfrac{\sqrt[3]{56}}{\sqrt[3]{7}}$

▼ Rationalize the denominator and simplify:

22. $\dfrac{\sqrt{7}}{\sqrt{3}}$ 23. $\dfrac{3}{\sqrt[3]{4}}$

24. $\dfrac{\sqrt{3}}{\sqrt{8x}}$ 25. $\dfrac{3}{\sqrt{5} - \sqrt{2}}$

▼ Write in exponential form:

26. $\sqrt[3]{xy^2}$ 27. $(\sqrt{a})^5$

▼ Simplify as much as possible:

28. $-25^{1/2}$ 29. $(-25)^{1/2}$

30. $(-27)^{1/3}$ 31. $\left(\dfrac{4}{25}\right)^{-1/2}$

32. $27^{2/3}$ 33. $9^{-1/2} + 8^{-2/3}$

▼ Simplify using the properties of exponents. Assume that all variables are positive.

34. $(x^{1/2})^{3/4}$ 35. $x^{1/4} \cdot x^{1/3}$

36. $(x^{1/2} \cdot y^{3/4})^{2/3}$ 37. $\dfrac{x^{2/3}}{x^{1/2}}$

▼ Simplify and write in imaginary form:

38. $\sqrt{-16}$ **39.** $\sqrt{-50}$

40. $-\sqrt{-24}$

▼ Write as i, -1, $-i$, or 1:

41. i^3 **42.** i^{11} **43.** i^{46}

▼ Combine as indicated:

44. $(3 + 4i) + (5 - 2i)$ **45.** $(7 - 2i) - (3 - 7i)$

46. $(4 + 3i) - (7 - 5i) + (2 + 4i)$

47. $4 + 3i - [(7 - 5i) + (2 + 4i)]$

▼ Multiply:

48. $-3i(2 - 7i)$ **49.** $(7 + 2i)(5 - 4i)$

50. $(6 - 5i)^2$ **51.** $(3 + 5i)(3 - 5i)$

▼ Divide and write the quotient in complex-number form:

52. $\dfrac{5 - i}{5i}$ **53.** $\dfrac{2 + 3i}{-3i}$

54. $\dfrac{-4}{4 - i}$ **55.** $\dfrac{6 + i}{-3 + i}$

56. Explain why $i^{21} = i$.

CRITICAL THINKING EXERCISES—CHAPTER 5

•••••• ▼

▼ There is an error in each of the following. Can you find it?

1. $\sqrt{x + 2y} = \sqrt{x} + \sqrt{2y}$

2. $\sqrt{13} = \sqrt{4 + 9} = \sqrt{4} + \sqrt{9} = 2 + 3 = 5$

3. $4^{2/3} = \sqrt{4^3} = \sqrt{64} = 8$

4. $(13 + 4i) - (5 - 2i) = 8 + 2i$

5. $(2 + 3i)(3 + 4i) = 6 + 12i^2 = 6 - 12 = -6$

6. $8\sqrt{5} - 3\sqrt{3} = 5\sqrt{2}$

7. $\sqrt{81} = \pm 9$

8. $(-4)^{1/2} = -2$

9. $x^{1/3} \cdot x^{1/2} = x^{1/6}$

10. $(2 + 3i)^2 = 4 + 9i^2 = 4 - 9 = -5$

11. $\dfrac{2 + 3i}{3i} = \dfrac{2 + \overset{1}{\cancel{3i}}}{\underset{1}{\cancel{3i}}} = \dfrac{2 + 1}{1} = 3$

12. The sum of two imaginary numbers is always an imaginary number.

13. The product of two complex numbers is always a real number.

14. Every complex number is a real number.

NAME

❙❙❙❙❙ CLASS

ACHIEVEMENT TEST—CHAPTER 5

• • • • • • • • ▼

1. Find the square roots of 81.

1. _____

▼ Evaluate the following if possible:

2. $\sqrt{64}$

2. _____

3. $\sqrt{-64}$

3. _____

4. $-\sqrt{64}$

4. _____

5. $\sqrt[3]{64}$

5. _____

6. $-\sqrt[3]{-64}$

6. _____

▼ Simplify:

7. $\sqrt{45}$

7. _____

8. $\sqrt{18x^3}$

8. _____

▼ Multiply and simplify:

9. $\sqrt{3x^3} \cdot \sqrt{6x}$

9. _____

10. $4\sqrt{5}(\sqrt{5} - 2\sqrt{15})$

10. _____

▼ Combine where possible:

11. $3\sqrt{32} - \sqrt{50}$ 11. _____

12. $\sqrt[3]{16} - 5\sqrt{8}$ 12. _____

▼ Divide and simplify:

13. $\dfrac{\sqrt[3]{162}}{\sqrt[3]{6}}$ 13. _____

14. Write the conjugate of $3\sqrt{2} + 6$.   14. _____

▼ Rationalize the denominator and simplify.

15. $\dfrac{\sqrt{6}}{\sqrt{2}}$ 15. _____

16. $\dfrac{\sqrt{5}}{\sqrt{12x}}$ 16. _____

17. $\dfrac{6}{\sqrt[3]{9}}$ 17. _____

18. $\dfrac{4}{\sqrt{11} - \sqrt{7}}$ 18. _____

▼ Write in exponential form.

19. $\sqrt[4]{b^7}$ 19. _____

▼ Simplify as much as possible.

20. $-49^{1/2}$

21. $(-36)^{1/2}$

22. $\left(\dfrac{16}{9}\right)^{-1/2}$

23. $81^{-3/4} - 81^{-1/2}$

24. $\left(x^{2/3}\right)^{3/8}$

25. $x^{2/3} \cdot x^{3/4}$

26. $\dfrac{x^{4/5}}{x^{2/3}}$

27. $\left(x^{3/4} \cdot y^{3/8}\right)^{1/3}$

28. $\sqrt{-44}$

29. $(-3 + 5i) - (1 - 2i)$

30. $(-3 + 5i)(1 - 2i)$

31. $\dfrac{-3 + 5i}{1 - 2i}$

20. _____

21. _____

22. _____

23. _____

24. _____

25. _____

26. _____

27. _____

28. _____

29. _____

30. _____

31. _____

Graphs and Functions

INTRODUCTION

The old adage that a picture is worth a thousand words may be more true in mathematics than anywhere else.

In previous chapters we have worked with equations in many ways. In this chapter we will investigate methods of illustrating equations graphically. To anyone studying mathematics or science, the concept of a mathematical function is very important. In this chapter, functions and their properties will be discussed.

CHAPTER 6—NUMBER KNOWLEDGE

The Golden Rectangle

Draw any rectangle that seems pleasing to your eye. Measure the length and width and use these measurements to form the ratio of the length to the width. In most cases this ratio will be close to what is called the *golden ratio,* and the rectangle will be called a *golden rectangle.*

$$\frac{\text{length}}{\text{width}} \approx \frac{1.6}{1}$$

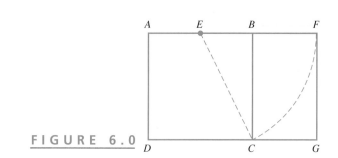

FIGURE 6.0

To construct a golden rectangle with a compass and a straightedge, begin with any square *ABCD,* as in Figure 6.0. Treating the midpoint of side *AB* (point *E*) as the center of a circle with radius *EC,* draw an arc, so that it intersects the extension of side *AB* at the point *F.* Complete the rectangle by drawing in *BF, FG,* and *CG.* The resulting rectangle, *AFGD,* is called a **golden rectangle;** *BFGC* is also a golden rectangle.

Many everyday objects around us are formed by rectangles using measurements that are close to golden rectangles: cereal boxes, books, 3 × 5 index cards, windows, bricks, desk and table tops, buildings.

Many famous artists such as Leonardo da Vinci, Albert Durer, and Pieter Mondriaan used the golden rectangle extensively in their paintings.

6.1　The Rectangular Coordinate System

In large cities, streets and avenues are often numbered to make it easy to find a particular place. In a similar way, finding points on a piece of paper is made convenient by a numbering system called a **rectangular coordinate system** (also called a **Cartesian coordinate system**).

In previous chapters we made extensive use of the number line. Each point on the number line has a specific number associated with it. Now let's draw another line that intersects the number line at an angle of 90 degrees (a right angle) and passes through the point zero. This point of intersection is called the **origin.** These two lines are called the **axes,** and they determine a flat surface called a **plane.** The horizontal line is called the **x-axis,** and the vertical line is called the **y-axis.** The y-axis is numbered using the same scale that is used on the number line. Starting with zero at the point of intersection (origin), we use positive numbers as we travel in an upward direction and negative numbers in the downward direction.

The *x*-axis and *y*-axis divide the plane into four regions called **quadrants.** Starting in the upper right quadrant, they are numbered I through IV as we move counterclockwise (See Figure 6.1).

There is a scheme for associating each point in the plane with a pair of numbers. For example, consider the **ordered pair** of numbers (4, −2). The first number, 4, is called the **x-coordinate** or **abscissa,** and it tells us how far to go to the left (−) or to the right (+). The second number, −2, is called the **y-coordinate** or the **ordinate,** and it tells us how far to go upward (+) or downward (−).

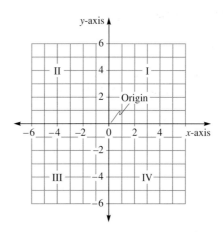

FIGURE 6.1. The Quadrants

So, in the case of $(4, -2)$, we start from the origin and travel 4 units to the right, since 4 is positive. From there, the second number, -2, tells us to go 2 units in a downward direction, since -2 is negative. Study the location of this point in Figure 6.2.

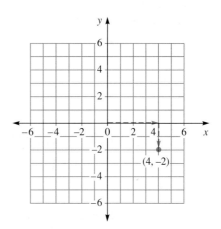

FIGURE 6.2.

EXAMPLE 1   Plot the point $(-3, 4)$.

SOLUTION   We start at the origin (see Figure 6.3). The *x*-coordinate is a negative number (-3), so we move 3 units to the left. From there, the *y*-coordinate (4) is positive, so we travel 4 units upward to find the location of the point $(-3, 4)$.   ∎

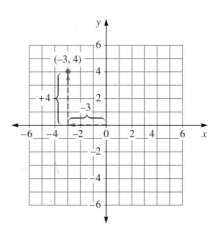

FIGURE 6.3.

The following examples are illustrated in Figure 6.4.

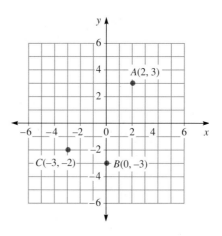

FIGURE 6.4.

EXAMPLE 2 **(a)** Locate the point $A(2, 3)$ on a rectangular coordinate system.

(b) Plot the point $B(0, -3)$ in the plane.

(c) Graph the point $C(-3, -2)$ in the plane.

SOLUTION **(a)** In the ordered pair $(2, 3)$, the x-coordinate (2) indicates that we move 2 units to the right of 0. From there the y-coordinate (3) directs us to go 3 units upward.

(b) In the ordered pair $(0, -3)$ the x-coordinate (0) tells us that we do not move right or left at all. The y-coordinate (-3) tells us to move downward 3 units.

(c) To graph the ordered pair $(-3, -2)$, we start at the origin, move 3 units to the left (-3), and then downward 2 units (-2). ▮▮

6.1   Exercises

▮ ▮ ▮ ▮ ▮

▼ On the graph provided, label the axes properly and locate the points.

1. **a.** $(1, 4)$
    b. $(3, 5)$
    c. $(4, -2)$
    d. $(1, -1)$
    e. $(-3, -3)$

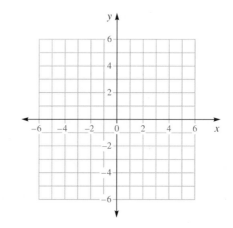

2. **a.** $(-3, 4)$
    b. $(-2, -5)$
    c. $(1, -4)$
    d. $(5, 5)$
    e. $(3, -3)$

4. **a.** $\left(\frac{1}{2}, \frac{1}{2}\right)$ (estimate)
    b. $(4.5, 3)$
    c. $\left(-1\frac{1}{2}, -2\frac{1}{2}\right)$ (estimate)
    d. $(-1.5, 2.5)$ (estimate)
    e. $(3.4, -1.8)$ (estimate)

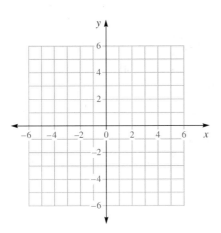

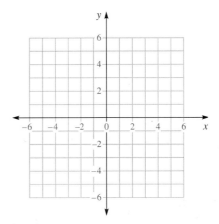

3. **a.** $(-4, 0)$
    b. $(0, 0)$
    c. $(0, 5)$
    d. $(-5, -5)$
    e. $\left(2\frac{1}{2}, 3\right)$ (estimate)

▼ In Exercise 5 give the ordered pair that represents each of the points in Figure 6.5.

5. A
    B
    C
    D
    E
    F
    G
    H

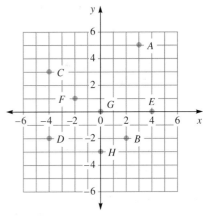

FIGURE 6.5.

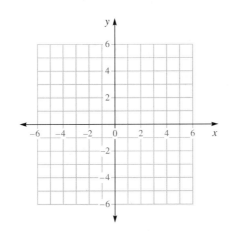

6. In which quadrant(s) is it possible for both coordinates of a point to be equal to each other?

6.2   Graphs of Linear Equations

Now that we know how to graph points in the plane, the next step is to graph *lines* in the plane.

The **graph of a linear equation with two variables x and y** is the set of points representing the ordered pairs (x, y) that are the solutions to the equation.

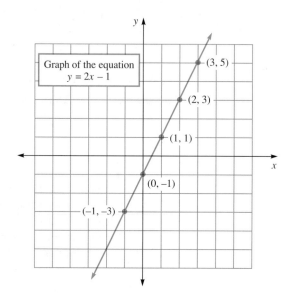

FIGURE 6.6.

Consider the equation $y = 2x - 1$ (see Figure 6.6). The points we are to graph are represented by the values for x and y that make the equation a true statement, that is, the solutions. We find these values by arbitrarily choosing a value for x and calculating a corresponding value for y. For example, if $x = 1$ we have

$$y = 2x - 1$$
$$y = 2 \cdot \boxed{1} - 1 \qquad \text{Substitute 1 for } x.$$
$$y = 2 - 1$$
$$y = 1$$

So the ordered pair $(x, y) = (1, 1)$ is a solution to the given equation. If $x = 0$, we have

$$y = 2x - 1$$
$$y = 2 \cdot \boxed{0} - 1 \qquad \text{Substitute 0 for } x.$$
$$y = 0 - 1$$
$$y = -1$$

So $(0, -1)$ is also a solution. Usually, we tabulate these values in a table such as the following:

Choose x	Calculate y	x	y	Ordered pairs
$x = 0$	$y = 2 \cdot 0 - 1 = 0 - 1 = -1$	0	-1	$(0, -1)$
$x = 1$	$y = 2 \cdot 1 - 1 = 2 - 1 = 1$	1	1	$(1, 1)$
$x = 2$	$y = 2 \cdot 2 - 1 = 4 - 1 = 3$	2	3	$(2, 3)$
$x = 3$	$y = 2 \cdot 3 - 1 = 6 - 1 = 5$	3	5	$(3, 5)$
$x = -1$	$y = 2 \cdot -1 - 1 = -2 - 1 = -3$	-1	-3	$(-1, -3)$

Remember that we picked any value for x that we wanted and calculated the value of y that corresponded to it. If we graph the five ordered pairs that we found, we notice an interesting fact.

When the points are connected, they all lie on a straight line. This is true of any **linear equation.** Although only two points are absolutely necessary to determine a straight line, we usually graph a third point as a check of our work.

EXAMPLE 1   Graph $x + y = 4$.

SOLUTION   It is easier in many cases to solve for y before making our substitution for x. Since we are free to choose any values we like for x, it is best to choose values that give easy calculations and will not run off the graph paper (see Figure 6.7 on next page).

$$x + y = 4$$
$$y = 4 - x$$

Choose x	Calculate y	x	y	Ordered pairs
0	$y = 4 - 0 = 4$	0	4	$(0, 4)$
2	$y = 4 - 2 = 2$	2	2	$(2, 2)$
4	$y = 4 - 4 = 0$	4	0	$(4, 0)$

Now we plot the points and connect them.

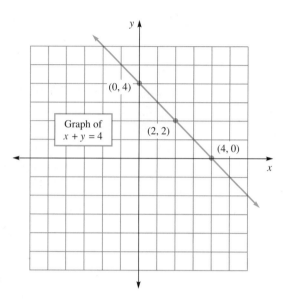

FIGURE 6.7.

Graphing Using x- and y-Intercepts

In many cases the easiest points to plot are the places where the line crosses the x- and y-axes; these are called the **x-intercept** and the **y-intercept.** The x-intercept occurs when y equals zero and the y-intercept is found by letting x equal zero.

> **To graph an equation using intercepts:**
>
> **1.** Find the y-intercept by letting $x = 0$ and solving for y;
>
> **2.** find the x-intercept by letting $y = 0$ and solving for x;
>
> **3.** plot one additional point as a check.

EXAMPLE 2   Using the intercept method, graph $2x - 5y = 10$.

SOLUTION $2x - 5y = 10$

Let $x = 0$: $2(0) - 5y = 10$ Replace x by 0.

                      $-5y = 10$ Divide both sides by −5.

                        $y = -2$ The y-intercept is (0, −2).

Let $y = 0$: $2x - 5(0) = 10$ Replace y by 0.

                      $2x = 10$ Divide both sides by 2.

                        $x = 5$ The x-intercept is (5, 0).

Let $y = -1$: $2x - 5(-1) = 10$ Replace y by -1.

$2x + 5 = 10$ Subtract 5 from both sides.

$2x = 5$ Divide both sides by 2.

$x = \dfrac{5}{2}$ or $2\frac{1}{2}$ Yields ordered pair $(2\frac{1}{2}, -1)$.

Plotting these ordered pairs in Figure 6.8 gives us the graph of the equation.   ▌▌

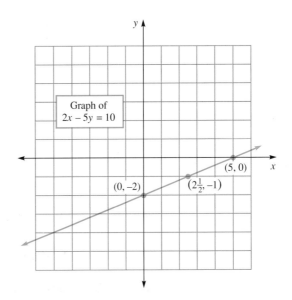

Graph of
$2x - 5y = 10$

$(5, 0)$

$(0, -2)$ $(2\frac{1}{2}, -1)$

FIGURE 6.8.

EXAMPLE 3   Graph $y = 4$.

SOLUTION   As we can immediately see, the x-term is missing. Think of this as having the coefficient of the x-term equal to 0:

$$y = 4 \quad \text{is the same as} \quad y = 0 \cdot x + 4$$

Now we make the table as before.

Choose x	Calculate y	x	y	Ordered pairs
0	$y = 0 \cdot x + 4 = 0(0) + 4 = 4$	0	4	$(0, 4)$
2	$y = 0 \cdot x + 4 = 0(2) + 4 = 4$	2	4	$(2, 4)$
-2	$y = 0 \cdot x + 4 = 0(-2) + 4 = 4$	-2	4	$(-2, 4)$

As you can see, no matter what value we choose for x, the value of y is always equal to 4. Graphing the ordered pairs yields a horizontal line on which the value of y is 4, no matter what the value of x. (See Figure 6.9.)

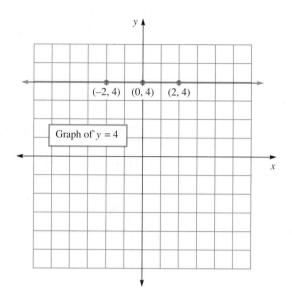

FIGURE 6.9.
▮▮

> The graph of an equation of the form $y = b$ (where b is some constant) is a
> **horizontal line** through all of the points where $y = b$. There is no
> x-intercept.

EXAMPLE 4   Graph $x = -2$.

SOLUTION   In this case the y-term is considered to have a zero coefficient, and we can say
that

$$x = -2 \quad \text{is the same as} \quad x + 0 \cdot y = -2.$$

We see that x is always equal to -2 regardless of the value of y. Some typical
ordered pairs would be $(-2, -2)$, $(-2, 4)$, and $(-2, 1)$. Graphing them gives us
a vertical line, and $x = -2$ at every point on this line (see Figure 6.10).

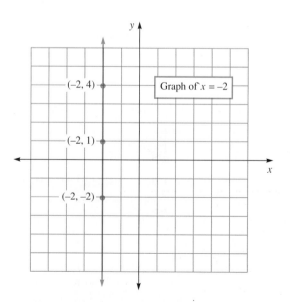

FIGURE 6.10.
▮▮

The graph of an equation of the form $x = c$ (where c is some constant) is a **vertical line** through all of the points where $x = c$. There is no y-intercept.

To graph a linear equation:

1. Choose an arbitrary value for one of the variables and calculate the corresponding value of the remaining variable. Repeat the procedure three times, yielding three ordered pairs. The x- and y-intercepts are often easiest to find.

2. Plot the three ordered pairs obtained in Step 1.

3. Draw a straight line through the points.

Vertical and horizontal lines:

1. The graph of the equation $x = c$ (where c is a constant) is a vertical line through all points where $x = c$.

2. The graph of the equation $y = b$ (where b is a constant) is a horizontal line through all points where $y = b$.

6.2   Exercises

▼ Graph each of the following equations.

1. $y = x - 5$

2. $x - y = 3$

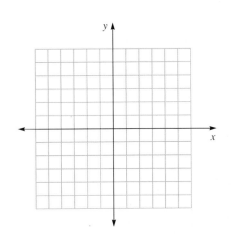

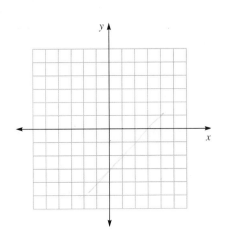

3. $3x + y = 4$

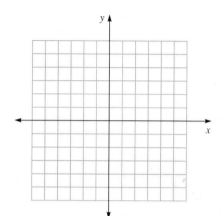

4. $y - x = 0$

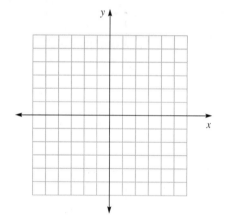

5. $y = 4x - 1$

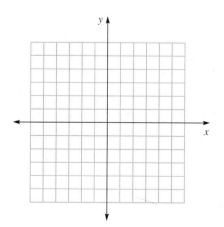

6. $2y + 3x = 6$

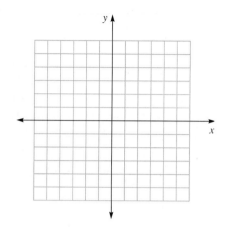

7. $y + 2x = 6$

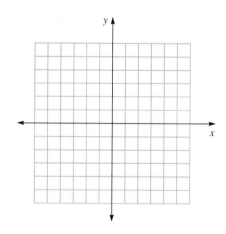

8. $4x - 2y = 12$

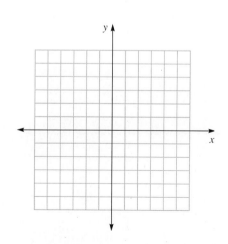

9. $5y = 2x - 5$

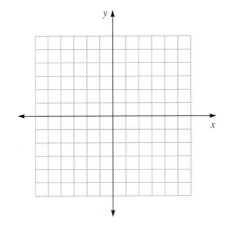

12. $2x - 6y = 12$

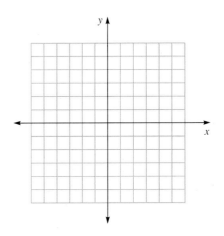

10. $2x + 3y = 1$

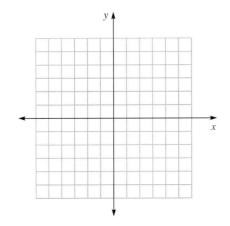

13. $3 - y = 2x$

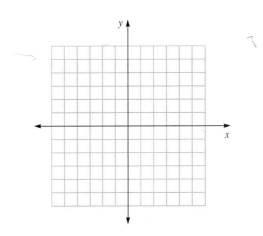

11. $3x + 4y = 12$

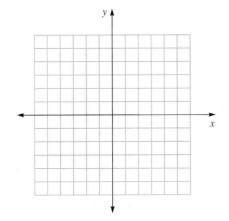

14. $y = \frac{1}{3}x + 2$

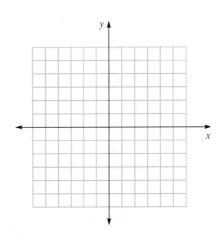

15. $y = -\frac{3}{4}x - 1$

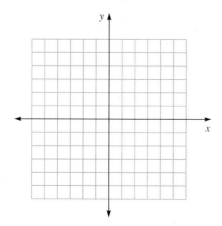

18. $x = -2$

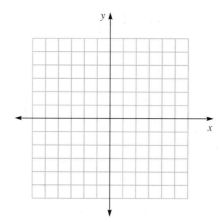

16. $y = -\frac{2}{5}x + 2$

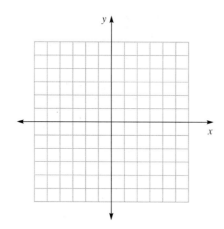

19. $y = 2$

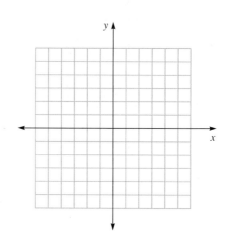

17. $x = -3$

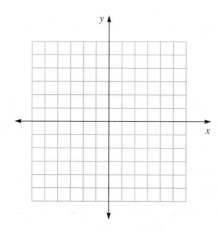

20. $y = -1$

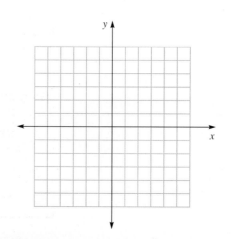

21. $x = 0$

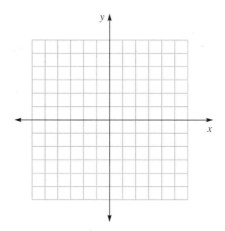

23. $x = \frac{7}{2}$

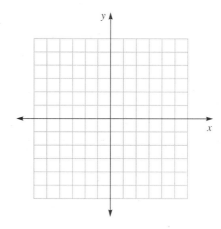

22. $y = 0$

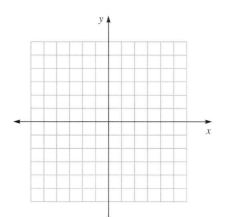

24. $y = -\frac{3}{2}$

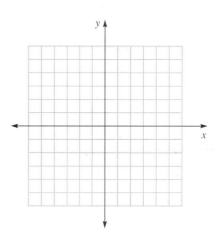

6.3   Graphing Curves

Graphing a straight line requires plotting only two points. (We plot a third as a check.) Graphing curves uses the same procedure except that more than two points need to be plotted.

> **To graph a curve:**
>
> **1.** make a table of ordered pairs by choosing values for one variable and calculating a corresponding value for the remaining variable;
>
> **2.** plot a sufficient number of points to determine the graph;
>
> **3.** connect the points with a smooth curve.

EXAMPLE 1   Graph $y = x^2 - 2$.

SOLUTION

Choose x	Calculate y	x	y	Ordered pairs
0	$y = (0)^2 - 2 = 0 - 2 = -2$	0	−2	(0, −2)
1	$y = (1)^2 - 2 = 1 - 2 = -1$	1	−1	(1, −1)
−1	$y = (-1)^2 - 2 = 1 - 2 = -1$	−1	−1	(−1, −1)
2	$y = (2)^2 - 2 = 4 - 2 = 2$	2	2	(2, 2)
−2	$y (-2)^2 - 2 = 4 - 2 = 2$	−2	2	(−2, 2)
3	$y = (3)^2 - 2 = 9 - 2 = 7$	3	7	(3, 7)
−3	$y = (-3)^2 - 2 = 9 - 2 = 7$	−3	7	(−3, 7)

We plot the ordered pairs (points) and connect the points using a smooth curve. (See Figure 6.11.)

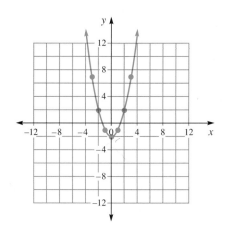

FIGURE 6.11.   ▮▮

EXAMPLE 2   Graph $y = x^2 + x - 3$.

SOLUTION

Choose x	Calculate y	x	y	Ordered pairs
0	$y = (0)^2 + 0 - 3 = -3$	0	−3	(0, −3)
1	$y = (1)^2 + 1 - 3 = -1$	1	−1	(1, −1)
−1	$y = (-1)^2 - 1 - 3 = -3$	−1	−3	(−1, −3)
2	$y = (2)^2 + 2 - 3 = 3$	2	3	(2, 3)
−2	$y = (-2)^2 - 2 - 3 = -1$	−2	−1	(−2, −1)
3	$y = (3)^2 + 3 - 3 = 9$	3	9	(3, 9)
−3	$y = (-3)^2 - 3 - 3 = 3$	−3	3	(−3, 3)
−4	$y = (-4)^2 - 4 - 3 = 9$	−4	9	(−4, 9)

Now we plot the points and connect them with a smooth curve. (See Figure 6.12.)

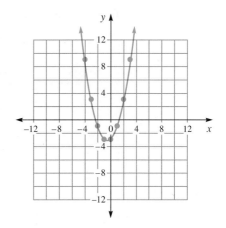

F I G U R E   6 . 1 2 .

6.3   Exercises

▼ Graph the following curves.

3. $y = x^2 + 3x + 1$

1. $y = x^2$

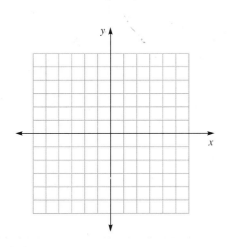

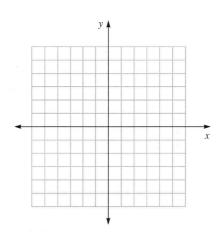

4. $y = -2 + x^2$

2. $y = x^2 - 4$

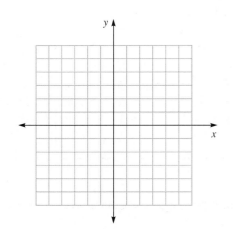

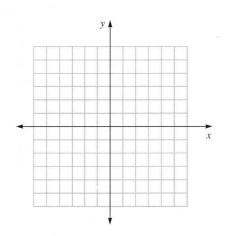

5. $y = x^2 + 2x$

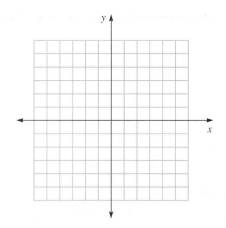

6. $y = -2x^2 + 3$

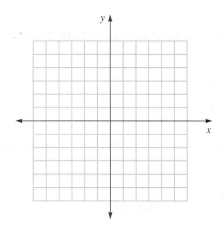

7. $y = x^2 + 2x - 3$

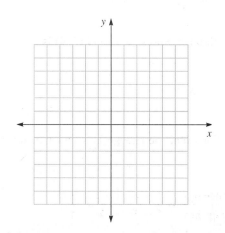

8. $y = x^3$

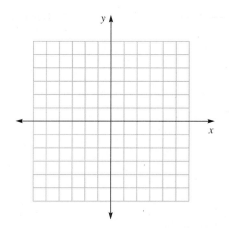

9. $y = -x^3$

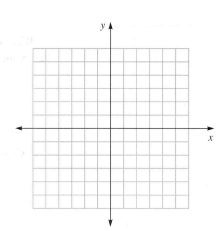

10. $y = x^3 - 2x$

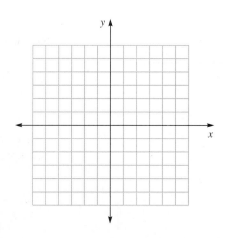

6.4   The Slope of a Line

▌ ▌ ▌ ▌ ▌

Consider the straight line drawn through the points $(1, 1)$ and $(5, 4)$. (See Figure 6.13.)

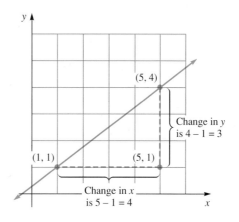

FIGURE 6.13.

We can calculate the change in both y and x as we go from the point $(1, 1)$ to the point $(5, 4)$

$$\text{change in } y \text{ is } 4 - 1 = 3$$

$$\text{change in } x \text{ is } 5 - 1 = 4$$

The **slope,** or *steepness,* of a line we denote by the letter m and define it as

$$\text{slope} = m = \frac{\text{change in } y}{\text{change in } x}$$

So in our example,

$$\text{slope} = m = \frac{\text{change in } y}{\text{change in } x} = \frac{4 - 1}{5 - 1} = \frac{3}{4}$$

We say that the slope of the line through points $(1, 1)$ and $(5, 4)$ is

$$m = \frac{3}{4}$$

It is interesting to note that if we find the change in y and x in the opposite direction, the result is the same:

$$\text{slope} = \frac{\text{change in } y}{\text{change in } y} = \frac{1 - 4}{1 - 5} = \frac{-3}{-4} = \frac{3}{4}$$

In more general terms, the definition of slope is the following:

> If (x_1, y_1) and (x_2, y_2) are any two points on a line, then **the slope of the line** is given by
>
> $$\text{slope} = m = \frac{\text{change in } y}{\text{change in } x} = \frac{y_2 - y_1}{x_2 - x_1}$$

EXAMPLE 1   Find the slope of the line drawn through the points $(-3, -3)$ and $(4, 2)$. (See Figure 6.14.)

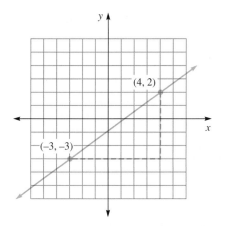

FIGURE 6.14.

SOLUTION   We let $(x_1, y_1) = (-3, -3)$ and $(x_2, y_2) = (4, 2)$. Then

$$\text{slope} = m = \frac{\text{change in } y}{\text{change in } x} = \frac{y_2 - y_1}{x_2 - x_1} = \frac{2 - (-3)}{4 - (-3)} = \frac{2 + 3}{4 + 3} = \frac{5}{7}$$

Remember, we could have found the changes in x and y in the reverse direction:

$$m = \frac{y_1 - y_2}{x_1 - x_2} = \frac{-3 - 2}{-3 - 4} = \frac{-5}{-7} = \frac{5}{7}$$

which is the same result as before.   ▮▮

STOP   Never put the y's in one order and the x's in another; that is,

$$m = \frac{y_2 - y_1}{x_1 - x_2} = \frac{2 - (-3)}{-3 - 4} = \frac{5}{-7} \text{ is not correct.}$$

EXAMPLE 2   Find the slope of the line containing the points $(-3, 2)$ and $(3, -1)$; see Figure 6.15.

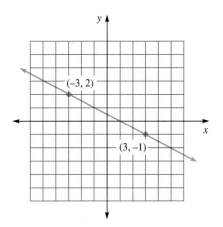

FIGURE 6.15.

SOLUTION   Let $(x_1, y_1) = (-3, 2)$ and $(x_2, y_2) = (3, 1)$. Then

$$m = \frac{y_2 - y_1}{x_2 - x_1}$$

$$= \frac{-1 - 2}{3 - (-3)} = \frac{-3}{6} = -\frac{1}{2}$$

EXAMPLE 3   Find the slope of the line containing the points $(-2, 2)$ and $(3, 2)$. (See Figure 6.16.)

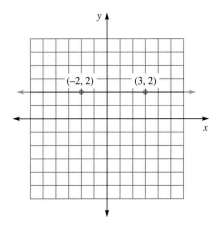

FIGURE 6.16.

SOLUTION $$m = \frac{y_2 - y_1}{x_2 - x_1} = \frac{2 - 2}{3 - (-2)} = \frac{0}{5} = 0$$

The slope is 0.

Note that *the slope of any horizontal line is 0:*

$$m = \frac{\text{vertical change}}{\text{horizontal change}} = \frac{0}{\text{horizontal change}} = 0$$

EXAMPLE 4   Find the slope of the line through the points $(-2, 3)$ and $(-2, -1)$; see Figure 6.17.

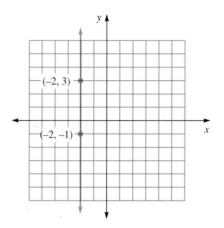

FIGURE 6.17.

SOLUTION

$$m = \frac{y_2 - y_1}{x_2 - x_1} = \frac{-1 - 3}{-2 - (-2)} = \frac{-4}{0}$$

The slope is *undefined* since we have 0 in the denominator.   ▮▮

Note that *the slope of any vertical line is undefined,* since

$$m = \frac{\text{vertical change}}{\text{horizontal change}} = \frac{\text{vertical change}}{0} \qquad \text{which is undefined}$$

The slope of the line in Example 1 is a positive number $(+\frac{5}{7})$, and the line slants upward as we move from left to right (see Figure 6.14). In Example 2, the slope of the line is negative $(-\frac{1}{2})$, and the line slants downward as we move from left to right (see Figure 6.15).

We know quite a bit about the nature of a line just by looking at the number that represents its slope. This is summarized in the following chart:

Slope	Graph of Line
positive	slants upward as it goes from left to right
negative	slants downward as it goes from left to right
0	horizontal line
undefined	vertical line

Slopes of Parallel and Perpendicular Lines

• • • • •

You can tell whether two lines are parallel (never intersect) or perpendicular (intersect at right angles) by looking at their slopes.

> **Given two different nonvertical lines with slopes m_1 and m_2:**
>
> **1.** If $m_1 = m_2$, the lines are parallel.
>
> **2.** If $m_1 = -\dfrac{1}{m_2}$, the lines are perpendicular. In other words, the slopes are negative reciprocals of each other (or $m_1 \cdot m_2 = -1$).

EXAMPLE 5 Show that the line thorough the points $(-1, 1)$ and $(2, 7)$ is parallel to the line thorugh the points $(0, -1)$ and $(3, 5)$. The two lines are pictured in Figure 6.18.

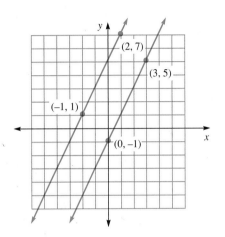

FIGURE 6.18.

SOLUTION

$$m_1 = \frac{7 - 1}{2 - (-1)} = \frac{6}{3} = 2$$

Equal slopes indicate that the lines are parallel.

$$m_2 = \frac{5 - (-1)}{3 - 0} = \frac{6}{3} = 2$$

EXAMPLE 6 Show that the line through the points $(4, 3)$ and $(-2, 0)$ is perpendicular to the line through the points $(0, 5)$ and $(4, -3)$; see Figure 6.19.

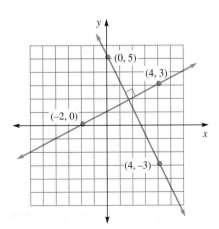

FIGURE 6.19.

SOLUTION $m_1 = \dfrac{3 - 0}{4 - (-2)} = \dfrac{3}{6} = \dfrac{1}{2}$

Slopes are negative reciprocals of each
other, so the lines are perpendicular.

$m_2 = \dfrac{5 - (-3)}{0 - 4} = \dfrac{8}{-4} = -2$ ▮ ▮

6.4   Exercises

▮ ▮ ▮ ▮ ▮

▼ Find the slope of the line through the given pair
of points.

1. (2, 3) and (6, 5)

2. (0, 0) and (2, 3)

3. (3, 4) and (1, 5)

4. (1, 2) and (−1, 3)

5. (4, 6) and (1, 3)

6. (−2, 1) and (−4, −1)

7. (−3, −2) and (−5, −6)

8. (2, −1) and (2, 5)

9. (−3, 4) and (−3, 7)

10. (−5, 2) and (−5, −6)

11. (3, −2) and (7, −2)

12. (4, 4) and (−3, 4)

13. (7, −1) and (−5, −1)

14. (4, 4) and (−2, −2)

15. (−3, −5) and (−6, −10)

16. $(-2, 4)$ and $(0, 6)$

17. $(-7, 0)$ and $(4, 0)$

18. $(0, -6)$ and $(0, 5)$

19. $(4, \frac{1}{2})$ and $(3, 2\frac{1}{2})$

20. $(2\frac{1}{2}, 3)$ and $(6, -2\frac{1}{2})$

▼ Find the slope of each equation by first finding two points on the line.

21. $x + y = 6$

22. $x - y = 7$

23. $2x + 3y = 6$

24. $4x - 3y = 12$

25. $y = 4x + 1$

26. $y = -2x - 3$

27. $y = \frac{1}{4}x + 3$

28. $y = \frac{1}{3}x - 2$

29. What is the slope of *any* line that is parallel to the line passing through the points $(-2, 1)$ and $(4, -2)$?

30. What is the slope of *any* line that is perpendicular to the line passing through the points $(-1, 4)$ and $(-2, -3)$?

31. Is the line through the points $(2, -1)$ and $(4, -2)$ parallel to the line through the points $(-2, -2)$ and $(2, -4)$?

32. Is the line through the points $(-5, -2)$ and $(-3, 6)$ parallel to the line through the points $(2, -1)$ and $(3, 4)$?

33. Is the line through the points $(4, -1)$ and $(-2, -3)$ perpendicular to the line through the points $(1, 2)$ and $(3, -2)$?

34. Is the line through the points $(-2, -1)$ and $(4, 1)$ perpendicular to the line through the points $(2, -4)$ and $(-1, 5)$?

35. Use slopes to determine whether the points $P(0, -4)$, $Q(-1, -7)$, and $R(2, 2)$ are collinear (lie on the same straight line).

36. Use slopes to determine whether the triangle with vertices $P(2, -3)$, $Q(3, 1)$, and $R(-5, 3)$ is a right triangle.

 37. Why is the slope of a horizontal line zero?

 38. Why is the slope of a vertical line undefined?

 39. What do you know about all lines that have positive slopes?

 40. What do you know about all lines that have negative slopes?

6.5   The Equation of a Line

The Point–Slope Form of the Equation of a Line

• • • • •

Consider a line having slope m and containing a known point with coordinates (x_1, y_1). If we let (x, y) by *any* other point on the line (see Figure 6.20), we can find the slope as follows:

$$\text{slope} = \frac{y - y_1}{x - x_1} = m$$

$$\cancel{(x - x_1)}\frac{(y - y_1)}{\cancel{(x - x_1)}} = m\,(x - x_1) \qquad \text{Multiply both sides by } x - x_1.$$

$$y - y_1 = m(x - x_1)$$

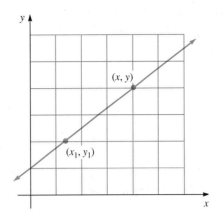

FIGURE 6.20.

The point–slope form of the equation of a line:

$$y - y_1 = m(x - x_1)$$

where m = slope and (x_1, y_1) is a known point on the line.

Equations are frequently written in a form called the **general form** of the equation of a straight line.

The general form of an equation of a line:

$$Ax + By + C = 0$$

where A, B, and C are integers and A and B are not both equal to 0. The coefficient A should be nonnegative.

EXAMPLE 1 Find the equation of the line passing through the point $(2, -1)$ and having a slope of 3. Write the equation in general form.

SOLUTION

$$y - y_1 = m(x - x_1) \qquad \text{Point–slope form}$$

$$y - \boxed{(-1)} = \boxed{3}\left(x - \boxed{2}\right) \qquad \text{Substitute } x_1 = 2, \ y_1 = -1, \text{ and } m = 3.$$

$$y + 1 = 3x - 6$$

$$3x - y - 7 = 0 \qquad \text{General form of the equation} \qquad ❙❙$$

EXAMPLE 2 Find the equation in general form of the line that has a slope of -5 and passes through the point $(0, -4)$.

SOLUTION

$$y - y_1 = m(x - x_1) \qquad \text{Point–slope form}$$

$$y - \boxed{(-4)} = \boxed{(-5)}\left(x - \boxed{0}\right) \qquad x_1 = 0, \ y_1 = -4, \text{ and } m = -5$$

$$y + 4 = -5x$$

$$5x + y + 4 = 0 \qquad \text{General form} \qquad ❙❙$$

To find the equation of a line when we are given *two* points on the line, we find the slope of the line as before and then use this slope and either of the two points in the point–slope formula.

EXAMPLE 3   Find the equation of the line through the points $(1, -2)$ and $(3, 4)$.

SOLUTION   The slope is

$$m = \frac{y_2 - y_1}{x_2 - x_1} = \frac{4 - (-2)}{3 - 1} = \frac{6}{2} = 3$$

Now we use the point–slope formula and *either* of the given points to find the equation. We use the point $(1, -2)$.

$$y - y_1 = m(x - x_1) \qquad \text{Point-slope form}$$
$$y - (-2) = 3(x - 1) \qquad x = 1, y = -2, \text{ and } m = 3$$
$$y + 2 = 3x - 3$$
$$3x - y - 5 = 0 \qquad \text{General form}$$

We could have also used the other point, $(3, 4)$:

$$y - y_1 = m(x - x_1) \qquad \text{Point–slope form}$$
$$y - 4 = 3(x - 3) \qquad x = 3, y = 4, \text{ and } m = 3$$
$$y - 4 = 3x - 9$$
$$3x - y - 5 = 0 \qquad \text{This is the same result.} \qquad ▪▪$$

EXAMPLE 4   Find the equation of the horizontal line that passes through the point $(2, 4)$.

SOLUTION   A horizontal line has a slope equal to 0.

$$y - y_1 = m(x - x_1) \qquad \text{Point–slope form}$$
$$y - 4 = 0(x - 2) \qquad x_1 = 2, y_1 = 4, \text{ and } m = 0$$
$$y - 4 = 0$$
$$y = 4 \qquad ▪▪$$

EXAMPLE 5   Find the equation of the vertical line that passes through the point $(1, -5)$.

SOLUTION   Since all vertical lines have an undefined slope, we cannot use the point–slope form. Even though we know that m is undefined, it would make no sense to substitute the word *undefined* for m in the equation $y - y_1 = m(x - x_1)$. However, we can write the equation directly if we realize that all points on any vertical line have the same x-coordinate. In this case, since the line passes through the point $(1, -5)$, x is always equal to 1 and the equation is $x = 1$.

A quick sketch of the situation is particularly helpful in this case: See Figure 6.21.

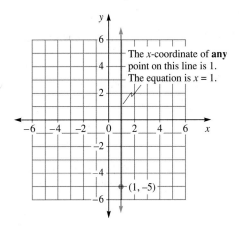

FIGURE 6.21.

The Slope–Intercept Form of the Equation of a Line

• • • • •

The point where a line crosses the y-axis is called the **y-intercept.** Since every point on the y-axis has an x-coordinate equal to zero, the y-intercept will be of the form $(0, b)$. Now let's find the equation of a line with y-intercept $(0, b)$ and slope m, as pictured in Figure 6.22.

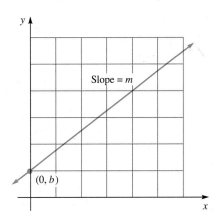

FIGURE 6.22.

$$y - y_1 = m(x - x_1)$$

$$y - \boxed{b} = m\left(x - \boxed{0}\right) \qquad \text{Substitute } x_1 = 0, \ y_1 = b.$$

$$y - b = mx$$

$$y = \boxed{m}\,x + \boxed{b}$$

Slope

y-intercept

$m = \text{slope}$

$b = y\text{-intercept}$

> ### The slope–intercept form of the equation of a line:
>
> $$y = mx + b$$
>
> where m = slope and b = y-intercept.

This form for the equation of a line is particularly useful because the slope and y-intercept can immediately be seen.

EXAMPLE 6   Write the equation, in slope–intercept form, of a line that crosses the y-axis at the point $(0, 3)$ and has a slope of $\frac{1}{2}$.

SOLUTION   Since $m = \frac{1}{2}$ and $b = 3$, we can write the equation immediately:

$$y = mx + b \qquad \text{Slope–intercept form}$$

$$y = \frac{1}{2}x + 3 \qquad \text{Substitute } m = \frac{1}{2},\ b = 3.$$

▮▮

EXAMPLE 7   Find the slope and y-intercept of the line with equation

$$3x - 2y + 8 = 0$$

SOLUTION   We must first put the given equation in slope–intercept form, $y = mx + b$, by solving for y:

$$3x - 2y + 8 = 0$$

$$-2y = -3x - 8 \qquad \text{Add } -3x \text{ and } -8 \text{ to both sides.}$$

$$y = \frac{-3x}{-2} - \frac{8}{-2} \qquad \text{Divide both sides by } -2.$$

$$y = \underset{\underset{m}{\uparrow}}{\boxed{\frac{3}{2}}}x + \underset{\underset{b}{\uparrow}}{\boxed{4}} \qquad \text{Slope–intercept form}$$

Therefore $m = \frac{3}{2}$ and the y-intercept is $(0, 4)$.

▮▮

Graphing an Equation in Slope–Intercept Form

• • • • •

The slope–intercept form of the equation of a line is particularly useful when graphing. Both the slope and the y-intercept are easy to identify.

E X A M P L E 8 Graph $y = \dfrac{3}{4}x + 1$.

S O L U T I O N By inspection, we see that the slope is

Numerator is the vertical change.

$$m = \dfrac{3}{4}$$

Denominator is the horizontal change.

The y-intercept is $b = 1$, therefore the graph crosses the y-axis at $(0, 1)$. Because the slope is $m = \frac{3}{4}$, if we start at the point $(0, 1)$ and move *up* 3 units (vertical change) and to the *right* 4 units (horizontal change), we will arrive at a new point, $(4, 4)$, that is also on the graph. See Figure 6.23.

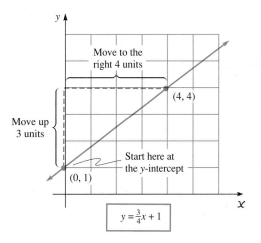

F I G U R E 6 . 2 3 .

$$y = \tfrac{3}{4}x + 1$$

To graph a line using slope and intercept:

1. Write the equation in slope–intercept form, $y = mx + b$.

2. Plot the y-intercept, $(0, b)$.

3. From the y-intercept, using the slope, mark off the vertical change (the numerator of the slope) and the horizontal change (the denominator of the slope).

4. This new point will be on the line. Draw a line through the two points to complete the graph.

E X A M P L E 9 Graph $y = -\dfrac{2}{3}x - 2$.

S O L U T I O N We first locate the y-intercept on the graph $(0, -2)$. The slope is $m = -\frac{2}{3}$, which can be written in either of two ways:

down 2 units

$$m = -\dfrac{2}{3} = \dfrac{-2}{3}$$

right 3 units

or

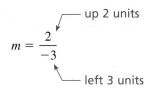

As you can see in Figure 6.24, writing m as either $\dfrac{-2}{3}$ or $\dfrac{2}{-3}$, will get us to a point on the line.

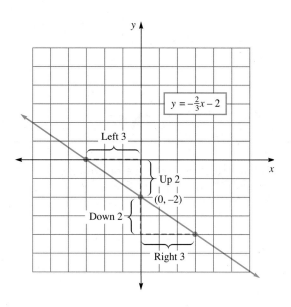

FIGURE 6.24.   ▪▪

EXAMPLE 10   Graph $3x + 2y = 4$, using the slope–intercept method.

SOLUTION   We start by putting the equation in slope–intercept form.

$$3x + 2y = 4$$
$$2y = -3x + 4 \qquad \text{Add } -3x \text{ to both sides.}$$
$$y = -\frac{3}{2}x + 2 \qquad \text{Divide both sides by 2.}$$

The slope is $-\dfrac{3}{2}$ and the y-intercept is 2 or the point $(0, 2)$. Now we plot the y-intercept, $(0, 2)$. Writing the slope $-\dfrac{3}{2}$ as $\dfrac{-3}{2}$, from the y-intercept $(0, 2)$ we move 3 units downward and 2 units to the right. This point must also be on the line. We draw a line through the two points and we have the graph of the given equation. See Figure 6.25.

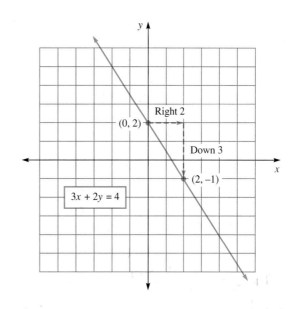

FIGURE 6.25.

6.5   Exercises

▼ Write the equation of the line with the given slope and passing through the given point. Give your answer in slope–intercept form.

1. $(1, 3)$; $m = 2$ **2.** $(-1, 4)$; $m = 4$

3. $(-3, -2)$; $m = -3$ **4.** $(4, -5)$; $m = -5$

5. $(-2, -7)$; $m = 1$ **6.** $(-6, 3)$; $m = \frac{2}{3}$

7. $(-1, -4)$; $m = \frac{3}{4}$ **8.** $(2, 0)$; $m = 0$

9. $(0, 4)$; $m = 0$ **10.** $(2, -5)$; $m = -\frac{3}{2}$

▼ Write the equation in slope–intercept form for the line containing the two given points.

11. $(0, 0)$ and $(3, 4)$ **12.** $(1, 3)$ and $(5, 6)$

13. $(2, 3)$ and $(1, 4)$ **14.** $(5 -2)$ and $(2, -2)$

15. $(0, 3)$ and $(6, 0)$ **16.** $(2, 0)$ and $(2, -6)$

17. $(4, -2)$ and $(-1, -8)$ **18.** $(3, 0)$ and $(4, 0)$

19. $(-5, 2)$ and $(-5, 5)$ **20.** $(-5, 3)$ and $(-8, -6)$

▼ Find the slope and *y*-intercept of the line having the given equation.

21. $y = 4x - 3$

22. $y = \frac{1}{4}x - 3$

23. $y = -\frac{3}{4}x + 2$

24. $y = -3x + \frac{1}{4}$

25. $4x - 3y + 6 = 0$

26. $3x - 5y - 12 = 0$

27. $x - 3y + 6 = 0$

28. $3x - 5 = -2y$

29. $\frac{3}{5}x - \frac{1}{2}y + \frac{1}{2} = 0$

30. $\frac{1}{2}x + \frac{2}{3}y - 2 = 0$

31. $y = -4$

32. $y = \frac{2}{3}$

33. $x = -\frac{1}{2}$

34. $x = 2$

35. Write the equation of the line with slope $\frac{2}{3}$ and *y*-intercept $(0, 4)$.

36. Write the equation of the line with slope $-\frac{3}{5}$ and *y*-intercept $(0, -2)$.

37. Write the equation of the line that crosses the *y*-axis at $(0, -2)$ and has a slope of -4.

38. Write the equation of the line that crosses the *y*-axis at $(0, 0)$ and has a slope of -3.

39. Write the equation of the line through the point $(2, -3)$ that has no *y*-intercept.

40. Write the equation of the line through the point $(4, 0)$ that does not cross the *y*-axis.

41. Write the equation of the horizontal line containing the point $(-2, 3)$

42. Write the equation of the horizontal line containing the point $(2, 1)$.

43. Write the equation of the vertical line containing the point $(-2, -3)$.

44. Write the equation of the vertical line containing the point $(0, 4)$

45. Write the equation of *any* line that is parallel to $y = \frac{2}{3}x - 4$.

46. Write the equation of *any* line that is parallel to $y = -3x + \frac{2}{3}$.

47. Write the equation of *any* line that is perpendicular to $y = \frac{2}{3}x - 4$.

48. Write the equation of *any* line that is perpendicular to $y = -3x + \frac{2}{3}$.

49. Write the equation of *any* line that is parallel to $2x - 3y + 5 = 0$.

50. Write the equation of *any* line that is perpendicular to $2x - 3y + 5 = 0$.

51. Identify the lines represented by the following pairs of equations as parallel, perpendicular, or neither.

a. $y = -\frac{2}{3}x + 4$ and $y = \frac{3}{2}x - 5$

b. $y = 6x - 2$ and $y = 4x + 2$

c. $y = \frac{1}{2}x - 5$ and $y = \frac{1}{2}x + 6$

d. $3x - y - 2 = 0$ and $x + 3y = -6$

▼ Graph each of the following lines using the slope–intercept method.

52. $y = \frac{2}{3}x - 3$

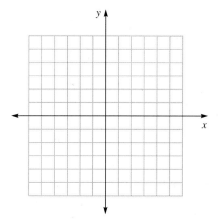

53. $y = \frac{3}{2}x - 1$

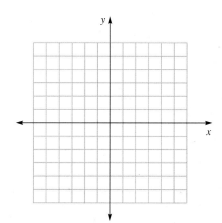

54. $y = x - 2$

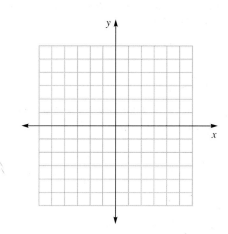

55. $y = x + 3$

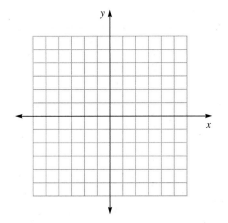

58. $y = -\frac{3}{4}x + 2$

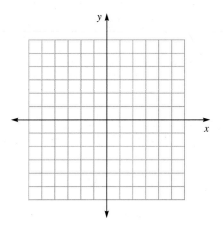

56. $y = 3x + 1$

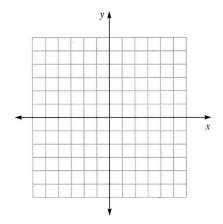

59. $y = -\frac{4}{3}x + 1$

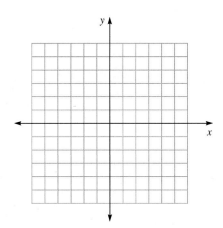

57. $y = -2x$

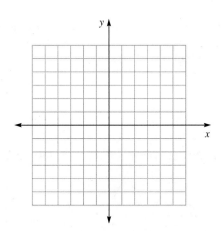

60. $3x + 2y = 4$

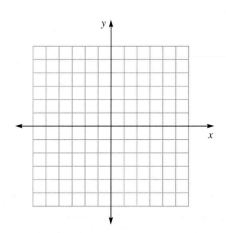

61. $4x - 3y = 6$

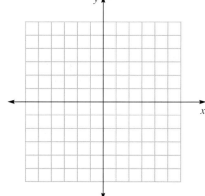

63. $4x + y + 3 = 0$

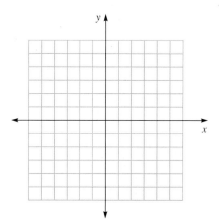

62. $2x - y + 1 = 0$

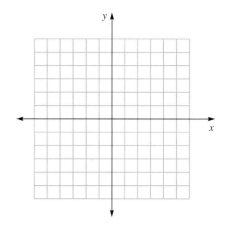

 64. How can you tell, without graphing, whether two lines are parallel, perpendicular, or neither?

6.6   The Distance and Midpoint Formulas

I I I I I

The Distance Formula

• • • • •

The distance between two points can be found by using the Pythagorean theorem, which says that the sum of the squares of the two legs of a right triangle is equal to the square of the hypotenuse; see Figure 6.26.

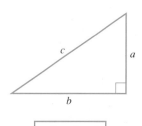

FIGURE 6.26.

$$a^2 + b^2 = c^2$$

The line segment joining any two points $P(x_1, y_1)$ and $Q(x_2, y_2)$ can be thought of as the hypotenuse of a right triangle; see Figure 6.27.

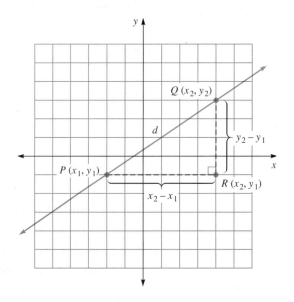

FIGURE 6.27.

According to the Pythagorean theorem,

$$d^2 = (x_2 - x_1)^2 + (y_2 - y_1)^2$$

or

$$d = \sqrt{(x_2 - x_1)^2 + (y_2 - y_1)^2}$$

Distance formula:

The distance between two points (x_1, y_1) and (x_2, y_2) is given by

$$d = \sqrt{(x_2 - x_1)^2 + (y_2 - y_1)^2}$$

EXAMPLE 1   Find the distance between (3, 7) and (5, 3).

SOLUTION $d = \sqrt{(x_2 - x_1)^2 + (y_2 - y_1)^2}$

$\qquad = \sqrt{(5 - 3)^2 + (3 - 7)^2}$

$\qquad = \sqrt{(2)^2 + (-4)^2}$

$\qquad = \sqrt{4 + 16}$

$\qquad = \sqrt{20} = 2\sqrt{5}$

Using a calculator to find $\sqrt{5}$, we have a rational approximation to our answer of $d \approx 2(2.236) = 4.472$.   ▪▪

EXAMPLE 2   Find x if the distance from $(x, 2)$ to $(2, 6)$ is 5.

SOLUTION
$$d = \sqrt{(x_2 - x_1)^2 + (y_2 - y_1)^2}$$
$$5 = \sqrt{(x - 2)^2 + (2 - 6)^2}$$
$$5^2 = (x - 2)^2 + (-4)^2$$
$$25 = x^2 - 4x + 4 + 16$$
$$0 = x^2 - 4x - 5$$
$$0 = (x - 5)(x + 1)$$
$$x = 5 \quad \text{or} \quad x = -1$$

There are two solutions, 5 and -1, which indicates that there are two points, (5, 2) and $(-1, 2)$, that are 5 units from the point (2, 6).   ▌▌

The Midpoint Formula

• • • • •

The coordinates of the midpoint of the line segment joining two points are found by averaging the x-coordinates and averaging the y-coordinates of the two points.

> **Midpoint Formula:**
>
> The midpoint $(\overline{x}, \overline{y})$ of the line segment joining the points $(x_1\ y_1)$ and (x_2, y_2) is given by
>
> $$(\overline{x}, \overline{y}) = \left(\frac{x_1 + x_2}{2}, \frac{y_1 + y_2}{2} \right)$$

EXAMPLE 3   Find the midpoint of the line segment joining the points $(-2, -4)$ and $(6, 6)$.

SOLUTION
$$(\overline{x}, \overline{y}) = \left(\frac{x_1 + x_2}{2}, \frac{y_1 + y_2}{2} \right)$$
$$= \left(\frac{-2 + 6}{2}, \frac{-4 + 6}{2} \right)$$
$$= \left(\frac{4}{2}, \frac{2}{2} \right)$$
$$= (2, 1)$$

The midpoint is (2, 1).   ▌▌

EXAMPLE 4   Show that the line segment joining $P(0, 0)$ and $Q(4, 6)$ and the line segment joining $R(-2, 4)$ and $S(6, 2)$ *bisect* each other (divide each other in half).

SOLUTION   If the midpoint of PQ is the same as the midpoint of RS, then the two line segments bisect each other. Figure 6.28 illustrates the situation.

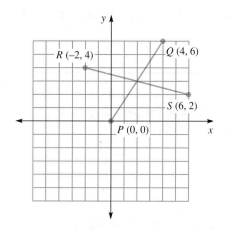

FIGURE 6.28.

$$\text{midpoint of } PQ = \left(\frac{0+4}{2}, \frac{0+6}{2}\right) = \left(\frac{4}{2}, \frac{6}{2}\right) = (2, 3)$$

$$\text{midpoint of } RS = \left(\frac{-2+6}{2}, \frac{4+2}{2}\right) = \left(\frac{4}{2}, \frac{6}{2}\right) = (2, 3)$$

Since the midpoints are the same, the line segments do bisect each other.   ▮ ▮

6.6   Exercises

▮ ▮ ▮ ▮ ▮

▼ Find the distance between each pair of points.

1. $(5, 8)$ and $(2, 4)$ **2.** $(-3, 7)$ and $(3, 1)$

3. $(0, -5)$ and $(12, 0)$ **4.** $(2, 1)$ and $(4, 7)$

5. $(0, -3)$ and $(7, 1)$ **6.** $(-3, 4)$ and $(2, -1)$

7. $(4, -1)$ and $(-2, 1)$ **8.** $(7, -4)$ and $(-1, -1)$

▼ Find the midpoint of the line segment joining each pair of points.

9. $(6, 2)$ and $(2, 4)$ **10.** $(-4, 2)$ and $(6, 8)$

11. $(-1, -3)$ and $(-3, 7)$ **12.** $(4, -5)$ and $(6, 5)$

13. $(2, 0)$ and $(-9, 6)$ **14.** $(3, -5)$ and $(4, -6)$

15. $(-2, 2)$ and $(2, -2)$ 16. $(3, -7)$ and $(-6, -2)$

21. Given the right triangle in Figure 6.29, show that the midpoint of the hypotenuse is equidistant from all three vertices of the triangle.

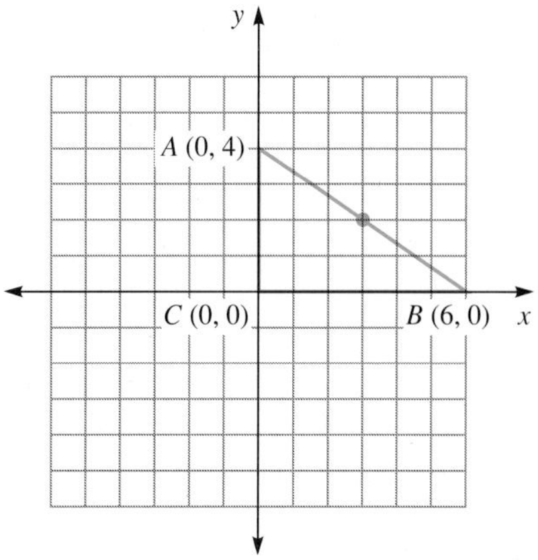

17. Find the lengths of the sides of the triangle with vertices at $(-2, -10)$, $(6, 5)$, and $(-2, 11)$.

18. Use the distance formula to show that the points $(0, 0)$, $(8, 0)$, and $(4, 4\sqrt{3})$ are the vertices of an equilateral triangle (all sides equal).

FIGURE 6.29.

19. Show that the points $(5, 0)$, $(8, -3)$, and $(2, -3)$ are the vertices of an isosceles triangle (two sides equals).

22. Find the points on the x-axis that are 5 units from the point $(1, 4)$.

20. Use the Pythagorean theorem to show that the points $(1, -3)$, $(-1, 2)$, and $(9, 6)$ are vertices of a right triangle.

23. Show that the line segment joining $A(4, 4)$ and $B(3, 2)$ and the line segment joining $C(6, -3)$ and $D(1, 9)$ bisect each other.

6.7   Functions

The concept of a function is very important to anyone who studies mathematics, physics, engineering, computer science, or any other science that deals with quantitative relationships.

A **function** deals with a correspondence between two sets. Objects from one set are matched up with objects of another set. Consider the case of four students and the grades they received on a mathematics exam:

$$\text{Ann} \rightarrow 81$$

$$\text{Bill} \rightarrow 90$$

$$\text{Charlie} \rightarrow 85$$

$$\text{David} \rightarrow 70$$

For each student on the left there is *exactly one* exam score on the right. It is this characteristic that determines a function from a correspondence between two sets. The set of students (the first set) is called the **domain** of the function and the set of exam scores (the second set) is called the **range** of the function.

> **Definition:**   A **function** is a correspondence between two sets, called the **domain** and **range,** such that each element in the domain corresponds to exactly one element in the range.

Suppose we look at the same students and their grades on a history exam.

$$\text{Ann} \rightarrow 80$$
$$\text{Bill} \rightarrow 80$$
$$\text{Charlie} \rightarrow 68$$
$$\text{David} \rightarrow 91$$

Even though each student received only one grade on their exam, you can see that both Ann and Bill are paired with the number 80. We normally do not list elements of a set more than once, and we can illustrate the function as follows:

Domain	Range
Ann ⟍	
Bill ↗ 80	
Charlie → 68	
David → 91	

This *is* a function since each student (in the domain) corresponds to exactly one grade (in the range).

The following situation would not be possible and does not meet the definition of a function:

$$\text{Ann} \rightarrow 80$$
$$\text{Bill} \rightarrow 85$$
$$\text{Charlie} \rightarrow 60$$
$$\rightarrow 75$$

Charlie can't receive more than one grade on the same exam. In this case each student (in the domain) does not correspond to *exactly one* grade (in the range), since Charlie corresponds to *more than one* grade. This is *not* a function.

In mathematics, the domain and range of functions usually consist of numbers and can be written as a collection of ordered pairs. Our example would look like this:

{(Ann, 81), (Bill, 90), (Charlie, 85), (David, 70)} is a function.

{(Ann, 80), (Bill, 80), Charlie, 68), (David, 91)} is a function.

{(Ann, 80), (Bill, 85), (Charlie, 60), (Charlie, 75)} is not a function.

This gives us the following alternate definition of a function.

> **Definition:** A **function** is a set of ordered pairs no two of which have the same first element. The **domain** is the set of first elements and the **range** is the set of second elements of the ordered pairs.

EXAMPLE 1   For each of the following sets, determine if it is a function and, if so, give the domain and range.

(a) {(3, 4), (5, 6), (1, 5)} *is* a function, since all the ordered pairs have different first elements.
Domain: {3, 5, 1}; range: {4, 6, 5}

(b) {(1, 2), (2, 2), (3, 2)} *is* a function, since all the ordered pairs have different first elements.
Domain: {1, 2, 3}; range: {2}

(c) {(1, 2), (2, 3), (1, 4)} *is not* a function, since two ordered pairs, (1, 2) and (1, 4), have the same first element. ❚❚

Vertical Line Test

• • • • •

If instead of a set of ordered pairs we are given an equation, we can look at the graph of the equation to determine whether it is a function. If two ordered pairs have the same first coordinate (not a function), they will both lie on a vertical line. This is shown in Figure 6.30.

The graph does not represent a function, since it contains two ordered pairs with the same first element, (x, y_1) and (x, y_2).

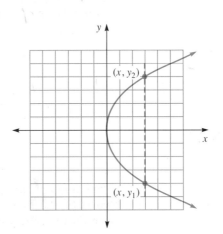

FIGURE 6.30.

Vertical line test:

If it is possible to draw a vertical line in such a way that it intersects a graph in more than one point, then the graph does not represent a function.

EXAMPLE 2   Apply the vertical line test to Figure 6.31(a) and (b) to see if they represent functions.   ▮▮

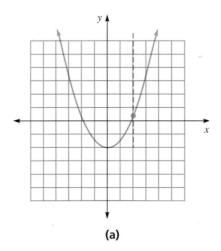
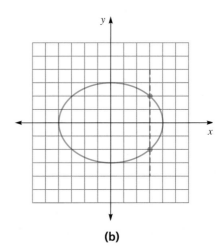

FIGURE 6.31.

(a)   (b)

No vertical line intersects the graph in more than one point. It *is* a function.

A vertical line *does* intersect the graph in more than one point. It *is not* a function.

EXAMPLE 3   Graph the equation $x = -y^2 + 1$ and apply the vertical line test to see if it describes a function.

SOLUTION   We begin by making a table of values:

x	y
1	0
0	1
0	-1
-3	2
-3	-2
-8	3
-8	-3

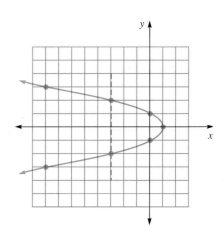

FIGURE 6.32.

In Figure 6.32, we *can* draw a vertical line in such a way that it intersects the graph in more than one point, so $x = -y^2 + 1$ does not describe a function.   ▮▮

6.7   Exercises

▮ ▮ ▮ ▮ ▮

▼ Indicate which of the following sets are functions and for those that are, give the domain and range.

1. {(1, 2), (3, 4), (5, 6)}

2. {(2, 1), (4, 3), (6, 5)}

3. {(−2, 0), (4, 2)}

4. {(−3, 5), (−3, 0)}

5. {(−1, 6), (2, 3), (−1, 7)}

6. {(4, 6), (2, 3), (−1, 6)}

7. {(−2, 3), (4, −1), (0, 3)}

8. {(0, 0)}

9. {(−1, 4), (6, −2), (0, 1), (−1, 0)}

10. {(0, 0), (0, 1), (1, 0)}

▼ Which of the following graphs represent functions?

11.

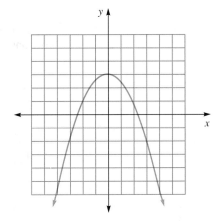

12.

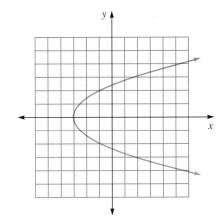

13.

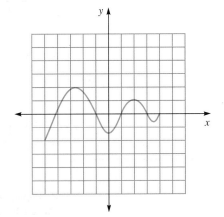

14.

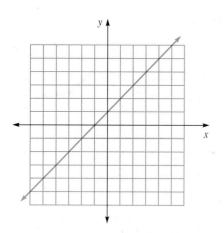

17.

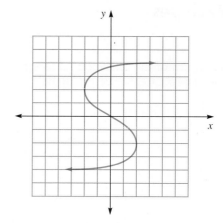

15.

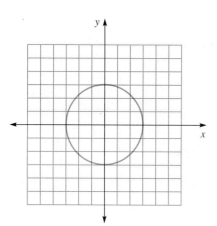

18.

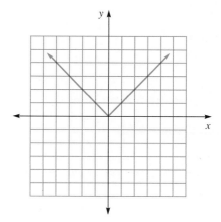

16.

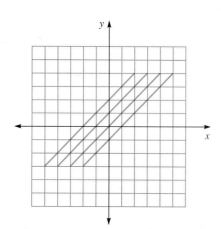

▼ Graph each of the following equations and
determine which represent functions.

19. $y = x^2$

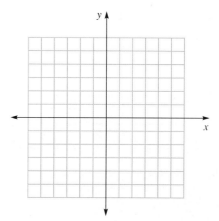

20. $x = y^2$

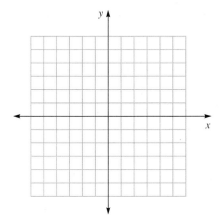

21. $x^2 + y^2 = 4$

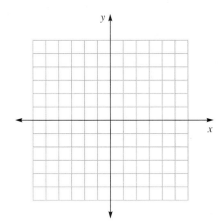

22. $y = x^2 - x + 2$

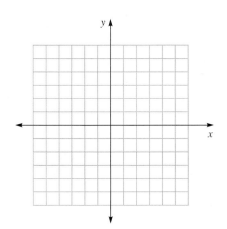

23. $y = 3x - 4$

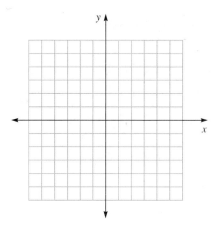

24. $y^2 = 2x + 1$

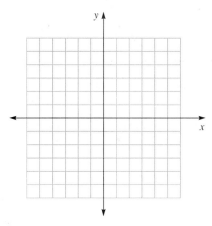

 25. Why does the vertical line test work?

6.8   Functional Notation

▮ ▮ ▮ ▮ ▮

In mathematics it is desirable to express all statements as briefly and accurately as possible.

Given the function $y = 2x^2 + 1$, consider the following two statements:

1. "Find the value of y, if $y = 2x^2 + 1$, and x is equal to 3."

2. "If $f(x) = 2x^2 + 1$, find $f(3)$."

Both statements say the same thing, only the second statement is shorter, and $f(x)$ replaces y in the equation.

$$f(x) = 2x^2 + 1 \text{ is read "} f \text{ of } x \text{ equals } 2x^2 + 1.\text{"}$$

The variable x has the role of a placeholder. If we used empty parentheses instead of x to describe the function, we would have

$$f(\ ) = 2(\ )^2 + 1$$

Therefore, to evaluate $f(3)$, we simply put 3 in each set of parentheses:

$$f(3) = 2(3)^2 + 1$$
$$= 2 \cdot 9 + 1$$
$$= 18 + 1$$
$$= 19$$

EXAMPLE 1   If $g(x) = 3x - 5$, evaluate: **(a)** $g(2)$;   **(b)** $g(-2)$;   **(c)** $g(0)$; **(d)** $g(a)$;   **(e)** $g(a + 2)$.

SOLUTION $g(x) = 3x - 5$

(a) $g(2) = 3(2) - 5 = 6 - 5 = 1$

(b) $g(-2) = 3(-2) - 5 = -6 - 5 = -11$

(c) $g(0) = 3(0) - 5 = 0 - 5 = -5$

(d) $g(a) = 3(a) - 5 = 3a - 5$

(e) $g(a + 2) = 3(a + 2) - 5 = 3a + 6 - 5 = 3a + 1$ ▮▮

The Domain of a Function

• • • • •

If a function is stated as an equation in terms of x, the domain is the set of all possible replacements for x. If the domain of a function is not specifically stated, we may assume that it consists of all real numbers for which the function is defined. There are two main areas where care must be taken.

> **The domain must *not* contain values of *x* that:**
>
> 1. yield 0 in any denominator;
> 2. yield a square root of a negative number.

E X A M P L E 2   Find the domain of the function $y = \dfrac{1}{x - 5}$.

S O L U T I O N   The domain contains all real numbers except $x = 5$, since $x = 5$ produces zero in the denominator.   ▌▌

E X A M P L E 3   Find the domain of the function $y = \dfrac{-6}{2x^2 - 5x - 3}$.

S O L U T I O N   The values of x for which the denominator is 0 are the solutions to the equation $2x^2 - 5x - 3 = 0$.

$$2x^2 - 5x - 3 = 0$$

$$(2x + 1)(x - 3) = 0$$

$$2x + 1 = 0 \quad \bigg| \quad x - 3 = 0$$

$$x = -\tfrac{1}{2} \quad \bigg| \quad x = 3$$

Therefore the domain contains all real numbers except $x = -\tfrac{1}{2}$ and $x = 3$.   ▌▌

E X A M P L E 4   Find the domain of the function $f(x) = \sqrt{x - 2}$.

S O L U T I O N   Since the square root of a negative number is not allowed, we must restrict the domain to values of x for which $x - 2 \geq 0$.

$$x - 2 \geq 0$$

$$x \geq 2$$

The domain consists of all x such that $x \geq 2$.   ▌▌

E X A M P L E 5   Find the domain of $g(x) = 2x^2 + 1$.

S O L U T I O N   The function g is defined for all values of x, so the domain consists of the set of real numbers.   ▌▌

6.8 Exercises

▮ ▮ ▮ ▮ ▮

1. If $f(x) = 2x + 5$, find:
   a. $f(2)$
   b. $f(-3)$
   c. $f(0)$
   d. $f(\frac{1}{2})$

2. If $g(x) = x^2 + 2x - 1$, find:
   a. $g(2)$
   b. $g(1)$
   c. $g(b)$
   d. $g(a + b)$

3. If $F(z) = z^2 - 2z + 4$, find:
   a. $F(3)$
   b. $F(-3)$
   c. $F(\frac{1}{2})$
   d. $F(a + b)$

4. If $f(x) = \dfrac{x - 2}{x + 2}$, find:
   a. $f(2)$
   b. $f(-2)$
   c. $f(\frac{1}{2})$
   d. x if $f(x) = 5$

▼ Find the domains of the following functions.

5. $y = 3x + 2$

6. $y = 2x^2$

7. $f(x) = -2x - 2$

8. $f(x) = \dfrac{6}{x}$

9. $g(x) = \dfrac{x + 1}{x - 1}$

10. $F(x) = \dfrac{6}{x^2}$

11. $f(x) = \dfrac{3}{2x + 1}$

12. $f(x) = \dfrac{5}{3x - 1}$

13. $h(x) = \dfrac{2x + 2}{x - 5}$

14. $h(x) = \dfrac{3x - 1}{x + 7}$

15. $f(x) = \dfrac{3}{x^2 - 2x - 3}$

16. $g(x) = \dfrac{-3}{x^2 - 9}$

17. $h(x) = \dfrac{x - 1}{x^2 + 3x - 4}$

18. $f(x) = \dfrac{2x + 2}{3x^2 - 11x - 4}$

19. $y = \dfrac{2x - 5}{2x^2 - x - 3}$

20. $p(x) = \dfrac{3x + 4}{2x^2 - 11x + 12}$

21. $y = \sqrt{x - 4}$

22. $y = \sqrt{x + 5}$

23. $f(x) = \sqrt{x - 8}$

24. $f(x) = \sqrt{x + 1}$

25. $f(x) = \sqrt{2x - 1}$

26. $f(x) = \sqrt{3x - 2}$

27. $g(x) = \sqrt{5x + 1}$

28. $g(x) = \sqrt{2x - 5}$

29. $h(x) = \sqrt{1 - 4x}$

30. $h(x) = \sqrt{2 - x}$

31. $y = \sqrt{1 - 3x}$

32. $y = \sqrt{2 - 5x}$

33. $f(x) = \dfrac{-7}{\sqrt{x - 5}}$

34. $f(x) = \dfrac{3}{\sqrt{2x + 1}}$

35. A repairperson charges \$35 for a house call plus \$20 per hour for each hour that the repair takes. If x is the number of hours for the call, then the cost function is $f(x) = 20x + 35$.

 a. Find $f(3)$.

 b. How much will it cost if the work takes 3 hours?

 c. If your bill is \$115, how many hours did the work take?

36. The function $V(s) = s^3$ gives the volume of a cube with side equal to s.

 a. Find the volume of a cube with a side of length 4.

 b. Find $V(4)$.

 c. Find $V(2)$

 d. Does it make sense to ask what $V(-3)$ is? Explain your answer.

6.9   The Inverse of a Function

▌ ▌ ▌ ▌ ▌

Consider the function

$$f = \{(1, 3), (3, 7), (4, 9), (-1, -1)\}$$

The **inverse of f,** denoted f^{-1}, is formed by interchanging the x- and y-values of the ordered pairs.

$$f = \{(1, 3), (3, 7), (4, 9), (-1, -1)\}$$
$$f^{-1} = \{(3, 1), (7, 3), (9, 4), (-1, -1)\}$$

When a function is defined by an equation instead of a set of ordered pairs, we find the inverse by interchanging x and y in the equation and then solving for y in terms of x.

EXAMPLE 1   Find the inverse of $f(x) = 2x + 1$.

SOLUTION

$$f(x) = 2x + 1$$

$$y = 2x + 1 \qquad \text{Write } f(x) \text{ as } y.$$

$$x = 2y + 1 \qquad \text{Interchange } x \text{ and } y.$$

$$x - 1 = 2y \qquad \text{Solve for } y \text{ in terms of } x.$$

$$y = \frac{x - 1}{2} \qquad \text{This can also be written as } f^{-1}(x) = \frac{x - 1}{2}.$$

We have $f(x) = y = 2x + 1$ and $f^{-1}(x) = y = \dfrac{x - 1}{2}$.   ▌ ▌

> **The find the inverse of a function, do the following:**
>
> 1. If f is written as a set of ordered pairs, interchange the x- and y-values.
>
> 2. If f is written in equation form:
>   a. write $f(x)$ as y;
>   b. interchange x and y;
>   c. solve for y;
>   d. write $f^{-1}(x)$ for y.

EXAMPLE 2   Graph $f(x) = 2x + 1$ and $f^{-1}(x) = \dfrac{x-1}{2}$ (from Example 1) on the same set of axes.

SOLUTION   Note in Figure 6.33 that the graphs of $f(x)$ and $f^{-1}(x)$ are symmetrical about the line $y = x$. This is because the inverse was obtained from the function by exchanging x and y in the equation. If (x, y) is on the graph of the function, then (y, x) will be on the graph of its inverse; and (x, y) and (y, x) are symmetrical about the line $y = x$.   ❚❚

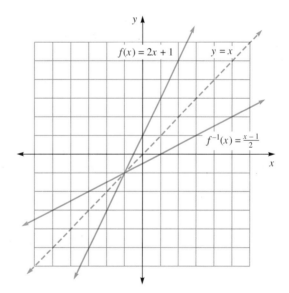

FIGURE 6.33.

EXAMPLE 3   Find the inverse of $y = x^2 - 3$ and graph both the function and its inverse on the same set of axes.

SOLUTION

$$y = x^2 - 3$$

$$x = y^2 - 3 \qquad \text{Interchange } x \text{ and } y.$$

$$y^2 = x + 3 \qquad \text{Solve for } y.$$

$$y = \pm\sqrt{x + 3} \qquad \text{Don't forget } \pm.$$

We graph this equation in Figure 6.34.

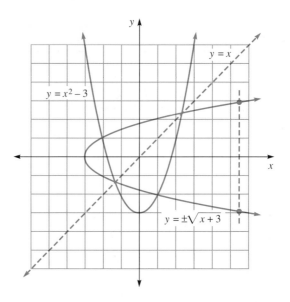

FIGURE 6.34.

Note that the graph of $y = \pm\sqrt{x + 3}$ does not represent a function, since a vertical line intersects the graph in more than one point. This shows that we do not always obtain a function when we attempt to find the inverse of a given function.   ▮▮

One-to-One Functions

• • • • •

A function with an inverse that is also a function is called a **one-to-one function.** Each value of x is paired with exactly one value of y and each value of y is paired with exactly one value of x.

One method for identifying a one-to-one function is by using the **horizontal line test.**

Horizontal Line Test:

If every *horizontal* line intersects the graph of the given function in no more than one point, then it is a one-to-one function and its inverse will also be a function.

In Example 2, both $f(x) = 2x + 1$ and $f^{-1}(x) = \dfrac{x - 1}{2}$ have graphs that are straight lines and so both represent functions. Therefore $f(x) = 2x + 1$ is a one-to-one function. In Example 3, a horizontal line intersects the graph of

$f(x) = x^2 - 3$ in more than one point and, as we discovered, its inverse is not a function. Therefore $f(x) = x^2 - 3$ is not a one-to-one function.

6.9   Exercises

▼ Find the inverse of each of the following functions. State whether or not the inverse is a function.

1. $f = \{(1, 3), (2, 7), (4, -2)\}$

2. $f = \{(-1, 0), (0, 6), (-2, -2)\}$

3. $f = \{(1, 3), (2, 3), (3, 3)\}$

4. $f = \{(4, 1), (6, 2), (8, 1)\}$

5. $f = \{(0, 0), (1, 1)\}$

6. $f = \{(0, 6), (6, 0)\}$

▼ For each of the following functions f, write the equation describing f^{-1}. Draw the graphs of f and f^{-1} on the same set of axes.

7. $f(x) = 3x - 2$

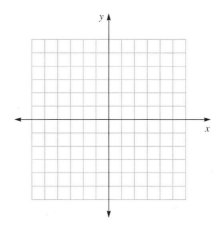

8. $f(x) = \frac{1}{2}x + 3$

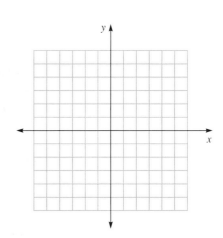

9. $f(x) = \frac{2}{3}x - 2$

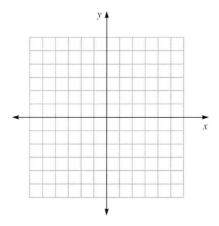

10. $f(x) = -\frac{1}{2}x + 2$

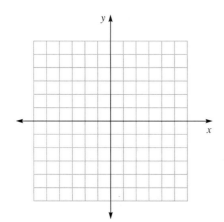

11. $2x + 3y = 3$

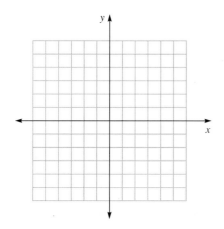

12. $-3x + 4y = -8$

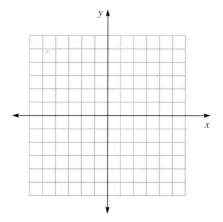

▼ For each of the following functions, write the equation describing its inverse. Graph both the function and the inverse on the same set of axes. Tell whether or not the inverse represents a one-to-one function.

13. $f(x) = -\frac{3}{4}x - 2$

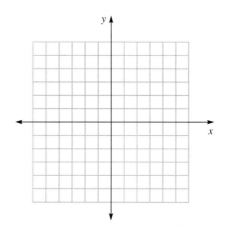

14. $f(x) = \frac{1}{3}x + 1$

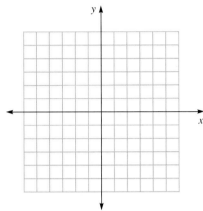

15. $f(x) = x^2 - 4$

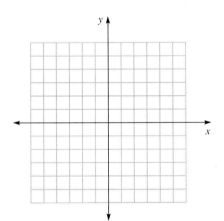

16. $f(x) = \frac{1}{2}x^2$

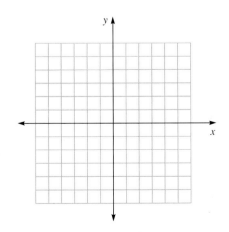

17. $f(x) = x^3$

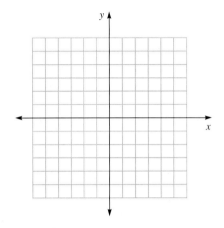

18. $f(x) = \frac{1}{2}x^3 + 1$

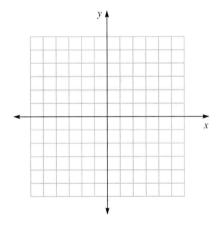

SUMMARY—CHAPTER 6

EXAMPLES

In an ordered pair (x, y):
 x is called the **x-coordinate** or **abscissa**.
 y is called the **y-coordinate** or **ordinate**.
The x-intercept is the point where a graph crosses the x-axis.
The y-intercept is the point where a graph crosses the y-axis.

To graph a linear equation:

1. Choose an arbitrary value for one of the variables and calculate the corresponding value for the remaining variable. Repeat this procedure three times, yielding three ordered pairs. The x- and y-intercepts are often easiest to find.

2. Plot the three ordered pairs obtained in Step 1.

3. Draw a straight line through the points.

Graph $y = 3x - 1$.

x	y
0	-1
$\frac{1}{3}$	0
2	5

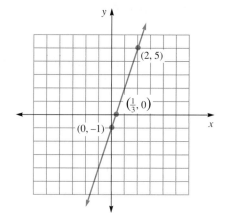

Vertical and horizontal lines:

1. The graph of the equation $x = c$ (where c is a constant) is a vertical line through all points where $x = c$.

2. The graph of the equation $y = b$ (where b is a constant) is a horizontal line through all points where $y = b$.

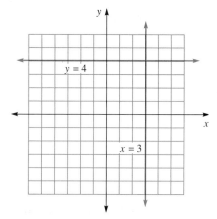

To find the slope of a line:
If (x_1, y_1) and (x_2, y_2) are any two points on a line, then the slope m of the line is given by

$$m = \frac{\text{change in } y}{\text{change in } x} = \frac{y_2 - y_1}{x_2 - x_1}$$

Find the slope of the line connecting the points $(2, -3)$ and $(4, 1)$.

$$m = \frac{-3 - 1}{2 - 4} = \frac{-4}{-2} = 2$$

EXAMPLES

The slope of any *horizontal line* is 0.

The slope of any *vertical line* is undefined.

Slope	Graph of line
positive	slants upward as it moves from left to right
negative	slants downward as it moves from left to right
0	horizontal line
undefined	vertical line

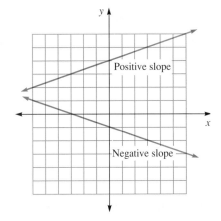

Slopes of **parallel lines** are the same.

$y = \boxed{3}\,x + 6$ and
$y = \boxed{3}\,x - 5$ are parallel.

Slopes of **perpendicular lines** are negative reciprocals of each other.

$y = \boxed{2}\,x - 1$ and
$y = \boxed{-\frac{1}{2}}\,x + 7$ are perpendicular.

The **point–slope form** of the equation of a line is

$$y - y_1 = m(x - x_1)$$

where m = slope and (x_1, y_1) is a known point on the line.

Find the equation of the line through points $(3, -2)$ and $(5, 2)$.

$$m = \frac{y_2 - y_1}{x_2 - x_1} = \frac{-2 - 2}{3 - 5} = \frac{-4}{-2} = 2$$

The **general form** of an equation of a line is

$$Ax + By + C = 0$$

where A, B, and C are integers and A and B are not both 0.

The **slope–intercept form** of the equation of a line is

$$y = mx + b$$

where m is the slope and b is the y-intercept.

$$y - y_1 = m(x - x_1)$$
$$y - 2 = 2(x - 5) \qquad \text{Point–slope form}$$
$$y - 2 = 2x - 10$$
$$y = 2x - 8 \qquad \text{Slope–intercept form}$$
$$2x - y - 8 = 0 \qquad \text{General form}$$

To graph a curve:

1. Make a table of ordered pairs by choosing values for one variable and calculating corresponding values for the remaining variable.

2. Plot a sufficient number of points to determine the graph.

3. Connect the points with a smooth curve.

Graph $y = 2x - 5$.

Choose x	Calculate y	Ordered pairs
0	$y = 2 \cdot 0^2 - 5 = -5$	$(0, -5)$
1	$y = 2 \cdot 1^2 - 5 = -3$	$(1, -3)$
2	$y = 2 \cdot 2^2 - 5 = 3$	$(2, 3)$
-1	$y = 2 \cdot (-1)^2 - 5 = -3$	$(-1, -3)$
-2	$y = 2 \cdot (-2)^2 - 5 = 3$	$(-2, 3)$

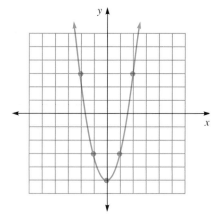

The distance formula: The distance between two points (x_1, y_1) and (x_2, y_2) is given by

$$d = \sqrt{(x_2 - x_1)^2 + (y_2 - y_1)^2}$$

The distance from $(3, 3)$ to $(2, 4)$ is

$$d = \sqrt{(2 - 3)^2 + (4 - 3)^2}$$
$$= \sqrt{1 + 1} = \sqrt{2} \approx 1.414$$

The midpoint formula: The midpoint $(\bar{x}, \bar{y})$ of the line segment joining the points (x_1, y_1) and (x_2, y_2) is $\left(\dfrac{x_1 + x_2}{2}, \dfrac{y_1 + y_2}{2} \right)$.

The midpoint of $(3, 1)$ and $(2, 4)$ is

$$\left(\frac{3 + 2}{2}, \frac{1 + 4}{2} \right) = \left(\frac{5}{2}, \frac{5}{2} \right)$$

A **function** is a correspondence between two sets, called the **domain** and the **range,** such that each element in the domain corresponds to exactly one element in the range.

DOMAIN		RANGE
2	$\rightarrow$	3
3	$\rightarrow$	5
-4	$\rightarrow$	6

A **function** is a set of ordered pairs, no two of which have the same first element. The **domain** is the set of first elements and the **range** is the set of second elements of the ordered pairs.

$f = \{(2, 3), (3, 5), (-4, 6)\}$ is a function.
Domain $= \{2, 3, -4\}$
Range $= \{3, 5, 6\}$
$g = \{(1, 4), (-2, 7), (-2, 8)\}$ is not a function.

The vertical line test: If all vertical lines intersect a graph in at most one point, then the graph represents a function.

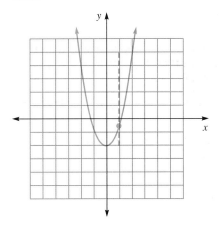

The graph pictured here represents a function.

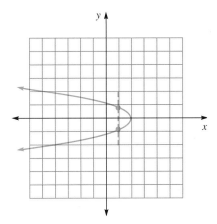

The graph pictured here does *not* represent a function.

If $f(x) = 3x^2 - 4$ then $f(2) = 3 \cdot 2^2 - 4 = 8$.

Functional notation: $f(x)$ is read "*f* of *x*." The notations $f(x)$ and y have the same meaning.

The domain of a function must not contain values that:

1. yield 0 in any denominator;

Let $f(x) = \dfrac{1}{x + 2}$. Then the domain of f is all real numbers except $x = -2$.

2. yield a square root of a negative number.

Let $g(x) = \sqrt{x + 2}$. Then the domain of g is all real numbers such that $x \geq -2$.

Given a function f, to find its **inverse f^{-1}** when it exists:

1. If f is written as a set of ordered pairs, interchange the x- and y-values.

Let $f = \{(2, 3), (4, 1)\}$. Then $f^{-1} = \{(3, 2), (1, 4)\}$.

2. If f is written in equation form:
   a. Write $f(x)$ as y.
   b. Interchange the variables x and y.
   c. Solve for y in terms of x.
   d. Write y as $f^{-1}(x)$.

Let $f(x) = y = 3x + 1$. Then $x = 3y + 1$
$$x - 1 = 3y$$
$$f^{-1}(x) = y = \frac{x - 1}{3}$$

A **one-to-one function** is a function with an inverse that is also a function.

The horizontal line test: A function is one-to-one if every horizontal line intersects the graph in at most one point.

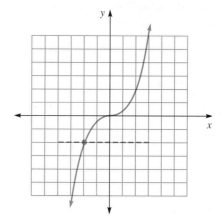

This graph represents a one-to-one function.

REVIEW EXERCISES—CHAPTER 6

▼ Graph each of the given equations.

1. $y = x + 1$

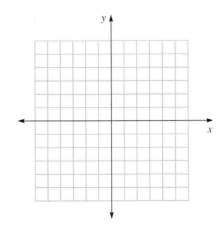

2. $2x + y = -1$

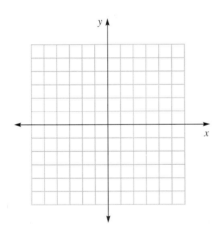

3. $x = -4$

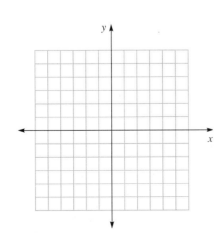

4. $y = -1$

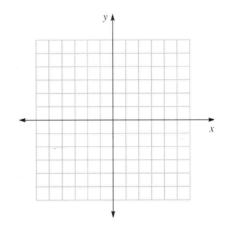

5. $y = -x^2$

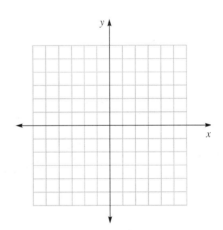

6. $y = x^2 - 2x - 1$

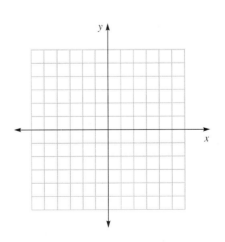

7. What is the slope of *any* line that is parallel to the line passing through the points $(2, -4)$ and $(-6, 6)$?

8. What is the slope of *any* line that is perpendicular to the line passing through the points $(2, -4)$ and $(-6, 6)$?

▼ Find the slope of the line segment connecting the given points.

9. $(-1, 3)$ and $(-6, -2)$ 10. $(-2, 4)$ and $(-2, -6)$

11. $(4, 3)$ and $(-6, 3)$

▼ Write the equation of the line that passes through the given point and has the given slope.

12. $(2, 1)$; $m = 4$ 13. $(3, -2)$; $m = -\frac{2}{3}$

▼ Write the equation of the line containing the two given points.

14. $(4, -1)$ and $(3, 6)$ 15. $(7, -4)$ and $(5, 1)$

▼ Find the slope and *y*-intercept of the line having the given equation.

16. $3x - 2y = 1$ 17. $4x + \frac{1}{2}y = 3$

18. Graph $y = \frac{1}{3}x - 2$ using the slope and *y*-intercept.

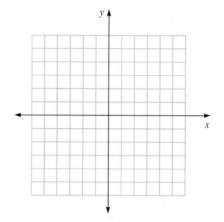

19. Find the distance between the points $(2, -4)$ and $(5, 1)$.

20. Find the midpoint of the line segment joining the points $(2, 4)$ and $(-2, 6)$.

▼ Which of the following sets are functions? Give the domain and range of any that are.

21. $\{(2, 4), (6, -8), (0, -8)\}$

22. $\{(-3, -4), (-6, -4), (-6, 1)\}$

▼ Which of the following graphs represent functions? Which are one-to-one functions:

23.

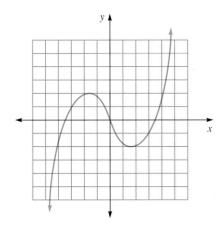

24.

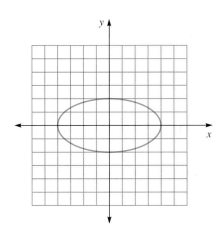

▼ If $f(x) = 2x^2 - x - 1$, find the following:

25. $f(1)$ **26.** $f(-1)$

27. $f(0)$ **28.** $f(b)$

29. $f(b + 1)$

▼ Find the domain of the following functions.

30. $y = 2x^2 - 1$

31. $y = \dfrac{x^2 + 2x - 1}{x^2 - 4}$

32. $y = \sqrt{x + 4}$

▼ Find the inverse of each of the following functions and state whether or not the inverse is a function.

33. $f = \{(2, 3), (4, -1), (6, 3)\}$

34. $f = \{(1, 1), (2, 2), (-4, -4)\}$

35. If $y = f(x) = 3x + 1$, find f^{-1} and graph f and f^{-1} on the same set of axes. Is f^{-1} a function? Is f a one-to-one function?

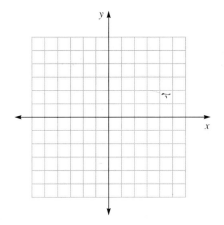

CRITICAL THINKING EXERCISES—CHAPTER 6

● ● ● ● ● ● ● ▼

▼ There is an error in each of the following. Can you find it?

1. The graph of the equation $y = 2$ is parallel to the y-axis.

2. The graph of the equation $x = -2$ has zero slope.

3. The slope of the graph of $y + 2x = -6$ is 2.

4. The y-intercept of the graph of $3y = 4x + 5$ is 5.

5. The set of ordered pairs $\{(1, 3), (2, 3), (3, 3)\}$ is not a function because every ordered pair has the same second element.

6. The lines given by $y = \frac{2}{3}x + 4$ and $y = \frac{3}{2}x + 1$ are perpendicular.

7. The domain of $y = \dfrac{3}{x^2 - 9}$ is the set of all real numbers except $x = 3$, since the denominator is equal to zero when $x = 3$.

8. The domain of $f(x) = \dfrac{3}{\sqrt{x - 1}}$ is the set of all real numbers such that $x \geq 1$.

9. The x-intercept occurs where $x = 0$.

10. The distance between two points is negative whenever the line segment connecting them slants downward as it goes from left to right.

NAME   **I I I I I CLASS**

ACHIEVEMENT TEST—CHAPTER 6

▼   Graph each of the following equations.

1. $y = -2x + 1$

1.

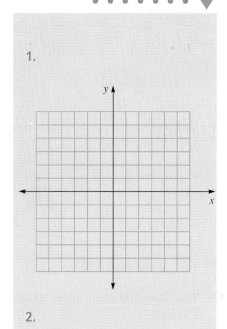

2. $y = \frac{1}{2}x + 3$

2.

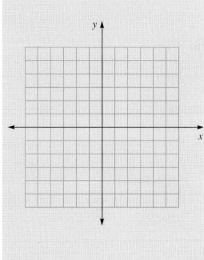

▼   Find the slope of the line containing the given points.

3. $(-2, 3)$ and $(4, -1)$

3. _____

4. $(3, -2)$ and $(5, -2)$

4. _____

5. $(-1, 4)$ and $(-1, -3)$

5. _____

▼ Write the equation of the line with the given slope that passes through the given point. Write the equation in slope–intercept form.

6. $(2, -3)$; $m = 4$ 6. _____

7. $(-1, -6)$; $m = \frac{1}{2}$ 7. _____

8. $(2, 1)$; m is undefined   8. _____

▼ Write the equation, in slope–intercept form, of the line containing the two given points.

9. $(-3, 2)$ and $(4, -1)$ 9. _____

10. $(-2, -3)$ and $(-2, 5)$ 10. _____

11. $(4, -1)$ and $(-2, -1)$ 11. _____

▼ Find the slope and y-intercept of the lines with the following equations:

12. $y = -5x - 3$ 12. _____

13. $y = \frac{3}{4}x + 3$ 13. _____

14. $y = 3$ 14. _____

15. $x = -1$ 15. _____

16. Find the distance between the points $(4, -5)$ and $(1, 0)$.   16. _____

17. Find the midpoint of the line segment joining the points $(-2, 0)$ and $(-8, 6)$.   17. _____

▼ Which of the following graphs represent functions?

18.

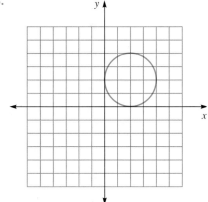

18. _____

19.

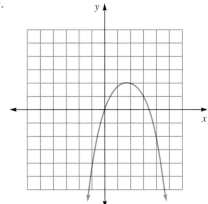

19. _____

▼ If $f(x) = x^2 - 4x + 1$, find the following:

20. $f(2)$

20. _____

21. $f(a + b)$

21. _____

▼ Find the domain of the following functions.

22. $y = \dfrac{x}{x - 3}$

22. _____

23. $y = \sqrt{x - 1}$

23. _____

24. a. Find the inverse of $\{(2, -1), (3, 2), (4, -1)\}$.
    b. Is the inverse a function?

24. _____

_____

25. Given $f(x) = 2x - 1$:

   a. find f^{-1};

   b. graph f and f^{-1} on the same set of axes.

   c. Is f^{-1} a function?

   d. Is f a one-to-one function?

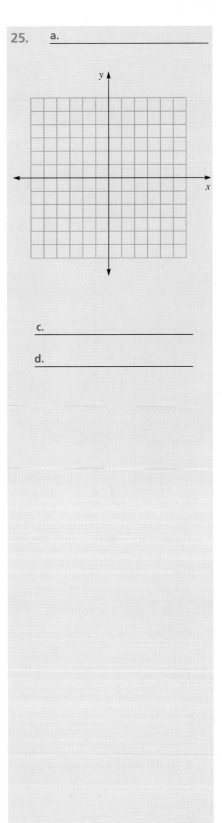

25. a.

c.

d.

NAME **▮▮▮▮▮ CLASS**

CUMULATIVE REVIEW—CHAPTERS 4, 5, 6

• • • • • • • • ▼

1. For what value of the variable is the fraction $\dfrac{x-7}{4x-2}$ undefined?

1. _____

2. Reduce to lowest terms:

   a. $\dfrac{-18x^4yz^5}{-3xy^4z^5}$ **b.** $\dfrac{6-2x}{2x-6}$ **c.** $\dfrac{2x^2-7x-4}{x^2-9x+20}$

2. _____

   _____

   _____

   _____

3. Multiply: $\dfrac{3x-3}{x^2-1} \cdot \dfrac{2x^2+7x+5}{12x+30}$.

3. _____

4. Subtract: $\dfrac{3x+5}{x^2+x-12} - \dfrac{2}{x+4}$.

4. _____

5. Simplify the complex fraction: $\dfrac{\dfrac{5}{a-b}+\dfrac{3}{a+b}}{\dfrac{4}{a}}$.

5. _____

6. Solve for x: $\dfrac{3x}{x-5} - \dfrac{3}{5-x} = 1$.

6. _____

7. Dotty and Doug can mow the lawn, working together, in 2 hours. It takes Doug 4 hours to do it, working alone. How long will it take Dotty to mow the lawn if she works by herself?

7. _____

▼ Evaluate:

8. $\sqrt{121}$

8. _____

9. $\sqrt[3]{125}$

9. _____

10. $\sqrt[5]{-32}$

10. _____

▼ Simplify:

11. $\sqrt{96}$

11. _____

12. $\sqrt[3]{48}$

12. _____

13. $\sqrt{24x^9y}$

13. _____

▼ Multiply and simplify:

14. $\sqrt{10} \cdot \sqrt{15}$

14. _____

15. $(\sqrt{3} - 2)(2\sqrt{3} + 5)$

15. _____

▼ Combine where possible:

16. $\sqrt{45} + \sqrt{20}$

16. _____

17. $\sqrt{8} - \sqrt{18}$

17. _____

▼ Rationalize the denominator and simplify:

18. $\dfrac{6}{\sqrt{2}}$

18. _____

19. $\dfrac{4}{\sqrt{8}}$

19. _____

20. $\dfrac{4}{\sqrt{3} - 2}$

20. _____

▼ Simplify:

21. $64^{2/3}$

21. _____

22. $\left(x^{1/4}\right)^{2/3}$

22. _____

23. $\dfrac{x^{2/3}}{x^{1/4}}$

23. _____

24. Write $\sqrt{-27}$ as an imaginary number and simplify.

24. _____

25. Add: $(2 + 3i) + (4 - 5i)$.

25. _____

26. Subtract: $(5 - 3i) - (2 - 4i)$.

26. _____

27. Multiply: $(2i + 1)(3i - 2)$.

27. _____

28. Divide: $\dfrac{3 + 5i}{4 - i}$

28. _____

29. Graph $2x + 3y = 12$.

29.

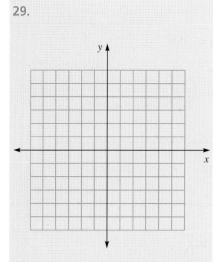

30. Graph $y = 2x^2 - 3x + 1$.

30.

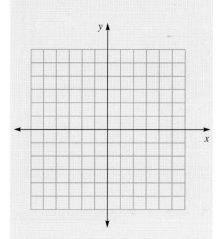

31. Write the equation of the line containing the points $(2, -3)$ and $(4, -9)$.

31. _____

32. Write the equation of *any* line perpendicular to the line $y = \frac{3}{4}x - 2$.

32. _____

33. Find the distance between the points $(-4, -1)$ and $(5, 2)$.

33. _____

34. If $f(x) = x^2 - 4x + 7$, find the following:
   a. $f(-1)$;
   b. $f(a + 1)$.

34. _____

_____

35. Find the inverse of $g(x) = \frac{1}{3}x + 2$.

35. _____

36. Find the domain of $y = \dfrac{3}{x^2 - 4}$.

36. _____

Quadratic Equations

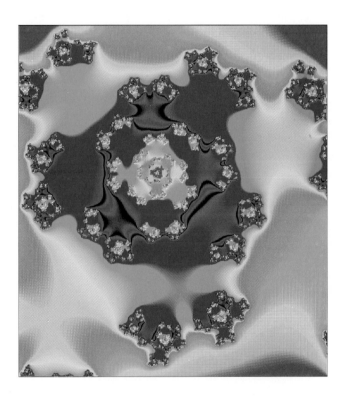

INTRODUCTION

In Chapter 7 we introduce a new type of equation called a **quadratic equation.** We will learn several methods of solving them. On some occasions our solutions will turn out to be complex numbers, which we investigated previously. We will also graph some quadratic equations and study some of their applications.

CHAPTER 7—NUMBER KNOWLEDGE

The Pythagorean Theorem and the Pythagoreans

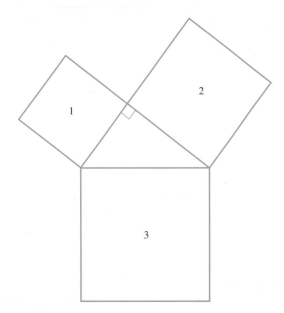

F I G U R E   7 . 1 .

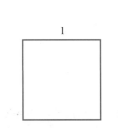 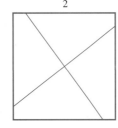

F I G U R E   7 . 2 .

Cut out the red squares numbered 1 and 2 in Figure 7.2 and verify that they are the same size as squares 1 and 2 of Figure 7.1. Now cut out the squares of Figure 7.2, and cut square 2 on the lines into 4 pieces and see if you can arrange square 1 and the pieces of square 2 together in such a way as to cover square 3 exactly. (The arrangement is given on the next page if you run out of patience.)

This suggests that the sum of the areas of squares 1 and 2 is equal to the area of square 3. This is known as the Pythagorean theorem and was discovered in the sixth century B.C. by a group of Greek scholars known as the Pythagoreans and headed by a man named Pythagorus. In addition to mathematics, they also studied music, astronomy, and philosophy. They were a secret society

and for this reason much of what they accomplished was not recorded.

They believed that numbers had mystical powers and meanings. For example, even numbers were considered feminine and odd numbers masculine. Since $2 + 3 = 5$, the number 5 represented marriage, since it is the sum of the first feminine and masculine numbers. The number 6 was considered a perfect number, since it is the sum of its proper divisors ($1 + 2 + 3 = 6$).

Returning to the theorem that bears the Pythagoreans' name, we see that they discovered that if they constructed squares on each of the sides of a right triangle, the sum of the areas of the squares on the two shorter sides (the *legs*) was equal to the area of the square on the longer side (the *hypotenuse*): see Figure 7.3.

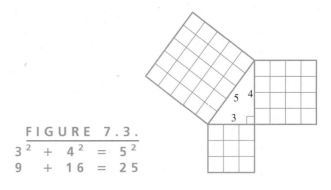

F I G U R E   7 . 3 .

$3^2 + 4^2 = 5^2$
$9 + 16 = 25$

Stated formally, we have

The Pythagorean Theorem

Given any right triangle with legs a and b and hypotenuse c,

$$a^2 + b^2 = c^2$$

Here is a simple but elegant geometric proof of the Pythagorean theorem. We are given two squares, each having length $a + b$ on a side and therefore of equal area. Each of the squares has been divided into squares and right triangles in a different fashion, as shown in Figures 7.4 and 7.5.

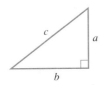

FIGURE 7.4.

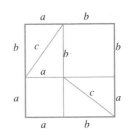
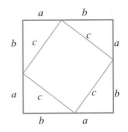

FIGURE 7.5.

From each of the squares remove four triangles with sides a, b, and c. In Figure 7.4, the area that remains consists of two squares, one of area a^2 and the other b^2. In Figure 7.5, a square of area c^2 remains. We conclude that

$$a^2 + b^2 = c^2$$

where a, b, and c are the sides of a right triangle.

7.1   Quadratic Equations

▮ ▮ ▮ ▮ ▮

The Standard Form of a Quadratic Equation

• • • • •

A **quadratic equation** is any equation that can be written in the form

$$ax^2 + bx + c = 0$$

where a, b, and c are real numbers and $a \neq 0$. This form is called the **standard form** of a quadratic equation. In this chapter we will require that $a > 0$ and that a, b, and c be integers.

E X A M P L E   1   Write $-2x = 3 - 9x^2$ in standard form, and identify a, b, and c.

SOLUTION $\qquad -2x = 3 - 9x^2 \qquad$ Add $9x^2$ and -3 to both sides.

$\qquad \left.\begin{array}{l} 9x^2 - 2x - 3 = 0 \\[4pt] a = 9,\ b = -2,\ c = -3 \end{array}\right\} \qquad$ In standard form   ▮▮

Answer to the puzzle on the previous page.

Solving Quadratic Equations by Factoring

· · · · ·

What do we know about two numbers with a product of zero? A little thought should convince you that at least *one* of them must be zero. This is known as **the rule of zero products** and is stated as follows:

The rule of zero products:

If $a \cdot b = 0$, then either $a = 0$, $b = 0$, or both $a = 0$ *and* $b = 0$.

We can use this principle to solve any quadratic equation that is factorable by use of the following procedure:

To solve a quadratic equation by factoring:

1. put the equation in standard form, $ax^2 + bx + c = 0$;

2. factor the left side of the equation;

3. set each factor equal to zero and solve for x;

4. check each solution in the original equation.

EXAMPLE 2   Solve $x^2 - 2x - 8 = 0$ by factoring.

SOLUTION

$$x^2 - 2x - 8 = 0 \qquad \text{Already in standard form}$$

$$(x - 4)(x + 2) = 0 \qquad \text{Factor the left side.}$$

$$x - 4 = 0 \quad | \quad x + 2 = 0 \qquad \text{Set each factor equal to 0.}$$

$$x = 4 \quad | \quad x = -2 \qquad \text{Solve each equation for } x.$$

There are *two* solutions, $x = 4$ and $x = -2$.

Check

$x = 4$	$x = -2$
$x^2 - 2x - 8 = 0$	$x^2 - 2x - 8 = 0$
$(4)^2 - 2(4) - 8 \overset{?}{=} 0$	$(-2)^2 - 2(-2) - 8 \overset{?}{=} 0$
$16 - 8 - 8 \overset{?}{=} 0$	$4 + 4 - 8 \overset{?}{=} 0$
$0 = 0$ ✔	$0 = 0$ ✔

EXAMPLE 3   Solve $3y^2 = 2 - 5y$ by factoring.

SOLUTION

$$3y^2 = 2 - 5y \qquad \text{Put in standard form.}$$

$$3y^2 + 5y - 2 = 0 \qquad \text{Factor.}$$

$$(3y - 1)(y + 2) = 0 \qquad \text{Set each factor equal to 0.}$$

$$3y - 1 = 0 \quad \bigg| \quad y + 2 = 0 \qquad \text{Solve each equation for } y.$$

$$y = \frac{1}{3} \quad \bigg| \quad y = -2$$

The solutions are $y = \dfrac{1}{3}$ and $y = -2$

Check

$$y = \frac{1}{3} \qquad\qquad\qquad\qquad y = -2$$

$$3y^2 = 2 - 5y \qquad\qquad\qquad 3y^2 = 2 - 5y$$

$$3\left(\frac{1}{3}\right)^2 \overset{?}{=} 2 - 5\left(\frac{1}{3}\right) \qquad 3(-2)^2 \overset{?}{=} 2 - 5(-2)$$

$$3 \cdot \left(\frac{1}{9}\right) \overset{?}{=} 2 - \frac{5}{3} \qquad\qquad 3 \cdot 4 \overset{?}{=} 2 + 10$$

$$\frac{1}{3} \overset{?}{=} \frac{6}{3} - \frac{5}{3} \qquad\qquad\qquad 12 = 12$$

$$\frac{1}{3} = \frac{1}{3}$$

EXAMPLE 4   Solve $3x^2 - 6x = 0$ by factoring.

SOLUTION

$$3x^2 - 6x = 0 \qquad \text{Already in standard form}$$

$$3x(x - 2) = 0 \qquad \text{Factor.}$$

$$3x = 0 \quad \bigg| \quad x - 2 = 0 \qquad \text{Set each factor equal to 0.}$$

$$x = 0 \quad \bigg| \quad x = 2 \qquad \text{Solve for } x \text{ in each equation.}$$

Check

$$x = 0 \qquad\qquad\qquad\qquad x = 2$$

$$3x^2 - 6x = 0 \qquad\qquad\qquad 3x^2 - 6x = 0$$

$$3 \cdot 0^2 - 6 \cdot 0 \overset{?}{=} 0 \qquad\qquad 3 \cdot 2^2 - 6 \cdot 2 \overset{?}{=} 0$$

$$0 = 0 \qquad\qquad\qquad\qquad 12 - 12 = 0$$

$$0 = 0$$

STOP   Never divide both sides of a quadratic equation by an expression containing the variable. This can cause one of the solutions to get lost. This is illustrated in the following example.

EXAMPLE 5   Solve $2x^2 = 6x$.

CORRECT SOLUTION

$$2x^2 = 6x \qquad \text{Put in standard form.}$$
$$2x^2 - 6x = 0 \qquad \text{Factor.}$$
$$2x(x - 3) = 0 \qquad \text{Set each equation equal to 0.}$$
$$2x = 0 \quad \big| \quad x - 3 = 0 \qquad \text{Solve for } x.$$
$$x = 0 \quad \big| \quad x = 3$$

There are two solutions, $x = 0$ and $x = 3$.

INCORRECT SOLUTION

$$2x^2 = 6x$$
$$\frac{2x^2}{2x} = \frac{6x}{2x} \qquad \text{Divide both sides by } 2x \text{ (which you should not do!).}$$
$$x = 3 \qquad \text{Only one solution results.}$$

Dividing by $2x$ caused us to lose the solution $x = 0$.   ▮▮

EXAMPLE 6   Solve $9x^2 - 16 = 0$ by factoring.

SOLUTION

$$9x^2 - 16 = 0 \qquad \text{Already in standard form}$$
$$(3x - 4)(3x + 4) = 0 \qquad \text{Difference of two squares}$$
$$3x - 4 = 0 \quad \big| \quad 3x + 4 = 0 \qquad \text{Factor.}$$
$$3x = 4 \quad \big| \quad 3x = -4 \qquad \text{Solve for } x.$$
$$x = \frac{4}{3} \quad \big| \quad x = -\frac{4}{3}$$

The solutions are $x = \dfrac{4}{3}$ and $x = -\dfrac{4}{3}$.   ▮▮

**Solving Quadratic Equations
of the Form $ax^2 - c = 0$**

• • • • •

If $b = 0$, a quadratic equation of the form $ax^2 - c = 0$ results. The equation $9x^2 - 16 = 0$ is of the form $ax^2 - c = 0$. In Example 6 we solved this equation by factoring. Following is an alternate method for solving equations of this type.

To solve a quadratic equation of the form $ax^2 - c = 0$:

1. add c to both sides to obtain $ax^2 = c$:

2. divide both sides by a, yielding $x^2 = \dfrac{c}{a}$;

3. take the square root of both sides, obtaining $x = \pm\sqrt{\dfrac{c}{a}}$;

4. check the solution in the original equation.

EXAMPLE 7 Solve $9x^2 - 16 = 0$ by the method just described.

SOLUTION

$$9x^2 - 16 = 0 \qquad \text{Add 16 to both sides.}$$

$$9x^2 = 16 \qquad \text{Divide both sides by 9.}$$

$$x^2 = \frac{16}{9} \qquad \text{Take the square root of both sides. (Don't forget the $\pm$.)}$$

$$x = \pm\sqrt{\frac{16}{9}} = \pm\frac{4}{3}$$

The solutions are $x = \frac{4}{3}$ and $x = -\frac{4}{3}$, as in Example 6. ▌▐

EXAMPLE 8 Solve $3x^2 - 24 = 0$.

SOLUTION

$$3x^2 - 24 = 0 \qquad \text{Add 24 to both sides.}$$

$$3x^2 = 24 \qquad \text{Divide both sides by 3.}$$

$$x^2 = 8 \qquad \text{Take the square root of both sides.}$$

$$x = \pm\sqrt{8}$$

$$x = \pm\sqrt{4 \cdot 2} = \pm 2\sqrt{2} \qquad \text{Note that there are two solutions.}$$

Check

$$x = 2\sqrt{2} \qquad\qquad x = -2\sqrt{2}$$

$$3x^2 - 24 = 0 \qquad\qquad 3x^2 - 24 = 0$$

$$3\left(2\sqrt{2}\right)^2 - 24 \overset{?}{=} 0 \qquad\qquad 3\left(-2\sqrt{2}\right)^2 - 24 \overset{?}{=} 0$$

$$3 \cdot 8 - 24 \overset{?}{=} 0 \qquad\qquad 3 \cdot 8 - 24 = 0$$

$$24 - 24 = 0 \quad \text{✔} \qquad\qquad 24 - 24 = 0 \quad \text{✔} \qquad ▌▐$$

7.1 Exercises

▌▐▌▐▌

▼ Write each equation in standard form, and identify a, b, and c.

1. $4x^2 = -2 + 3x$

2. $5 - 4x + 7x^2 = 0$

3. $x = 3x^2 + 1$

4. $3 = 4x^2$

5. $3x^2 + 2x - 1 = 4x^2 + 5$

6. $-6x^2 + 2x - 1 = 3x^2 + 4x - 5$

7. $-2 - x^2 = 4x^2 + x$

8. $2x^2 - 3x - 1 = -3(x^2 + 4)$

▼ Solve and check.

9. $x^2 - x - 12 = 0$

10. $x^2 - 5x + 4 = 0$

11. $x^2 + 10x + 16 = 0$

12. $y^2 - 7y = 0$

13. $t^2 + t - 12 = 0$

14. $x^2 - 64 = 0$

15. $x^2 = 14x$

16. $2x^2 - 3x - 20 = 0$

17. $10x^2 - 13x + 4 = 0$

18. $3x^2 = 12x$

19. $4x^2 + 4x - 3 = 0$

20. $5x^2 - 14x - 3 = 0$

21. $3a^2 + 5a - 2 = 0$

22. $12x^2 - 11x + 2 = 0$

23. $4x^2 - 25 = 0$

24. $16x^2 - 9 = 0$

25. $y^2 - 18 = 0$

26. $2y^2 - 24 = 0$

27. $18x^2 = 2 - 9x$

28. $7x^2 = 15 + 32x$

29. $49t^2 = 24$

30. $\frac{2}{3}y^2 - \frac{27}{2} = 0$

31. Why must the equations $2x^2 + 5x - 12 = 0$ and $-2x^2 - 5x + 12 = 0$ have the same solutions?

7.2   Completing the Square

While solving quadratic equations by factoring, you may well have wondered what happens if the equation is not factorable. Obviously we need another method for solving quadratic equations like $x^2 - 4x - 7 = 0$, which is not factorable. The method we will learn is called *completing the square* and involves construction of perfect-square trinomials.

Examples of perfect-square trinomials are:

$$x^2 + 6x + 9 = (x + 3)^2$$

If we take $\frac{1}{2}$ of 6 and square it, we get $3^2 = 9$.

$$x^2 - 8x + 16 = (x - 4)^2$$

Take $\frac{1}{2}$ of -8 and square it, obtaining $+16$.

In any perfect-square trinomial, if we take one-half of the coefficient of the middle term and square it, we get the last term. The process of obtaining perfect-square trinomials is called **completing the square** and is illustrated in the following example.

EXAMPLE 1   Fill in the blanks so that perfect-square trinomials are obtained.

(a) $x^2 + 6x +$ ____ $= (x +$ ____ $)^2$

$$\left(\frac{1}{2} \cdot 6\right)^2 = 3^2 = 9$$

so $x^2 + 6x + 9 = (x + 3)^2$

(b) $x^2 - 10x +$ ____ $= (x +$ ____ $)^2$

$$\left[\frac{1}{2} \cdot (-10)\right]^2 = (-5)^2 = 25$$

$$x^2 - 10x + 25 = (x - 5)^2$$

(c) $x^2 + 3x +$ ____ $= (x +$ ____ $)^2$

$$\left(\frac{1}{2} \cdot 3\right)^2 = \left(\frac{3}{2}\right)^2 = \frac{9}{4}$$

$$x^2 + 3x + \frac{9}{4} = \left(x + \frac{3}{2}\right)^2$$

▮▮

We will now make use of completing the square to solve quadratic equations.

EXAMPLE 2   Solve $x^2 - 4x - 7 = 0$ by completing the square.

SOLUTION $x^2 - 4x - 7 = 0$ Add 7 to both sides.

$x^2 - 4x +$ ____ $= 7$ Notice the space that we left.

$x^2 - 4x +$ **4** $= 7 +$ **4**   Add the square of half the coefficient of x to both sides: $(-\frac{4}{2})^2 = (-2)^2 = 4$.

$(x - 2)^2 = 11$ Factor the left side, which is a perfect square trinomial

$x - 2 = \pm\sqrt{11}$ Take the square root of both sides.

$x = 2 \pm\sqrt{11}$ Add 2 to both sides.

The solutions are $x = 2 + \sqrt{11}$ and $x = 2 - \sqrt{11}$.

Check

$x = 2 + \sqrt{11}$	$x = 2 - \sqrt{11}$
$x^2 - 4x - 7 = 0$	$x^2 - 4x - 7 = 0$
$(2 + \sqrt{11})^2 - 4(2 + \sqrt{11}) - 7 \overset{?}{=} 0$	$(2 - \sqrt{11})^2 - 4(2 - \sqrt{11}) - 7 \overset{?}{=} 0$
$4 + 4\sqrt{11} + 11 - 8 - 4\sqrt{11} - 7 \overset{?}{=} 0$	$4 - 4\sqrt{11} + 11 - 8 + 4\sqrt{11} - 7 \overset{?}{=} 0$
$0 = 0$ ✔	$0 = 0$ ✔

▮▮

To complete the square, the coefficient of the x^2-term must be 1. If it is not, each term must be divided by the coefficient of the x^2-term before applying the procedure.

> **To solve a quadratic equation $ax^2 + bx + c = 0$ by completing the square:**
>
> 1. subtract the constant term c from both sides of the equation;
> 2. if $a \neq 1$, divide each term on both sides by a;
> 3. complete the square by adding the square of half the coefficient of x to both sides;
> 4. factor the left side of the equation (it will be a perfect-square trinomial);
> 5. take the square root of both sides of the equation and solve for x.

EXAMPLE 3 Solve $2x^2 + 3x - 2 = 0$ by completing the square.

SOLUTION

$2x^2 + 3x - 2 = 0$ Add 2 to both sides.

$2x^2 + 3x + \underline{\quad} = 2$ Divide each term by 2.

$x^2 + \dfrac{3}{2}x + \underline{\quad} = 1$ Take half of $\dfrac{3}{2}$ and square it.

$\left(\dfrac{1}{2} \cdot \dfrac{3}{2}\right)^2 = \left(\dfrac{3}{4}\right)^2 = \dfrac{9}{16}$ Add this result to both sides.

$x^2 + \dfrac{3}{2}x + \boxed{\dfrac{9}{16}} = 1 + \boxed{\dfrac{9}{16}}$ $1 + \dfrac{9}{16} = \dfrac{16}{16} + \dfrac{9}{16} = \dfrac{25}{16}$

$\left(x + \dfrac{3}{4}\right)^2 = \dfrac{25}{16}$ Take the square root of both sides.

$x + \dfrac{3}{4} = \pm\sqrt{\dfrac{25}{16}} = \pm\dfrac{5}{4}$ Subtract $\dfrac{3}{4}$ from both sides.

$x = -\dfrac{3}{4} \pm \dfrac{5}{4}$

The solutions are $x = \dfrac{-3 + 5}{4} = \dfrac{2}{4} = \dfrac{1}{2}$ and $x = \dfrac{-3 - 5}{4} = \dfrac{-8}{4} = -2$.

Check

$x = \dfrac{1}{2}$

$2x^2 + 3x - 2 = 0$

$2\left(\dfrac{1}{2}\right)^2 + 3 \cdot \dfrac{1}{2} - 2 \stackrel{?}{=} 0$

$2 \cdot \dfrac{1}{4} + \dfrac{3}{2} - 2 \stackrel{?}{=} 0$

$\dfrac{1}{2} + \dfrac{3}{2} - \dfrac{4}{2} \stackrel{?}{=} 0$

$0 = 0$ ✓

$x = -2$

$2x^2 + 3x - 2 = 0$

$2(-2)^2 + 3(-2) - 2 \stackrel{?}{=} 0$

$2(4) - 6 - 2 \stackrel{?}{=} 0$

$8 - 6 - 2 \stackrel{?}{=} 0$

$0 = 0$ ✓

7.2   Exercises

▮ ▮ ▮ ▮ ▮

▼ Fill in the blanks to make a perfect-square trinomial.

1. $x^2 + 4x + \underline{\quad} = (x + \underline{\quad})^2$

2. $x^2 - 10x + \underline{\quad} = (x - \underline{\quad})^2$

3. $y^2 - 8y + \underline{\quad} = (y - \underline{\quad})^2$

4. $y^2 + 3y + \underline{\quad} = (y + \underline{\quad})^2$

5. $x^2 - 7x + \underline{\quad} = (x - \underline{\quad})^2$

6. $x^2 - \dfrac{2}{3}x + \underline{\quad} = (x - \underline{\quad})^2$

7. $x^2 + \dfrac{1}{4}x + \underline{\quad} = (x + \underline{\quad})^2$

8. $x^2 - x + \underline{\quad} = (x - \underline{\quad})^2$

▼ Solve by completing the square:

9. $x^2 + 6x + 8 = 0$

10. $y^2 - 2y - 3 = 0$

11. $t^2 - 8t + 15 = 0$

12. $x^2 - 4x - 12 = 0$

13. $x^2 + 10x + 21 = 0$

14. $x^2 - 4x - 1 = 0$

15. $y^2 - 3y - 5 = 0$

16. $z^2 + 6z - 4 = 0$

17. $y^2 + y - 7 = 0$

18. $x^2 + 3x - 5 = 0$

19. $m^2 + 2m - 5 = 0$

20. $t^2 + 3t + 1 = 0$

21. $x^2 + 3x - 2 = 0$

22. $3y^2 + 2y - 6 = 0$

23. $2x^2 - 5x - 5 = 0$

24. $2y^2 - y - 7 = 0$

25. $6x^2 + 5x = 4$

26. $10x^2 + 23x = 5$

 27. What are the characteristics of a perfect-square trinomial?

7.3   The Quadratic Formula

Since we can use the process of completing the square on *any* quadratic equation, whether it is factorable or not, let's now apply this procedure on the standard form of the quadratic equation, $ax^2 + bx + c = 0$. In doing so, we will derive a formula known as the **quadratic formula.**

Derivation of the Quadratic Formula

We start with the standard form and complete the square:

$$ax^2 + bx + c = 0 \qquad \text{$a \neq 0$, standard form}$$

$$ax^2 + bx = -c \qquad \text{Add $-c$ to both sides.}$$

$$\frac{ax^2}{a} + \frac{bx}{a} = \frac{-c}{a} \qquad \text{Divide both sides by a.}$$

$$x^2 + \frac{b}{a}x + \frac{b^2}{4a^2} = \frac{-c}{a} + \frac{b^2}{4a^2} \qquad \text{Add $\left(\frac{1}{2} \cdot \frac{b}{a}\right)^2 = \frac{b^2}{4a^2}$ to both sides.}$$

$$\left(x + \frac{b}{2a}\right)^2 = \frac{-c}{a} + \frac{b^2}{4a^2} \qquad \text{Factor the left side.}$$

$$\left(x + \frac{b}{2a}\right)^2 = \frac{-4ac + b^2}{4a^2} = \frac{b^2 - 4ac}{4a^2} \qquad \text{Add the fractions on the right.}$$

$$x + \frac{b}{2a} = \pm\sqrt{\frac{b^2 - 4ac}{4a^2}} = \pm\frac{\sqrt{b^2 - 4ac}}{2a} \qquad \text{Take the square root of both sides.}$$

$$x = -\frac{b}{2a} \pm \frac{\sqrt{b^2 - 4ac}}{2a} \qquad \text{Add $-\frac{b}{2a}$ to both sides.}$$

$$x = \frac{-b \pm \sqrt{b^2 - 4ac}}{2a} \qquad \text{This is called the \textbf{quadratic formula.} (Note: $\pm$ gives \textit{two} solutions.)}$$

This formula must be memorized, since it provides solutions to *any* quadratic equation. The best approach to solving a quadratic equation is to try to factor it first. If, after a reasonable length of time, you are unable to factor it, solve it using the quadratic formula. The quadratic formula makes the tedious method of completing the square unnecessary when solving a quadratic equation.

To solve a quadratic equation using the quadratic formula

$$x = \frac{-b \pm \sqrt{b^2 - 4ac}}{2a}$$

---

1. write the given equation in standard form, $ax^2 + bx + c = 0$;

2. identify the constants a, b, and c;

3. substitute the values for a, b, and c into the quadratic formula;

4. evaluate to find the solutions.

E X A M P L E   1   Solve: $x^2 - 5x + 3 = 0$.

S O L U T I O N $x^2 - 5x + 3 = 0$ is in standard form as given, but it is not factorable. We see that

$$a = 1, \quad b = -5, \quad c = 3$$

We substitute these values into the quadratic formula:

$$x = \frac{-b \pm \sqrt{b^2 - 4ac}}{2a}$$

$$x = \frac{-(-5) \pm \sqrt{(-5)^2 - 4 \cdot 1 \cdot 3}}{2(1)} \qquad \text{Substitute for } a, b, \text{ and } c.$$

$$= \frac{5 \pm \sqrt{25 - 12}}{2}$$

$$= \frac{5 \pm \sqrt{13}}{2}$$

∎∎

STOP   The *entire numerator* in the quadratic formula is divided by $2a$:

$$x = -b \pm \frac{\sqrt{b^2 - 4ac}}{2a} \qquad \text{is not correct}$$

$$x = \frac{-b \pm \sqrt{b^2 - 4ac}}{2a} \qquad \text{is correct}$$

E X A M P L E   2   Solve $2x^2 - 5x - 3 = 0$ **(a)** by factoring; **(b)** by using the quadratic formula.

S O L U T I O N   **(a)**

$$2x^2 - 5x - 3 = 0$$

$$(2x + 1)(x - 3) = 0 \qquad \text{Factors}$$

$$2x + 1 = 0 \quad \bigg| \quad x - 3 = 0 \qquad \text{Set each factor equal to 0.}$$

$$2x = -1 \quad \bigg| \quad x = 3$$

$$x = -\frac{1}{2} \quad \bigg|$$

The solutions are $x = -\dfrac{1}{2}$ and $x = 3$.

(b) $2x^2 - 5x - 3 = 0$ Already in standard form

$a = 2, \quad b = -5, \quad c = -3$ Identify a, b, and c.

$x = \dfrac{-b \pm \sqrt{b^2 - 4ac}}{2a}$ Use the quadratic formula.

$= \dfrac{-(-5) \pm \sqrt{(-5)^2 - 4(2)(-3)}}{2(2)}$ Substitute for a, b, and c.

$= \dfrac{5 \pm \sqrt{25 + 24}}{4} = \dfrac{5 \pm \sqrt{49}}{4} = \dfrac{5 \pm 7}{4}$

$x = \dfrac{5 + 7}{4} = \dfrac{12}{4} = 3$ or $x = \dfrac{5 - 7}{4} = \dfrac{-2}{4} = -\dfrac{1}{2}.$

The solutions are $x = 3$ and $x = -\dfrac{1}{2}$, as before.   ▮▮

EXAMPLE 3   Solve: $x^2 - 2x + 7 = 0$.

SOLUTION   In this case we have $a = 1$, $b = -2$, and $c = 7$.

$x = \dfrac{-b \pm \sqrt{b^2 - 4ac}}{2a}$

$= \dfrac{-(-2) \pm \sqrt{(-2)^2 - 4 \cdot 1 \cdot 7}}{2 \cdot 1}$ Substitute for a, b, and c.

$= \dfrac{2 \pm \sqrt{-24}}{2}$ $\sqrt{-24} = \sqrt{4 \cdot (-1) \cdot 6} = 2i\sqrt{6}$

$= \dfrac{2 \pm 2i\sqrt{6}}{2}$ Simplify.

$= 1 \pm i\sqrt{6}$

The solutions are the complex numbers $1 + i\sqrt{6}$ and $1 - i\sqrt{6}$. This example shows that quadratic equations can have complex numbers as well as real numbers as solutions.

The Discriminant and Types of Solutions to Quadratic Equations

• • • • •

In the following example we will examine the different types of solutions obtained by solving three different quadratic equations using the quadratic formula.

EXAMPLE 4   **(a)** $x^2 - 3x - 4 = 0$

$$x = \frac{-b \pm \sqrt{b^2 - 4ac}}{2a}$$

$$= \frac{-(-3) \pm \sqrt{(-3)^2 - 4(1)(-4)}}{2(1)} \qquad a = 1, \ b = -3, \ c = -4$$

$$= \frac{3 \pm \sqrt{9 + 16}}{2} = \frac{3 \pm \sqrt{25}}{2} = \frac{3 \pm 5}{2}$$

$$x = \frac{3 + 5}{2} = \frac{8}{2} = 4 \quad \text{or} \quad x = \frac{3 - 5}{2} = \frac{-2}{2} = -1$$

We obtain two real-number solutions, 4 and -1.

(b) $x^2 - 4x + 4 = 0$

$$x = \frac{-b \pm \sqrt{b^2 - 4ac}}{2a}$$

$$= \frac{-(-4) \pm \sqrt{(-4)^2 - 4(1)(4)}}{2(1)} \qquad a = 1, \ b = -4, \ c = 4$$

$$= \frac{4 \pm \sqrt{16 - 16}}{2} = \frac{4 \pm \sqrt{0}}{2}$$

$$= \frac{4}{2} = \boxed{2}$$

We obtain one real-number solution, 2.

(c) $x^2 - 2x + 2 = 0$

$$x = \frac{-b \pm \sqrt{b^2 - 4ac}}{2a}$$

$$= \frac{-(-2) \pm \sqrt{(-2)^2 - 4(1)(2)}}{2(1)} \qquad a = 1, \ b = -2, \ c = 2$$

$$= \frac{2 \pm \sqrt{4 - 8}}{2} = \frac{2 \pm \sqrt{-4}}{2}$$

$$= \frac{2 \pm 2i}{2} = 1 \pm i$$

We obtain two complex conjugate solutions, $1 + i$ and $1 - i$,   ▌▐

In each case, the type of solution was determined by the expression $b^2 - 4ac$ under the radical in the quadratic formula. This number is called the **discriminant** of the equation $ax^2 + bx + c = 0$. Examine the following table.

Equation	$b^2 - 4ac$	Solutions	Type of Solutions
$x^2 - 3x - 4 = 0$	25	$4, -1$	two real numbers
$x^2 - 4x + 4 = 0$	0	2	one real number
$x^2 - 2x + 2 = 0$	-4	$1 + i, 1 - i$	two complex numbers

These results are summarized by the following:

The quadratic equation $ax^2 + bx + c = 0$ has:

1. two real-numbers solutions if $b^2 - 4ac > 0$;

2. one real-number solution if $b^2 - 4ac = 0$;

3. two complex-number solutions if $b^2 - 4ac < 0$.

EXAMPLE 5   Determine the type of solutions each of the following equations has by examining the discriminant.

(a) $3x^2 - 2x - 1 = 0$

$$a = 3, \quad b = -2, \quad c = -1$$
$$b^2 - 4ac = (-2)^2 - 4(3)(-1)$$
$$= 4 + 12$$
$$= 16 > 0$$

There are two real-number solutions.

(b) $x^2 - 6x + 9 = 0$

$$a = 1, \quad b = -6, \quad c = 9$$
$$b^2 - 4ac = (-6)^2 - 4(1)(9)$$
$$= 36 - 36$$
$$= 0$$

There is one real-number solution.

(c) $x^2 + x + 5 = 0$

$$a = 1, \quad b = 1, \quad c = 5$$
$$b^2 - 4ac = (1)^2 - 4(1)(5)$$
$$= 1 - 20$$
$$= -19 < 0$$

There are two complex-number solutions.   ■ ■

7.3   Exercises

▼ Solve, using the quadratic formula:

1. $x^2 + 5x + 4 = 0$

2. $y^2 + 6y - 8 = 0$

3. $2y^2 - 8y + 3 = 0$

4. $t^2 + 2t - 4 = 0$

5. $x^2 + 6x + 3 = 0$

6. $3x^2 - 8x + 1 = 0$

7. $x^2 + x - 1 = 0$

8. $x^2 - 5x - 3 = 0$

9. $3x^2 - 2x + 1 = 0$

10. $x^2 + 3x + 5 = 0$

11. $x^2 - 2x + 5 = 0$

17. $2x^2 - 2x = -1$

12. $x^2 - 2x - 1 = 0$

18. $x^2 - 3x = -3$

13. $x^2 - 2x - 6 = 0$

19. $x^2 + 10x = -40$

14. $x^2 - 3x + 3 = 0$

20. $2x^2 + 8x = -3$

15. $t^2 + 4t = 1$

21. $3x^2 + 7 = 2x$

16. $2x^2 + 5x = -1$

22. $3x^2 - 6x = 2x^2 + 3$

23. $x(x + 3) = -5$

24. $(3y - 1)^2 + 6y = 0$

25. $x^2 + 2(3x - 2) = 0$

26. $x(x + 2) + 3(x - 2) = -1$

27. $(x + 1)^2 = 3(1 - x)$

28. $(x + 3)(x + 4) = 15$

▼ Do not solve, but use the discriminant to determine the number and type of solutions for each of the following equations.

29. $x^2 + 3x + 1 = 0$

30. $x^2 - 6x - 2 = 0$

31. $x^2 - 10x + 25 = 0$

32. $x^2 + 6x + 9 = 0$

33. $x^2 + 4x + 5 = 0$

34. $x^2 + 5x + 9 = 0$

35. $3x^2 + 2x - 1 = 0$

36. $2x^2 + 4x + 3 = 0$

37. $6x^2 - 2x + 3 = 0$

38. $3x^2 - 4 = 0$

39. $3x^2 + 4 = 0$

40. $4x^2 - 12x + 9 = 0$

▼ Find a value of k so that each of the following has one real-number solution.

41. $x^2 - kx + 36 = 0$

42. $9x^2 + 30x + k = 0$

43. $kx^2 - 4x = -1$

7.4   Equations Resulting in Quadratic Equations

I I I I I

Equations Containing Fractions

• • • • •

So far, all equations that we have considered that contain fractions have resulted in *linear equations*. Now we will discuss equations that contain fractions and lead to **quadratic equations.** The procedure we use may result in equations with solutions that are not solutions to the original equations. Therefore, *all* apparent solutions must be checked in the *original* equation.

EXAMPLE 1   Solve: $\dfrac{3}{x+1} - \dfrac{2}{x-1} = -1$.

SOLUTION   We multiply by the LCD $= (x+1)(x-1)$ to clear of fractions:

$$(x+1)(x-1)\left[\frac{3}{x+1} - \frac{2}{x-1}\right] = (x+1)(x-1)\,(-1)$$

$$(x+1)(x-1)\frac{3}{(x+1)} - (x+1)(x-1)\frac{2}{(x-1)} = (x+1)(x-1)(-1)$$

$$3x - 3 - 2x - 2 = -x^2 + 1$$

$$x - 5 = -x^2 + 1$$

$$x^2 + x - 6 = 0 \qquad \text{Write in standard form.}$$

$$(x+3)(x-2) = 0 \qquad \text{Factor.}$$

$$\begin{array}{c|c} x + 3 = 0 & x - 2 = 0 \\ x = -3 & x = 2 \qquad \text{Solve for } x. \end{array}$$

The solutions are $x = -3$ and $x = 2$.

Check

$x = -3$	$x = 2$

$$\frac{3}{x+1} - \frac{2}{x-1} = -1 \qquad\qquad \frac{3}{x+1} - \frac{2}{x-1} = -1$$

$$\frac{3}{(-3)+1} - \frac{2}{(-3)-1} \stackrel{?}{=} -1 \qquad\qquad \frac{3}{2+1} - \frac{2}{2-1} \stackrel{?}{=} -1$$

$$\frac{3}{-2} - \frac{2}{-4} \stackrel{?}{=} -1 \qquad\qquad \frac{3}{3} - \frac{2}{1} \stackrel{?}{=} -1$$

$$\frac{6}{-4} - \frac{2}{-4} \stackrel{?}{=} -1 \qquad\qquad 1 - 2 \stackrel{?}{=} -1$$

$$\frac{4}{-4} \stackrel{?}{=} -1 \qquad\qquad\qquad -1 = -1 \quad \blacktriangleright$$

$$-1 = -1 \quad \blacktriangleright$$

Both solutions check.

I I

EXAMPLE 2 Solve: $\dfrac{y^2}{y-2} = \dfrac{4}{y-2}$.

SOLUTION We multiply both sides of the equation by $y - 2$, the LCD.

$$(y-2) \cdot \dfrac{y^2}{(y-2)} = (y-2) \cdot \dfrac{4}{(y-2)}$$

$$y^2 = 4$$

$$y = \pm 2 \qquad \text{Take the square root of both sides.}$$

The apparent solutions are $y = -2$ and $y = 2$.

Check

$$
\begin{array}{c|c}
y = -2 & y = 2 \\[4pt]
\dfrac{y^2}{y-2} = \dfrac{4}{y-2} & \dfrac{y^2}{y-2} = \dfrac{4}{y-2} \\[10pt]
\dfrac{(-2)^2}{(-2)-2} \stackrel{?}{=} \dfrac{4}{(-2)-2} & \dfrac{(2)^2}{2-2} \stackrel{?}{=} \dfrac{4}{2-2} \\[10pt]
\dfrac{4}{-4} = \dfrac{4}{-4} & \dfrac{4}{0} \stackrel{?}{=} \dfrac{4}{0} \\[10pt]
-1 = -1 \quad \checkmark &
\end{array}
$$

Since $\dfrac{4}{0}$ is undefined, $y = 2$ is not a solution. Therefore $y = -2$ is the only solution. This example illustrates the importance of checking all of your results in the original equation. Here $y = 2$ is called an *extraneous solution*. ▌▌

Equations Quadratic in Form

• • • • •

The equation $x^4 - 3x^2 - 4 = 0$ is quadratic in x^2, since we can write it

$$(x^2)^2 - 3(x^2) - 4 = 0$$

You may find it easier to make a temporary substitution for x^2, say $y = x^2$, to obtain

$$y^2 - 3y - 4 = 0$$

This is solved by factoring or by using the quadratic formula:

$$y^2 - 3y - 4 = 0$$

$$(y - 4)(y + 1) = 0$$

$$
\begin{array}{c|c}
y - 4 = 0 & y + 1 = 0 \\
y = 4 & y = -1
\end{array}
$$

The solutions are $y = 4$ and $y = -1$.

We have solutions in terms of y, but our original equation is in terms of x. Now go back to our temporary substitution, $y = x^2$, which gives us

$$x^2 = 4 \qquad \text{and} \qquad x^2 = -1$$
$$x = \pm 2 \qquad \text{and} \qquad x = \pm \sqrt{-1} = \pm i$$

There are four solutions: 2, -2, i, and $-i$.

EXAMPLE 3   Solve: $2x^4 - 5x^2 - 3 = 0$.

SOLUTION

$$2x^4 - 5x^2 - 3 = 0$$
$$2(x^2)^2 - 5(x)^2 - 3 = 0 \qquad \text{Substitute } y = x^2.$$
$$2y^2 - 5y - 3 = 0 \qquad \text{Factor.}$$
$$(y - 3)(2y + 1) = 0$$

$$y - 3 = 0 \qquad \bigm| \qquad 2y + 1 = 0 \qquad \text{Solve for } y.$$

$$y = 3 \qquad \bigm| \qquad y = -\frac{1}{2}$$

Since $y = x^2$, we go back to our original variables; x:

$$x^2 = 3 \qquad \text{and} \qquad x^2 = -\frac{1}{2} \qquad\qquad \text{Solve for } x.$$

$$x = \pm\sqrt{3} \qquad \text{and} \qquad x = \pm\sqrt{-\frac{1}{2}} = \pm\frac{i\sqrt{2}}{2}$$

The solutions are $\sqrt{3}$, $-\sqrt{3}$, $\dfrac{i\sqrt{2}}{2}$, and $\dfrac{-i\sqrt{2}}{2}$.   ■ ■

EXAMPLE 4   Solve: $(x - 1)^2 - 6(x - 1) + 8 = 0$.

SOLUTION

$$(\,x - 1\,)^2 - 6(\,x - 1\,) + 8 = 0 \qquad \text{Quadratic in } (x - 1).$$
$$y^2 - 6y + 8 = 0 \qquad \text{Let } y = (x - 1).$$
$$(y - 4)(y - 2) = 0 \qquad \text{Factor.}$$

$$y - 4 = 0 \qquad \bigm| \qquad y - 2 = 0$$
$$y = 4 \qquad \bigm| \qquad y = 2$$

Since $y = x - 1$,

$$x - 1 = 4 \qquad \text{and} \qquad x - 1 = 2 \qquad \text{Solve for } x.$$
$$x = 5 \qquad \text{and} \qquad x = 3$$

The solutions are 5 and 3.   ■ ■

Radical Equations Containing Square Roots

Radical equations are generally solved by isolating a radical on one side of the equation and then squaring both sides of the equation. Some of the equations that result are quadratic equations.

This squaring process sometimes produces apparent solutions that don't check in the original equation. For example, the equation $x = 5$ has only one solution, the number 5. Squaring both sides ($x^2 = 25$) produces a new equation that has two solutions, 5 and -5.

Therefore, *all* solutions must be checked in the *original equation*.

EXAMPLE 5   Solve: $\sqrt{x + 12} = x$.

SOLUTION   Square both sides and solve for x.:

$$\sqrt{x + 12} = x \qquad \text{Square both sides.}$$
$$(\sqrt{x + 12})^2 = x^2$$
$$x + 12 = x^2 \qquad \text{Put in standard form.}$$
$$0 = x^2 - x - 12$$
$$0 = (x - 4)(x + 3) \qquad \text{Factor.}$$

$$
\begin{array}{c|c}
x - 4 = 0 & x + 3 = 0 \\
x = 4 & x = -3
\end{array}
$$

The apparent solutions are $x = 4$ and $x = -3$.

$$
\begin{array}{l|l}
\textbf{Check} \quad x = 4 & x = -3 \\
\sqrt{x + 12} = x & \sqrt{x + 12} = x \\
\sqrt{4 + 12} \stackrel{?}{=} 4 & \sqrt{(-3) + 12} \stackrel{?}{=} -3 \\
\sqrt{16} \stackrel{?}{=} 4 & \sqrt{9} \stackrel{?}{=} -3 \qquad \sqrt{9} = +3 \text{ only.} \\
4 = 4 \quad \text{Checks.} & 3 \neq -3 \qquad \text{Does not check!}
\end{array}
$$

Therefore the *only* solution is $x = 4$.   ▮▮

STOP   When solving any equation containing a radical or a variable in a denominator, it is absolutely essential that every apparent solution be checked in the original equation. As you can see from Examples 2 and 5, some apparent solutions may not be solutions at all. These are called *extraneous solutions* or *extraneous roots*.

EXAMPLE 6   Solve: $\sqrt{y-5} - y + 7 = 0$.

SOLUTION

$\sqrt{y-5} = y - 7$	Isolate the radical.
$(\sqrt{y-5})^2 = (y-7)^2$	Square both sides.
$y - 5 = y^2 - 14y + 49$	Remember the middle term on the right.
$0 = y^2 - 15y + 54$	Factor.
$0 = (y-9)(y-6)$	Solve for y.

$$y - 9 = 0 \quad \bigg| \quad y - 6 = 0$$
$$y = 9 \quad \bigg| \quad y = 6$$

The apparent solutions are $y = 9$ and $y = 6$.

Check

$$
\begin{array}{c|c}
y = 9 & y = 6 \\
\sqrt{y-5} = y - 7 & \sqrt{y-5} = y - 7 \\
\sqrt{9-5} \overset{?}{=} 9 - 7 & \sqrt{6-5} \overset{?}{=} 6 - 7 \\
\sqrt{4} \overset{?}{=} 2 & \sqrt{1} \overset{?}{=} -1 \\
2 = 2 \quad \checkmark & 1 \neq -1 \quad \text{Does not check.}
\end{array}
$$

The *only* solution is $y = 9$.   ▮▮

EXAMPLE 7   Solve: $\sqrt{3x+1} - \sqrt{x-4} = 3$.

SOLUTION

$\sqrt{3x+1} = 3 + \sqrt{x-4}$	Isolate one radical on the left.
$(\sqrt{3x+1})^2 = (3 + \sqrt{x-4})^2$	Square both sides.
$3x + 1 = 9 + 6\sqrt{x-4} + x - 4$	Don't forget the middle term, $6\sqrt{x-4}$.
$2x - 4 = 6\sqrt{x-4}$	Isolate the radical on the right.
$x - 2 = 3\sqrt{x-4}$	Divide both sides by 2 to simplify.
$(x-2)^2 = (3\sqrt{x-4})^2$	Square both sides.
$x^2 - 4x + 4 = 9(x-4)$	
$x^2 - 4x + 4 = 9x - 36$	Simplify to put in standard form.
$x^2 - 13x + 40 = 0$	Factor.
$(x-8)(x-5) = 0$	

$$x - 8 = 0 \quad \bigg| \quad x - 5 = 0$$
$$x = 8 \quad \bigg| \quad x = 5$$

17. $2x^4 - 9x^2 = -4$ **18.** $3x^2 + 18 = 29x$

▼ Solve, making sure to check all apparent solutions.

23. $\sqrt{x^2 + 7} = 4$ **24.** $\sqrt{5x - 6} = x$

19. $(x - 2)^2 + 3(x - 2) + 2 = 0$

25. $\sqrt{x - 3} = x - 9$ **26.** $\sqrt{5x - 1} = 3 - x$

20. $(3x - 1)^2 - 4(3x - 1) + 4 = 0$

27. $x + 2 = \sqrt{x + 2}$ **28.** $\sqrt{-4x - 2} = 2x + 1$

21. $(x - 5)^2 + 5(x - 5) + 6 = 0$

29. $\sqrt{2x + 7} = x + 4$ **30.** $x - 1 = \sqrt{2x + 1}$

31. $\sqrt{3x + 1} - \sqrt{x - 4} = 3$

22. $(x + 3)^2 - 2(x + 3) - 8 = 0$

32. $\sqrt{x + 3} + 4 = \sqrt{5x + 31}$

33. $\sqrt{2x} + \sqrt{x+1} = 1$

34. $\sqrt{2-3x} + \sqrt{3x+3} = 1$

35. $\sqrt{2x+1} = 2 + \sqrt{5-x}$

36. $\sqrt{x} - \sqrt{2x+4} = 2$

 37. Why must all solutions be checked in the original equation when solving a radical equation?

 38. Consider the equation $x = 2$. When we square both sides, we obtain the equation $x^2 = 4$, which has two solutions, $+2$ and -2. How is this possible?

 39. Can you square both numerator and denominator of a fraction to obtain an equivalent fraction?

7.5 Applications of Quadratic Equations

Many application problems result in quadratic equations. We must be careful that all solutions make sense in the original problem.

EXAMPLE 1   The product of two consecutive even whole numbers is 168. Find the numbers.

SOLUTION   Let $x = $ the first even whole number.

Then $x + 2 = $ the next even whole number.

Since their product is 168, we have the following equation:

$$x(x + 2) = 168$$
$$x^2 + 2x = 168 \qquad \text{Put in standard form.}$$
$$x^2 + 2x - 168 = 0 \qquad \text{Factor.}$$
$$(x + 14)(x - 12) = 0$$

$$
\begin{array}{c|c}
x + 14 = 0 & x - 12 = 0 \\
x = -14 & x = 12
\end{array}
$$

Since our problem asks for *whole* numbers, -14 is not a solution (whole numbers are not negative). The two numbers are therefore 12 and 14 (x and $x + 2$).

Check   12 and 14 are consecutive even whole numbers, and $12 \cdot 14 = 168$.   ▮ ▮

EXAMPLE 2 The height of a triangle is 3 cm greater than the base. If the area of the triangle is 54 cm^2, find the base and the height of the triangle.

SOLUTION   Let b = the base of the triangle.

Then $b + 3$ = the height of the triangle.

See Figure 7.6 for an illustration.

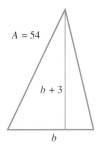

FIGURE 7.6.

$$\frac{1}{2} \cdot \text{base} \cdot \text{height} = \text{area}$$

$$\frac{1}{2} \cdot b \cdot (b + 3) = A = 54 \qquad \text{Multiply by 2 to clear of fractions}$$

$$b(b + 3) = 108$$

$$b^2 + 3b = 108 \qquad \text{Put in standard form.}$$

$$b^2 + 3b - 108 = 0 \qquad \text{Factor.}$$

$$(b + 12)(b - 9) = 0$$

$$b + 12 = 0 \quad \bigg| \quad b - 9 = 0$$

$$b = -12 \quad \bigg| \quad b = 9$$

Since the base cannot be negative, our only solution is $b = 9$ cm, and the height will be $b + 3 = 9 + 3 = 12$ cm.
Check height = 12 cm is 3 cm more than $b = 9$ cm, and the area is $A = \frac{1}{2}bh = \frac{1}{2} \cdot 9 \cdot 12 = 54$ cm^2.   ▮▮

The **Pythagorean theorem** states that in any right triangle, the square of the longest side, the *hypotenuse, c,* is equal to the sum of the squares of the other two sides, *a* and *b.* (See the discussion on the Pythagorean theorem at the beginning of this chapter in "Number Knowledge.") stated as an equation;

$$a^2 + b^2 = c^2$$

EXAMPLE 3 The length of the hypotenuse of a right triangle is 5 inches. Find the lengths of the two sides if one side is one inch longer than the other.

SOLUTION   Let x = the length of the shorter leg.

Then $x + 1$ = the length of the longer leg.

Figure 7.7 is a diagram showing this relationship.

FIGURE 7.7.

According to the Pythagorean theorem, we have the following equation:

$$x^2 + (x + 1)^2 = 5^2 \qquad \text{Square } (x + 1) \text{ and } 5.$$

$$x^2 + x^2 + 2x + 1 = 25 \qquad \text{Combine like terms.}$$

$$2x^2 + 2x + 1 = 25 \qquad \text{Put in standard form.}$$

$$2x^2 + 2x - 24 = 0 \qquad \text{Divide both sides by 2.}$$

$$x^2 + x - 12 = 0 \qquad \text{Factor.}$$

$$(x - 3)(x + 4) = 0$$

$$x - 3 = 0 \quad | \quad x + 4 = 0$$

$$x = 3 \quad | \quad x = -4$$

The number -4 is not a solution, since it represents the length of the side of the triangle, which cannot be negative. Therefore, our only solution is $x = 3$ in., which represents the length of the shorter leg. The longer leg is $x + 1 = 3 + 1 = 4$ in.

Finally, $3^2 + 4^2 = 5^2$, since $9 + 16 = 25$, and our result checks.   ▮▮

EXAMPLE 4   A page in a book measures 7 in. by 9 in. There is a blank border of even width around the page that leaves 35 in.^{2} of print. How wide is the border?

SOLUTION   Let x = the width of the border; then $l = 9 - 2x$ = the length of the printed portion, and $u = 7 - 2x$ = the width of the printed portion. See Figure 7.8.

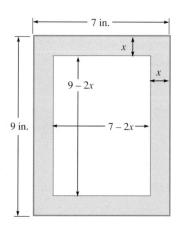

FIGURE 7.8.

$$\text{area of print} = l \cdot w = 35$$

$$(9 - 2x)(7 - 2x) = 35$$

$$63 - 32x + 4x^2 = 35$$

$$4x^2 - 32x + 28 = 0 \qquad \text{Divide by 4.}$$

$$x^2 - 8x + 7 = 0 \qquad \text{Factor.}$$

$$(x - 7)(x - 1) = 0$$

$$x - 7 = 0 \ \bigg| \ x - 1 = 0$$

$$x = 7 \ \bigg| \qquad x = 1$$

Check 7 cannot be a solution, since a border of 7 in. would result in no print on the page. If the border is 1 in., the length of the print will be $9 - 2(1) = 7$ in., the width will be $7 - 2(1) = 5$ in., and $7 \cdot 5 = 35$ in.^{2}, which is the area of the printed portion of the page. Therefore, the width of the border is indeed 1 in. ▮▮

EXAMPLE 5   A canoeist travels 4 miles upstream, turns around, and travels 4 miles back downstream in a total time of 3 hours. If the speed of the current is 1 mph, what is the speed of the canoe in still water?

SOLUTION   Let r = the rate of speed of the canoe in still water.

Then $r - 1$ = the rate of the canoe upstream

and $r + 1$ = the rate of the canoe downstream.

distance upstream = distance downstream = 4

Since rate · time = distance, or $r \cdot t = d$, dividing by r gives us $t = \dfrac{d}{r}$. In tabular form, it looks like this:

	Rate	Distance	Time = $\dfrac{\text{distance}}{\text{rate}}$
Upstream	$r - 1$	4	$\dfrac{4}{r - 1}$
Downstream	$r + 1$	4	$\dfrac{4}{r + 1}$

Realizing that the canoeist's time upstream plus his time downstream must equal his total time, 4 hours, we have this equation:

$$\text{time upstream} + \text{time downstream} = \text{total time}$$

$$\frac{4}{r - 1} + \frac{4}{r + 1} = 3$$

To solve, we multiply by the LCD $= (r - 1)(r + 1)$.

$$\cancel{(r - 1)}(r + 1) \cdot \frac{4}{\cancel{r - 1}} + (r - 1)\cancel{(r + 1)} \cdot \frac{4}{\cancel{r + 1}} = (r - 1)(r + 1) \cdot 3$$

$$4(r + 1) + 4(r - 1) = 3(r - 1)(r + 1)$$

$$4r + 4 + 4r - 4 = 3r^2 - 3$$

$$8r = 3r^2 - 3 \qquad \text{Put in}$$
$$\text{standard form.}$$

$$3r^2 - 8r - 3 = 0 \qquad \text{Factor.}$$

$$(3r + 1)(r - 3) = 0$$

$$3r + 1 = 0 \quad | \quad r - 3 = 0$$

$$r = -\frac{1}{3} \quad | \quad r = 3$$

Since the rate of the boat cannot be negative, we reject $r = -\frac{1}{3}$.

Check $r = 3$

$$\frac{4}{r - 1} + \frac{4}{r + 1} = 3$$

$$\frac{4}{3 - 1} + \frac{4}{3 + 1} \stackrel{?}{=} 3$$

$$\frac{4}{2} + \frac{4}{4} = 3 \quad \text{☑}$$

The speed of the canoe in still water is 3 mph. ▮▮

EXAMPLE 6   How many sides does a polygon have if it has 20 diagonals?

SOLUTION   The formula for finding the number of diagonals of an *n*-sided polygon is $\frac{n(n - 3)}{2}$. For example, as pictured in Figure 7.9, a 5-sided polygon (a pentagon) has

$$\frac{5(5 - 3)}{2} = \frac{5 \cdot 2}{2} = 5 \text{ diagonals}$$

FIGURE 7.9.

$$\frac{n(n-3)}{2} = 20 \qquad \text{Multiply both sides by 2.}$$

$$n(n-3) = 40$$

$$n^2 - 3n = 40 \qquad \text{Put in standard form.}$$

$$n^2 - 3n - 40 = 0 \qquad \text{Factor.}$$

$$(n-8)(n+5) = 0$$

$$n - 8 = 0 \quad | \quad n + 5 = 0$$

$$n = 8 \quad | \quad n = -5$$

We reject $n = -5$, so the polygon with 20 diagonals has 8 sides (an octagon). Draw an octagon and see if you can draw 20 diagonals. ▮▮

EXAMPLE 7   The sum of a number and its reciprocal is $2\frac{1}{6}$. Find the number.

SOLUTION   Let x = the desired number.

Then $\dfrac{1}{x}$ = the reciprocal of x.

Our equation is

$$x + \frac{1}{x} = 2\frac{1}{6} = \frac{13}{6}$$

$$6x \cdot x + 6x \cdot \frac{1}{x} = 6x \cdot \frac{13}{6} \qquad \text{Multiply by } 6x, \text{ the LCD.}$$

$$6x^2 + 6 = 13x \qquad \text{Put in standard form.}$$

$$6x^2 - 13x + 6 = 0 \qquad \text{Factor.}$$

$$(3x - 2)(2x - 3) = 0$$

$$3x - 2 = 0 \quad | \quad 2x - 3 = 0$$

$$3x = 2 \quad | \quad 2x = 3$$

$$x = \frac{2}{3} \quad | \quad x = \frac{3}{2}$$

Each solution is the reciprocal of the other, and since

$$\frac{2}{3} + \frac{3}{2} = \frac{4}{6} + \frac{9}{6} = \frac{13}{6} = 2\frac{1}{6}$$

both solutions check. The solutions are $\dfrac{2}{3}$ or $\dfrac{3}{2}$. ▮▮

7.5   Exercises

▮ ▮ ▮ ▮ ▮

1. The sum of the squares of two consecutive whole numbers is 113. Find the numbers.

2. Find a number such that the sum of the number and twice its square is 21.

3. The hypotenuse of a right triangle is 15 cm and one leg is 3 cm longer than the other. Find the length of each of the legs.

4. The hypotenuse of a right triangle is 1 cm greater than one of the legs. If the other leg is 5 cm, find the length of the other two sides.

5. The base of a triangle is 1 in. longer than its height. If the area of the triangle is 15 in.^{2}, find the base and height of the triangle.

6. The length of a rectangle is one more than three times the width. If the area is 52, find its dimensions.

7. The diagonal of a rectangle is 20. If the length of the rectangle is 4 more than its width, find the dimensions of the rectangle.

8. The product of two consecutive odd whole numbers is 143. Find the numbers.

9. The area of a square is equal to its perimeter. Find its dimensions.

10. A frame of equal width around a painting has outside dimensions of 22 in. by 16 in. If the area of the painting inside the frame is 247 in^2, how wide is the frame?

11. The page of a book measures 25 cm by 20 cm. There is a blank border of equal width on all sides surrounding the printed material. Find the width of the border if the area of the printed portion of the page is 336 cm^2.

12. A photograph measures 6 in. by 9 in. The frame around the photograph is of constant width and has an area equal to that of the photograph. Find the width of the frame.

13. A boat travels upstream a distance of 15 miles and downstream the same distance in a total time of 4 hours. If the speed of the current is 5 mph, what is the speed of the boat in still water?

14. A boat takes a 10-hour trip traveling 40 miles upstream and 40 miles downstream. If the current is moving at a rate of 3 mph, what is the speed of the boat in still water?

15. How many sides does a polygon have if it has 9 diagonals? (See Example 6.)

16. How many sides does a polygon have if it has 27 diagonals? (See Example 6.)

17. The sum of a number and its reciprocal is $\frac{25}{12}$. What is the number?

18. The sum of a number and its reciprocal is $3\frac{1}{24}$. What is the number?

19. The sum of a number and its reciprocal is 4. Find the number.

20. A box is 4 in. longer than it is wide. If the box is 5 in. high and has a volume of 300 in.³, find the length and width of the box.

21. A rectangular sheet of cardboard is three times as long as it is wide. If 3-in. squares are cut from the corners and the sides and ends are folded up, a box having a volume of 288 in.³ is formed. What are the length and width of the piece of cardboard?

22. Find the width x of an L-shaped corridor that measures 24 ft by 9 ft if the total floor area is 116 ft². See Figure 7.10.

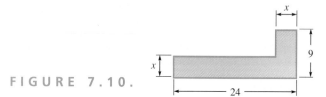

FIGURE 7.10.

23. A rectangular garden is 20 ft longer than it is wide. If the area is 300 ft², find the length and width of the garden.

24. Twice the square of a number is equal to 20 times the number. What is the number?

25. The length of a rectangular table is 2 ft more than its width. If the area of the table is 4 more than its perimeter, find the dimensions of the table.

26. The sum of the squares of two consecutive even integers is 100. Find the numbers.

27. It takes 480 square tiles to cover a floor. If square tiles that are 2 in. longer on a side are used, only 270 tiles are required to cover the same floor. Find the size of the smaller tiles.

28. If you subtract 10 from John's age and square the result, you will get a number that is five times as large as John's age. How old is John?

30. The sides of a right triangle are x, $x + 2$, and $x + 4$. Find the length of each side.

29. With the current moving at a rate of 3 mph, a boat travels 6 miles upstream and 12 miles downstream in a total time of 2 hours. What is the rate of speed of the boat in still water?

7.6   Parabolas

The graph of a quadratic equation $y = ax^2 + bx + c$ (where $a \neq 0$) forms a cup-shaped figure called a **parabola,** as pictured in Figure 7.11. Every parabola is symmetric about its *axis of symmetry,* which means that, if folded about this line, the two parts of the parabola would match. The lowest point [Figure 7.11(a)] or the highest point [Figure 7.11(b)] on the parabola, where the curve turns around, is called the **vertex.**

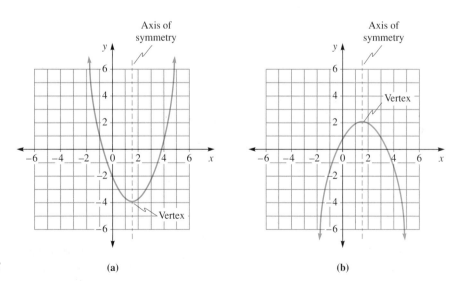

FIGURE 7.11.

(a)   (b)

There are many familiar applications of parabolas. Cross sections of automobile headlamps or backyard satellite dishes are parabolas. If a stone is thrown into the air, the path that it follows is a parabola.

Consider the graphs of the quadratic equations $y = x^2$ and $y = (x - 1)^2 - 3$. We will make a table of values for each equation and then graph the resulting ordered pairs.

$y = x^2$

x	y
-2	4
-1	1
0	0
1	1
2	4

$y = (x - 1)^2 - 3$

x	y
-2	6
-1	1
0	-2
1	-3
2	-2
3	1
4	6

The graphs are pictured in Figures 7.12(a) and (b):

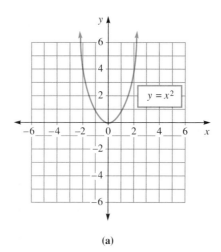
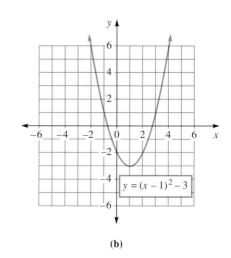

FIGURE 7.12.

(a)   (b)

Notice that the graph of $y = (x - 1)^2 - 3$ looks the same as the graph of $y = x^2$ except that it is shifted 1 unit to the right and 3 units downward. The vertex of $y = x^2$ is at $(0, 0)$, while the vertex of $y = (x - 1)^2 - 3$ is at $(1, -3)$. This can be generalized for quadratic equations of the form $y = a(x - h)^2 + k$. Its vertex will be at the point (h, k) and the vertical line through this point will be the axis of symmetry of the parabola.

> **The graph of $y = a(x - h)^2 + k$ is a parabola with vertex (h, k) and axis of symmetry $x = h$.** If $a > 0$ the parabola opens upward, while if $a < 0$ it opens downward.

Graphing Parabolas of the Form $y = a(x - h)^2 + k$

• • • • •

To graph a parabola of this form, we will make use of the vertex, which we just discovered is at the point (h, k). The x- and y-intercepts are also useful in graphing the parabola. The y-intercept, which is where the graph crosses the y-axis, is found by setting $x = 0$. To find any x-intercepts, we let $y = 0$.

EXAMPLE 1   Graph the parabola $y = (x - 1)^2 - 4$.

SOLUTION   1. Vertex: The vertex is at (h, k), which is $(1, -4)$.

2. The graph of the parabola opens *upward*, since the value of a is 1 (implied), which is positive.

3. y-intercept: We let $x = 0$:

$$y = (0 - 1)^2 - 4$$
$$y = -3$$

The y-intercept is $(0, -3)$.

4. x-intercepts: We let $y = 0$:

$$(x - 1)^2 - 4 = 0 \qquad \text{Substitute } y = 0.$$
$$x^2 - 2x + 1 - 4 = 0 \qquad \text{Square } x - 1.$$
$$x^2 - 2x - 3 = 0 \qquad \text{Collect like terms.}$$
$$(x - 3)(x + 1) = 0 \qquad \text{Factor.}$$
$$x = 3, -1 \qquad \text{Solve.}$$

The x-intercepts are $(3, 0)$ and $(-1, 0)$.

The graph is as pictured in Figure 7.13:

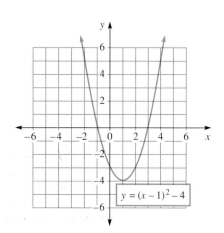

FIGURE 7.13.

EXAMPLE 2   Graph $y = \dfrac{1}{2}(x - 1)^2 + 2$.

SOLUTION   The vertex is at $(1, 2)$, which is in the first quadrant, and, since $a = \frac{1}{2}$, the parabola opens upward. Therefore the graph will not cross the x-axis and there are no x-intercepts. To find the y-intercept we set $x = 0$:

$$y = \frac{1}{2}(0 - 1)^2 + 2$$

$$= \frac{1}{2} \cdot 1 + 2 = \frac{5}{2}$$

The y-intercept is $(0, \frac{5}{2})$. For increased accuracy, we should plot an additional point. We let $x = -1$ and calculate y:

$$y = \frac{1}{2}(-1 - 1)^2 + 2 \qquad \text{Substitute } x = -1.$$

$$= \frac{1}{2} \cdot 4 + 2 \qquad\qquad (-2)^2 = 4$$

$$= 4 \qquad\qquad\qquad \text{The point is } (-1, 4).$$

Making use of symmetry around the axis of the parabola, which is the line $x = h$ or, in this case, $x = 1$, we complete our graph in Figure 7.14.

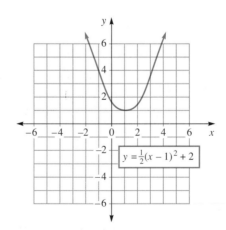

FIGURE 7.14.

Graphing Parabolas of the Form
$y = ax^2 + bx + c$

• • • • •

If we complete the square on the equation $y = ax^2 + bx + c$ in order to put it in the form $y = a(x - h)^2 + k$, we will be able to identify the vertex.

$$y = ax^2 + bx + c \quad (a \neq 0)$$

Divide both sides by a.

$$\frac{y}{a} = x^2 + \frac{b}{a}x + \frac{c}{a}$$

Subtract $\frac{c}{a}$ from both sides.

$$\frac{y}{a} - \frac{c}{a} = x^2 + \frac{b}{a}x$$

Complete the square in x.

$$\frac{y}{a} - \frac{c}{a} + \frac{b^2}{4a^2} = x^2 + \frac{b}{a}x + \frac{b^2}{4a^2}$$

Combine terms on the left side and factor the right side.

$$\frac{y}{a} + \frac{b^2 - 4ac}{4a^2} = \left(x + \frac{b}{2a}\right)^2$$

$$\frac{y}{a} = \left(x + \frac{b}{2a}\right)^2 - \frac{b^2 - 4ac}{4a^2}$$

Multiply both sides by a.

$$y = a\left(x + \frac{b}{2a}\right)^2 - \frac{b^2 - 4ac}{4a}$$

$$= a\underbrace{\left[x - \left(\frac{-b}{2a}\right)\right]^2}_{h} + \underbrace{\frac{4ac - b^2}{4a}}_{k}$$

In this final equation, if we let $h = \dfrac{-b}{2a}$ and $k = \dfrac{4ac - b^2}{4a}$, we have $y = a(x - h)^2 + k$. Therefore the vertex is at (h, k), or $\left(\dfrac{-b}{2a}, \dfrac{4ac - b^2}{4a}\right)$. It is generally easier to obtain the y-coordinate of the vertex simply by evaluating y when $x = \dfrac{-b}{2a}$ rather than memorizing the rather complicated expression for y.

To find the vertex of a parabola of the form $y = ax^2 + bx + c$:

1. find the x-coordinate of the vertex, $x = \dfrac{-b}{2a}$;

2. find the y-coordinate by substituting the value of x in the equation.

Using the Discriminant

Given a quadratic equation $ax^2 + bx + c = 0$, recall that if the discriminant $b^2 - 4ac$ is positive, there are two real-number solutions. Since $y = 0$ here, these real-number solutions are the x-intercepts of the graph. If $b^2 - 4ac = 0$, there is only one real solution, or one x-intercept, and if $b^2 - 4ac$ is negative, there are no real solutions, so there will be no x-intercepts. We summarize this as follows.

For a parabola of the form $y = ax^2 + bx + c$, calculate the discriminate if:

$b^2 - 4ac > 0$, there are two x-intercepts (two real solutions);
$b^2 - 4ac = 0$, there is one x-intercept (one real solution);
$b^2 - 4ac < 0$, there are no x-intercepts (no real solutions).
See Figure 7.15.

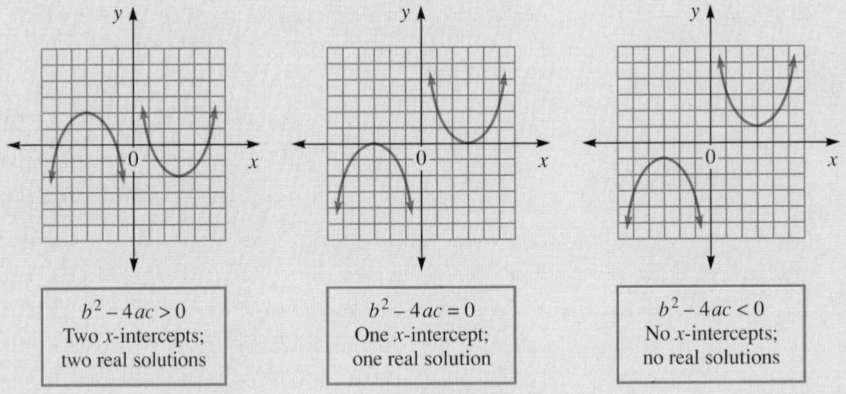

| $b^2 - 4ac > 0$ | $b^2 - 4ac = 0$ | $b^2 - 4ac < 0$ |
| Two x-intercepts; two real solutions | One x-intercept; one real solution | No x-intercepts; no real solutions |

FIGURE 7.15.

EXAMPLE 3   Graph the parabola $y = -x^2 + 2x + 4$.

SOLUTION   **1.** Vertex: The x-coordinate of the vertex is $x = \dfrac{-b}{2a} = \dfrac{-2}{-2} = 1$. The y-coordinate of the vertex is $y = -1^2 + 2 \cdot 1 + 4 = 5$. The vertex is at $(1, 5)$.

2. Since $a = -1$ (a is negative), the graph opens downward.

3. y-intercept: When $x = 0$ we have $y = 4$, so the y-intercept is $(0, 4)$.

4. x-intercepts: Letting $y = 0$ gives us

$$-x^2 + 2x + 4 = 0 \qquad \text{Multiply by } -1 \text{ on both sides.}$$

$$x^2 - 2x - 4 = 0$$

$$x = \frac{2 \pm \sqrt{4 - 4 \cdot 1 \cdot (-4)}}{2} \qquad \text{Use the quadratic formula to solve for } x.$$

$$= \frac{2 \pm \sqrt{20}}{2}$$

$$= \frac{2 \pm 2\sqrt{5}}{2} \qquad \text{Write } \sqrt{20} \text{ as } \sqrt{4 \cdot 5} = 2\sqrt{5}.$$

$$= 1 + \sqrt{5} \quad \text{and} \quad 1 - \sqrt{5} \qquad \text{Divide by 2.}$$

$$\approx 3.24 \quad \text{and} \quad -1.24 \qquad \text{Use a calculator to approximate each value of } x.$$

The x-intercepts are $(3.24, 0)$ and $(-1.24, 0)$

Now we complete the graph in Figure 7.16.

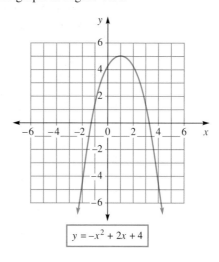

$$y = -x^2 + 2x + 4$$

FIGURE 7.16.

FIGURE 7.16.

EXAMPLE 4   Graph the parabola $y = x^2 + 4x + 4$.

SOLUTION   **1.** Vertex:

$$x = -\frac{b}{2a} = -\frac{4}{2} = -2$$

$$y = (-2)^2 + 4(-2) + 4 = 0$$

The vertex is at $(-2, 0)$.

2. Since $a = 1$, the graph opens upward.

3. y-intercept: When $x = 0$ we see that $y = 4$, so the y-intercept is $(0, 4)$.

4. x-intercepts: When $y = 0$, we have

$$x^2 + 4x + 4 = 0 \qquad \text{Let } y = 0.$$

$$(x + 2)(x + 2) = 0 \qquad \text{Factor.}$$

$$x = -2$$

There is only one x-intercept, at $(-2, 0)$, which happens also to be the vertex. See Figure 7.17.

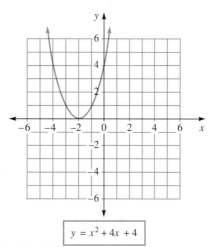

$$y = x^2 + 4x + 4$$

FIGURE 7.17.

7.6   Exercises

▼ Graph the following parabolas. Identify the vertex and any *x*-intercepts for each.

1. $y = 4x^2$

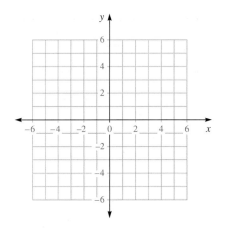

2. $y = 3x^2$

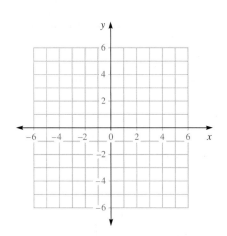

3. $y = -4x^2$

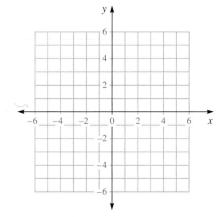

4. $y = -\frac{1}{2}x^2$

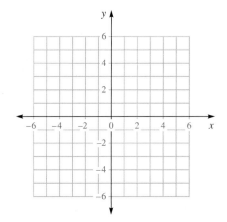

5. $y = (x - 3)^2 + 1$

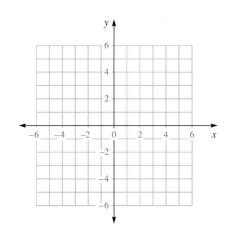

6. $y = (x - 2)^2 + 2$

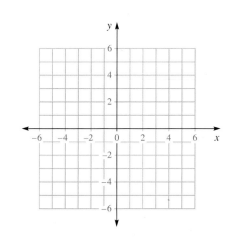

7. $y = 2(x - 3)^2$

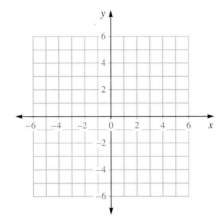

10. $y = -4(x - 2)^2$

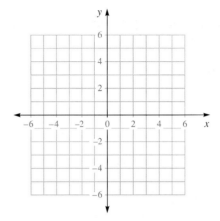

8. $y = (x + 1)^2$

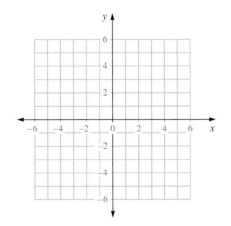

11. $y = (x + 2)^2 + 1$

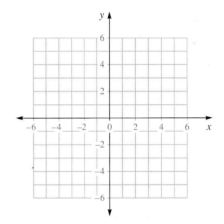

9. $y = -2(x - 1)^2$

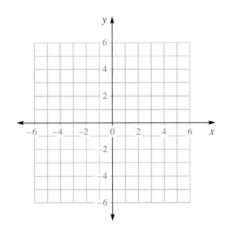

12. $y = -2(x - 1)^2 - 2$

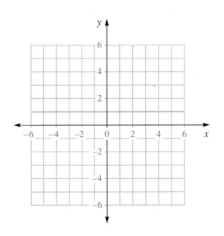

▼   Graph each of the following parabolas. Identify the vertex and any *x*-intercepts.

13. $y = x^2 + 4x - 5$

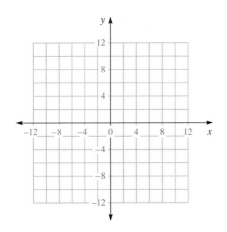

14. $y = x^2 - 2x - 3$

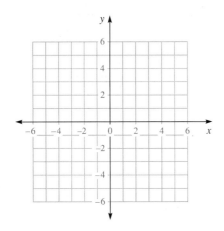

15. $y = -x^2 - 2x + 3$

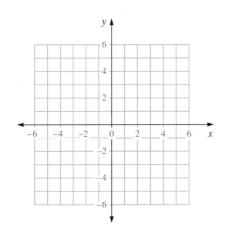

16. $y = -x^2 + 4x - 3$

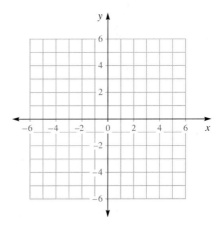

17. $y = -x^2 + 2x + 3$

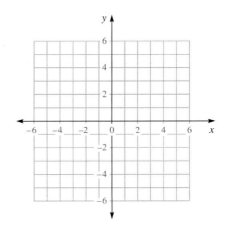

18. $y = x^2 + 4x + 3$

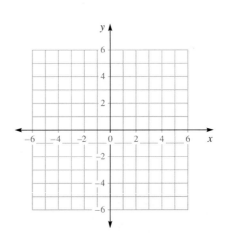

19. $y = -x^2 + 4x + 1$

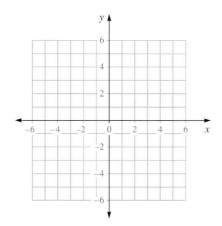

22. $y = 6 - x^2$

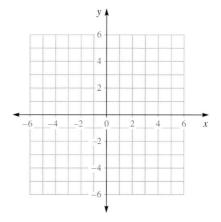

20. $y = -2x^2 + 2x + 1$

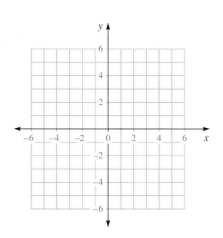

23. $y = 3x^2 + 4x + 2$

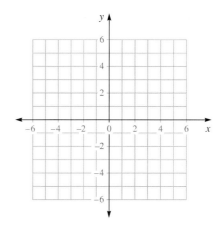

24. $y = 2x^2 + 4x + 3$

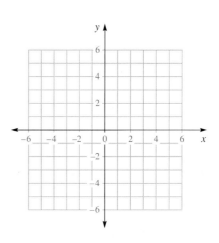

21. $y = 4 - x^2$

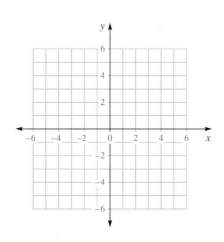

25. If the vertex of a parabola is in the first quadrant and the value of a in the general formula is positive, explain why there are no x-intercepts.

27. For each of the following, tell whether the parabola opens upward or downward and whether the graph is wider, narrower or the same as $y = x^2$.

    a. $y = x^2 + 3$

    b. $y = -2(x - 2)^2$

    c. $y = -2x^2$

    d. $y = \frac{1}{2}x^2 - 2$

    e. $y = -\frac{1}{3}(x + 1)^2$

26. Must all parabolas that open either upward or downward have at least one y-intercept? Might they have more than one?

7.7   Applications Involving Parabolas

Many practical problems involve finding the largest or the smallest value of something. Frequently we want to know the maximum profit or the minimum cost in business applications. The y-value of the vertex of a parabola gives the maximum or the minimum value of y, while the x-value of the vertex tells us *where* that maximum or minimum occurs.

EXAMPLE 1   Bryan Taylor wants to fence in a rectangular garden that has one side bordered by a stream. If he has 80 feet of fence, what is the maximum area he can enclose? See Figure 7.18.

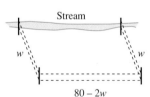

FIGURE 7.18.

SOLUTION   We are interested in finding the maximum area of the rectangular area when we use the formula $A = lw$. We let w be the width of the rectangle. Then the length l must be the 80 ft of fence available less the $2w$ that must be used up for the two sides, or $l = 80 - 2w$. The area A is

$$A = lw$$
$$= (80 - 2w)(w)$$
$$= 80w - 2w^2$$
$$= -2w^2 + 80w$$

Because the area A is represented by a quadratic equation with $a = -2 < 0$, its graph curves downward and so will be highest at its vertex. Therefore the area will be maximum when the graph is at its vertex. The w-coordinate of the vertex is at

$$w = \frac{-b}{2a} = \frac{-80}{2(-2)} = \frac{-80}{-4} = 20$$

To find the maximum area, we substitute this value for w back into our original equation and solve for A:

$$A = -2w^2 + 80w$$
$$= -2(20)^2 + 80(20) \qquad \text{Substitute } w = 20.$$
$$= -800 + 1600$$
$$= 800$$

The maximum area is 800 square feet. The graph of the parabola looks like Figure 7.19.

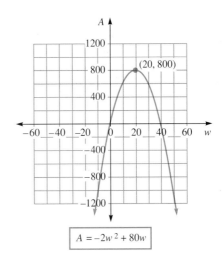

FIGURE 7.19.

$$\boxed{A = -2w^2 + 80w}$$

EXAMPLE 2   The total weekly cost to operate the Sun-Flow Swimming Pool Store is given by $C = x^2 - 20x + 500$, where x is the number of pools sold per week. How many pools should be sold per week in order to minimize costs?

SOLUTION   Since the coefficient of x^2 is positive, the parabola opens upward and the vertex will give us a *minimum* value for the cost C.

$$x = \frac{-b}{2a} = \frac{-(-20)}{2(1)} = \frac{20}{2} = 10$$

Ten pools should be sold per week to minimize costs.

7.7   Exercises

▮▮▮▮▮

▼ Solve:

1. A farmer has 100 ft of fence with which to enclose a rectangular garden. What is the maximum area that he can enclose? What are the dimensions of the garden?

2. A carpenter is building a rectangular room with a perimeter of 64 ft. What are the dimensions of the room that will give the maximum area? What is the maximum area?

3. What is the maximum product of two numbers with a sum of 32? What are the two numbers?

4. What is the maximum product of two numbers with a sum of 45? What are the two numbers?

5. What is the minimum product of two numbers with a difference of 6? What are the two numbers?

6. What is the minimum product of two numbers that differ by 4? What are the two numbers?

7. Joy Lungen runs a hot dog stand. Her total daily fixed cost to operate the business is given by $C = x^2 - 160x + 6440$, where x represents the number of hot dogs sold per day and C is in dollars. How many hot dogs must Joy sell to keep the cost at its lowest? What is this minimum cost?

8. In Renuka Khanna's Boutique, she finds that her profit on blouses is given by $P = -x^2 + 14x + 100$, where x is the number of blouses sold in a day and P represents the daily profit. How many blouses should Renuka sell in order to maximize her profit? What is the profit?

9. The total weekly profit P, on the sale of x cameras is given by $P = -2x^2 + 70x + 1900$. How many cameras must be sold in order to maximize the weekly profit? How much is this profit?

10. The Video Den finds that the profits P of the store are given by $P = -x^2 + 460x - 52770$, where P is the daily profit and x is the number of videos rented. Find the number of video rentals that will maximize the profit. What is the maximum profit?

11. The height h of a ball thrown upward with an initial velocity of 64 ft/sec is given by $h = -16t^2 + 64t$, where h is the height in feet after t seconds. Find the maximum height of the ball. How many seconds after the ball is thrown is the maximum height attained?

12. A stone is thrown upward from the top of a 96-foot-high building. The height, t seconds after it is thrown, is given by $h = -16t^2 + 80t + 96$. What is the maximum height that the stone attains? When does it reach its maximum height?

13. The sum of the length and width of a rectangle is 28 in. What is the length of the rectangle that will provide the maximum area? What is the maximum area?

14. Nature's Products roadside stand wants to fence in a rectangular parking lot using the highway as one of the boundaries. They have 200 ft of fencing available. What will the dimensions of the parking lot be so that the area will be maximum? What is the maximum area?

 15. How do the length and width of a rectangular-shaped field relate to each other if you wish to fence in a maximum area and there is fencing on all four sides? (Refer to the results of Exercises 1 and 2.) Answer the same question when fencing is needed on only three sides. (See Example 1 and Exercise 14.)

7.8   Conic Sections

The conic sections are formed by intersecting a cone with a plane. The resulting curves are *parabolas, circles, ellipses,* and *hyperbolas*. (See Figure 7.20.) The equations of these graphs are all second-degree equations in two variables.

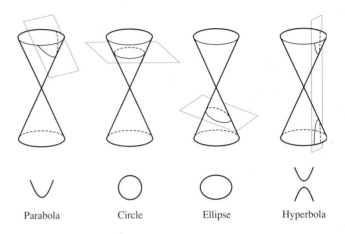

FIGURE 7.20.

Parabola   Circle   Ellipse   Hyperbola

We have already investigated parabolas in detail, including their graphs and applications. We now turn our attention to the other three conic sections.

Circles

Consider the graph of a circle with center at (h, k) and radius r; see Figure 7.21. The distance r from any point (x, y) on the circle to the center (h, k) can be found by using the distance formula:

$$r = \sqrt{(x - h)^2 + (y - k)^2}$$

Squaring both sides gives us $r^2 = (x - h)^2 + (y - k)^2$.

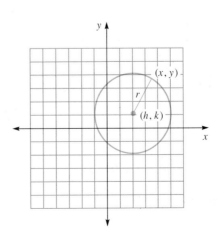

FIGURE 7.21.

> The graph of $(x - h)^2 + (y - k)^2 = r^2$ is a **circle with center at (h, k) and radius r.**

This equation allows us to graph a circle without graphing a great number of points.

EXAMPLE 1   Graph the circle $(x - 1)^2 + (y + 2)^2 = 9$.

SOLUTION   We write the equation in the form

$$(x - h)^2 + (y - k)^2 = r^2$$
$$(x - 1)^2 + [y - (-2)]^2 = 3^2$$

The center is at $(1, -2)$ and the radius is 3; see Figure 7.22.

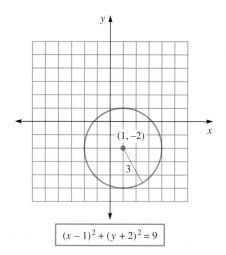

FIGURE 7.22.

$$\boxed{(x - 1)^2 + (y + 2)^2 = 9}$$

▮▮

EXAMPLE 2   Graph $x^2 + 2x + y^2 - 6y + 6 = 0$.

SOLUTION   We put the given equation into the form

$$(x - h)^2 + (y - k)^2 = r^2$$

by completing the square on x and y:

$$x^2 + 2x + y^2 - 6y + 6 = 0$$
$$x^2 + 2x + \underline{} + y^2 - 6y + \underline{} = -6$$
$$x^2 + 2x + 1 + y^2 - 6y + 9 = -6 + 1 + 9$$
$$(x + 1)^2 + (y - 3)^2 = 4$$
$$[x - (-1)]^2 + (y - 3)^2 = 2^2$$

The center is at $(-1, 3)$ and the radius is 2; see Figure 7.23.

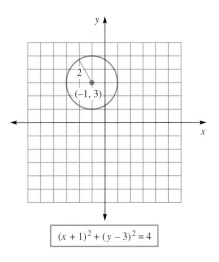

$$(x + 1)^2 + (y - 3)^2 = 4$$

FIGURE 7.23.

FIGURE 7.23.

Ellipses

• • • • •

We will consider only the graphs of ellipses that are centered at the origin.

EXAMPLE 3   Graph $\dfrac{x^2}{16} + \dfrac{y^2}{9} = 1$.

SOLUTION   Let's first find the x- and y-intercepts.

$$x\text{-intercept } (y = 0): \frac{x^2}{16} + \frac{0^2}{9} = 1$$

$$x^2 = 16$$

$$x = \pm 4$$

The x-intercepts are $(4, 0)$ and $(-4, 0)$.

$$y\text{-intercept } (x = 0): \frac{0^2}{16} + \frac{y^2}{9} = 1$$

$$y^2 = 9$$

$$y = \pm 3$$

The y-intercepts are (0, 3) and (0, −3). We now graph these points and connect them with a smooth curve, as pictured in Figure 7.24:

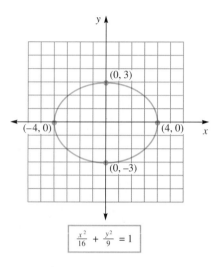

FIGURE 7.24.

$$\frac{x^2}{16} + \frac{y^2}{9} = 1$$

▌▌

Generalizing gives us the following:

> The graph of $\dfrac{x^2}{a^2} + \dfrac{y^2}{b^2} = 1$ is an **ellipse with x-intercepts (a, 0) and (−a, 0) and y-intercepts (0, b) and (0, −b).**

EXAMPLE 4   Graph $25x^2 + 4y^2 = 100$.

SOLUTION   We divide both sides of the equation by 100 to put it into standard form:

$$\frac{x^2}{a^2} + \frac{y^2}{b^2} = 1$$

$$\frac{25x^2}{100} + \frac{4y^2}{100} = \frac{100}{100}$$

$$\frac{x^2}{4} + \frac{y^2}{25} = 1$$

or

$$\frac{x^2}{2^2} + \frac{y^2}{5^2} = 1$$

The x-intercepts are (2, 0) and (−2, 0) and the y-intercepts are (0, 5) and (0, −5). We graph the ellipse in Figure 7.25.

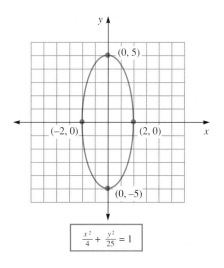

$$\frac{x^2}{4} + \frac{y^2}{25} = 1$$

FIGURE 7.25.

Hyperbolas

• • • • •

EXAMPLE 5   Graph $\dfrac{x^2}{9} - \dfrac{y^2}{4} = 1$.

SOLUTION   We solve for y in terms of x and tabulate some ordered pairs that satisfy the equation.

$$\frac{y^2}{4} = \frac{x^2}{9} - 1 = \frac{x^2 - 9}{9}$$

$$y^2 = \frac{4(x^2 - 9)}{9}$$

$$y = \pm\frac{2}{3}\sqrt{x^2 - 9}$$

Notice that we must have $x^2 - 9 \geq 0$, since we do not want square roots of negative numbers. Therefore, we need not consider any values for x between -3 and 3. The ordered pairs are graphed in Figure 7.26.

x	y
-5	$\pm\frac{8}{3} \approx \pm2.7$
-4	$\pm\frac{2}{3}\sqrt{7} \approx \pm1.8$
-3	0
3	0
4	$\pm\frac{2}{3}\sqrt{7} \approx \pm1.8$
5	$\pm\frac{8}{3} \approx \pm2.7$

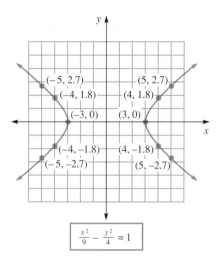

FIGURE 7.26.

We have obtained two separate branches of a curve called a **hyperbola.** Every hyperbola has a pair of **asymptotes** that are found by extending the diagonals of a rectangle that has sides parallel to the x- and y-axes and passes through the points $(3, 0)$, $(-3, 0)$, $(0, 2)$, and $(0, -2)$. These points come from the numbers written beneath the x- and y-terms of our original equation.

$$\frac{x^2}{9} - \frac{y^2}{4} = 1$$

$$\pm\sqrt{9} = \pm 3 \qquad\qquad \pm\sqrt{4} = \pm 2$$

See Figure 7.27 for a graphic interpretation of these ideas.

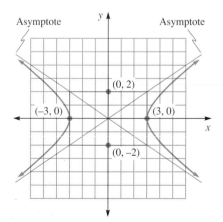

FIGURE 7.27.

To graph the equation of a hyperbola:

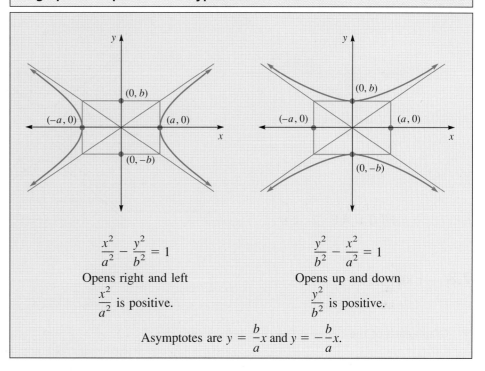

$$\frac{x^2}{a^2} - \frac{y^2}{b^2} = 1$$

Opens right and left

$\dfrac{x^2}{a^2}$ is positive.

$$\frac{y^2}{b^2} - \frac{x^2}{a^2} = 1$$

Opens up and down

$\dfrac{y^2}{b^2}$ is positive.

Asymptotes are $y = \dfrac{b}{a}x$ and $y = -\dfrac{b}{a}x$.

EXAMPLE 6   Graph $y^2 - x^2 = 4$.

SOLUTION   We must first divide both sides of the equation by 4 to get it into standard form. This gives us

$$\frac{y^2}{4} - \frac{x^2}{4} = 1$$

This hyperbola will open up and down, since $\dfrac{y^2}{4}$ is positive. We now find the x- and y-intercepts of the rectangle that will determine our asymptotes.

$$\frac{y^2}{4} - \frac{x^2}{4} = 1$$

$$b = \pm\sqrt{4} = \pm2 \qquad a = \pm\sqrt{4} = \pm2$$

The equations of the asymptotes are $y = \dfrac{b}{a}x$ and $y = -\dfrac{b}{a}x$:

$$y = \frac{2}{2}x \quad \text{or} \quad y = x \qquad \text{and} \qquad y = -\frac{2}{2}x \quad \text{or} \quad y = -x$$

Using the intercepts and asymptotes, we can graph the hyperbola in Figure 7.28.

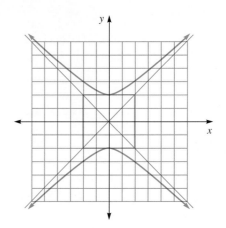

FIGURE 7.28.   ▮▮

7.8   Exercises

▮▮▮▮▮

▼ Graph each of the circles in Exercises 1–6.

2. $x^2 + y^2 = 9$

1. $x^2 + y^2 = 16$

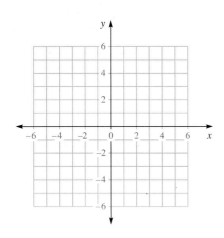

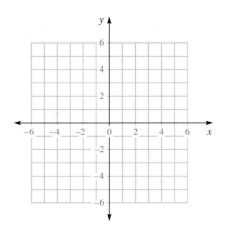

3. $(x - 2)^2 + (y - 1)^2 = 4$

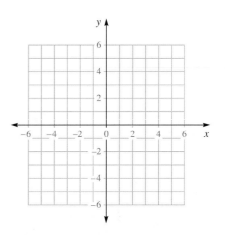

4. $(x + 3)^2 + (y - 2)^2 = 9$

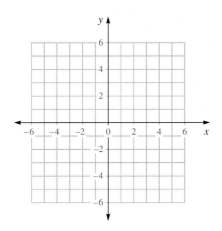

▼ Graph each of the ellipses in Exercises 7–12.

7. $\dfrac{x^2}{4} + \dfrac{y^2}{9} = 1$

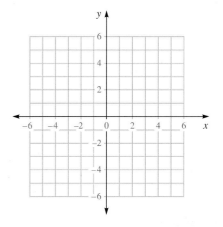

5. $x^2 - 2x + y^2 + 6y + 1 = 0$

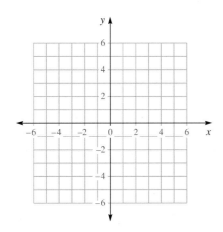

8. $\dfrac{x^2}{4} + y^2 = 1$

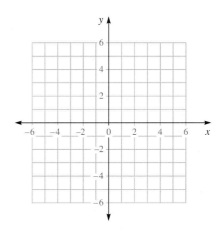

6. $x^2 + 4x + y^2 + 8y + 4 = 0$

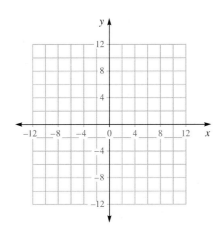

9. $\dfrac{x^2}{25} + \dfrac{y^2}{9} = 1$

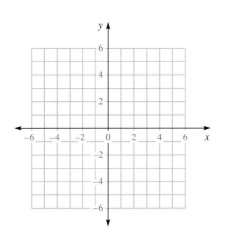

10. $\dfrac{x^2}{4} + \dfrac{y^2}{16} = 1$

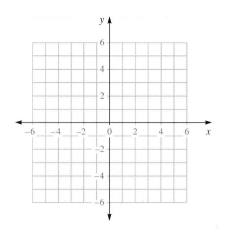

11. $9x^2 + y^2 = 9$

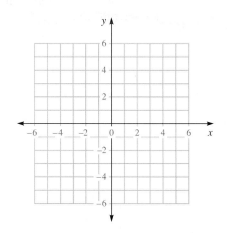

12. $4x^2 + 9y^2 = 36$

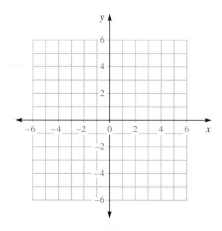

▼ Graph each of the hyperbolas in Exercises 13–18.

13. $\dfrac{x^2}{16} - \dfrac{y^2}{4} = 1$

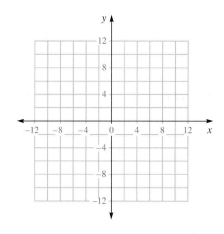

14. $\dfrac{x^2}{25} - \dfrac{y^2}{9} = 1$

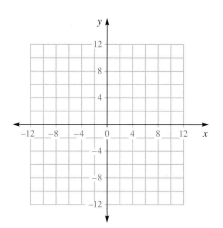

15. $\dfrac{y^2}{16} - \dfrac{x^2}{25} = 1$

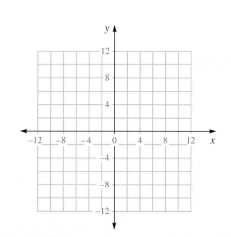

16. $x^2 - \dfrac{y^2}{4} = 1$

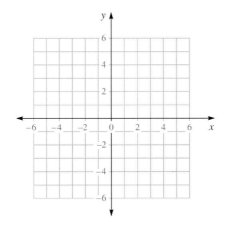

17. $9y^2 - x^2 = 9$

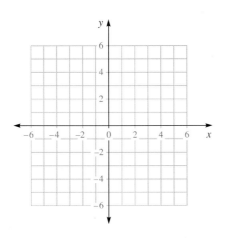

18. $4x^2 - 20y^2 = 20$

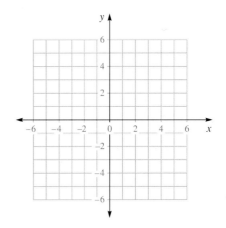

19. When graphing an ellipse of the form $\dfrac{x^2}{a^2} + \dfrac{y^2}{b^2} = 1$, what does the graph look like if $a^2 = b^2$?

SUMMARY—CHAPTER 7

●●●●●●● ▼

EXAMPLES

The **standard form of a quadratic equation** is

$$ax^2 + bx + c = 0$$

where *a, b,* and *c* are real numbers and $a > 0$.

$6x^2 - 8x + 1 = 0$ is in standard form.

To solve a quadratic equation by factoring:

1. Put the equation in standard form, $ax^2 + bx + c = 0$;

2. Factor the left side of the equation;

3. Set each factor equal to zero and solve for *x*;

Solve $x^2 - 7x = -10$ by factoring.

$$x^2 - 7x + 10 = 0$$
$$(x - 5)(x - 2) = 0$$

$$
\begin{array}{c|c}
x - 5 = 0 & x - 2 = 0 \\
x = 5 & x = 2
\end{array}
$$

Check

4. Check each solution in the original equation.

$$
\begin{array}{c|c}
5^2 - 7 \cdot 5 \overset{?}{=} -10 & 2^2 - 7 \cdot 2 \overset{?}{=} -10 \\
25 - 35 \overset{?}{=} -10 & 4 - 14 \overset{?}{=} -10 \\
-10 = -10 \quad ✓ & -10 = -10 \quad ✓
\end{array}
$$

To solve a quadratic equation of the form $ax^2 - c = 0$:

1. Add *c* to both sides to obtain $ax^2 = c$;

2. Divide both sides by *a*, yielding $x^2 = \dfrac{c}{a}$;

3. Take the square root of both sides, obtaining $x = \pm\sqrt{\dfrac{c}{a}}$;

Solve: $2x^2 - 18 = 0$.

$$2x^2 = 18$$
$$x^2 = 9$$
$$x = \pm\sqrt{9} = \pm 3$$

Check

4. Check the solutions in the original equation.

$$
\begin{array}{c|c}
2(3)^2 - 18 \overset{?}{=} 0 & 2(-3)^2 - 18 \overset{?}{=} 0 \\
2 \cdot 9 - 18 \overset{?}{=} 0 & 2 \cdot 9 - 18 \overset{?}{=} 0 \\
18 - 18 \overset{?}{=} 0 & 18 - 18 \overset{?}{=} 0 \\
0 = 0 \quad ✓ & 0 = 0 \quad ✓
\end{array}
$$

To solve a quadratic equation $ax^2 + bx + c = 0$ by completing the square:

1. Subtract *c* from both sides of the equation;

2. If $a \neq 1$, divide both sides by *a*;

3. Complete the square by adding $\left(\dfrac{b}{2a}\right)^2$ the square of half the coefficient of *x*, to both sides;

4. Factor the left side of the equation;

5. Take the square root of both sides and solve for *x*.

Solve $x^2 - 4x - 9 = 0$ by completing the square.

$$x^2 - 4x + \underline{\quad} = 9$$

Coefficient of x^2 is 1.

$$x^2 - 4x + 4 = 9 + 4$$
$$(x - 2)^2 = 13$$
$$x - 2 = \pm\sqrt{13}$$
$$x = 2 \pm\sqrt{13}$$

To solve a quadratic equation $ax^2 + bx + c = 0$ **using the quadratic formula:**

$$x = \frac{-b \pm \sqrt{b^2 - 4ac}}{2a}$$

EXAMPLES

Solve $3x^2 + 2x - 5 = 0$ using the quadratic formula.

1. Write the equation in standard form, $ax^2 + bx + c = 0$;

Already in standard form.

2. Identify the constants, a, b, and c;

$a = 3$, $b = 2$, $c = -5$

3. Substitute the values for a, b, and c into the quadratic formula.

$$\begin{aligned} x &= \frac{-2 \pm \sqrt{2^2 - 4 \cdot 3 \cdot (-5)}}{2 \cdot 3} \\ &= \frac{-2 \pm \sqrt{4 + 60}}{6} \\ &= \frac{-2 \pm \sqrt{64}}{6} = \frac{-2 \pm 8}{6} \end{aligned}$$

$$x = \frac{-2 + 8}{6} = \frac{6}{6} = 1$$

and

$$x = \frac{-2 - 8}{6} = \frac{-10}{6} = \frac{-5}{3}$$

The solutions are $x = 1$ and $x = -\frac{5}{3}$.

$b^2 - 4ac$ is called the **discriminant.**

The quadratic equation $ax^2 + bx + c = 0$ has:

1. Two real-number solutions if $b^2 - 4ac > 0$;

2. One real-number solution if $b^2 - 4ac = 0$;

3. Two complex-number solutions if $b^2 - 4ac < 0$

Given $2x^2 + 2x - 1 = 0$.

$$\begin{aligned} b^2 - 4ac &= 2^2 - 4 \cdot 2 \cdot (-1) \\ &= 4 + 8 = 12 > 0 \end{aligned}$$

There are 2 real solutions.

To solve an equation containing square roots:

1. Isolate a radical on one side of the equation.

2. Square both sides of the equation.

3. Simplify and solve the resulting equation. If a radical remains, isolate it and square both sides again.

4. Check all apparent solutions in the original equation.

Solve: $2x = \sqrt{5x - 1}$.

$$4x^2 = 5x - 1$$

$$4x^2 - 5x + 1 = 0$$

$$(4x - 1)(x - 1) = 0$$

$$4x - 1 = 0 \quad \bigg| \quad x - 1 = 0$$

$$x = \tfrac{1}{4} \quad \bigg| \quad x = 1$$

Check

$$2\left(\tfrac{1}{4}\right) \overset{?}{=} \sqrt{5 \cdot \tfrac{1}{4} - 1}$$

$$\tfrac{1}{2} \overset{?}{=} \sqrt{\tfrac{1}{4}}$$

$$\tfrac{1}{2} = \tfrac{1}{2} \quad \blacktriangleright$$

$$2(1) \overset{?}{=} \sqrt{5 \cdot 1 - 1}$$

$$2 \overset{?}{=} \sqrt{4}$$

$$2 = 2 \quad \blacktriangleright$$

The solutions are $\tfrac{1}{4}$ and 1.

The graph of $y = a(x - h)^2 + k$ is a **parabola with vertex (h, k) and axis of symmetry $x = h$.** If $a > 0$ the parabola opens upward, While if $a < 0$, it opens downward.

Given $y = 2(x - 1)^2 + 2$:
The vertex is $(1, 2)$.
The axis of symmetry is $x = 1$.
$a = 2$, so it opens upward.

The graph of $y = ax^2 + bx + c$ is a **parabola.** The x-coordinate of the vertex is $x = \dfrac{-b}{2a}$. The y-coordinate is found by substituting the value of x in the original equation.

Given $y = 2x^2 - 4x + 4$,
The x-coordinate of the vertex
is $x = \dfrac{-b}{2a} = \dfrac{4}{2 \cdot 2} = \dfrac{4}{4} = 1.$

The y-coordinate of the vertex
is $y = 2 \cdot 1^2 - 4 \cdot 1 + 4 = 2.$

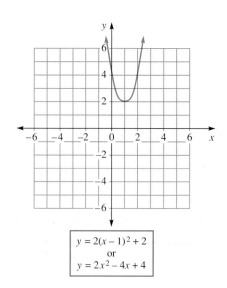

$$y = 2(x - 1)^2 + 2$$
$$\text{or}$$
$$y = 2x^2 - 4x + 4$$

The graph of $(x - h)^2 + (y - k)^2 = r^2$ is a **circle with a center at (h, k) and radius r.**

Graph $(x + 1)^2 + (y - 2)^2 = 9$.
The center is $(-1, 2)$.
The radius is 3.

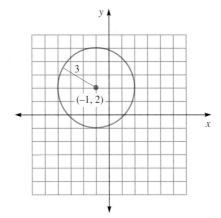

The graph of $\dfrac{x^2}{a^2} + \dfrac{y^2}{b^2} = 1$ is an **ellipse with**

x-intercepts (a, 0) and (−a, 0) and y-intercepts (0, b) and (0, −b).

Graph $\dfrac{x^2}{16} + \dfrac{y^2}{4} = 1$.

x-intercepts are $(4, 0)$ and $(-4, 0)$.
y-intercepts are $(0, 2)$ and $(0, -2)$.

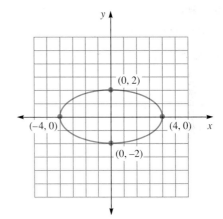

The graph of $\dfrac{x^2}{a^2} - \dfrac{y^2}{b^2} = 1$ is a **hyperbola opening right and left.**

The graph of $\dfrac{y^2}{b^2} - \dfrac{x^2}{a^2} = 1$ is a **hyperbola opening up and down.**

The equations of the **asymptotes** of a hyperbola with center at the origin are $y = \dfrac{b}{a}x$ **and** $y = -\dfrac{b}{a}x.$

Graph $\dfrac{x^2}{4} - y^2 = 1$.

x-intercepts of the rectangle are $(2, 0)$ and $(-2, 0)$.
y-intercepts of the rectangle are $(0, 1)$ and $(0, -1)$.

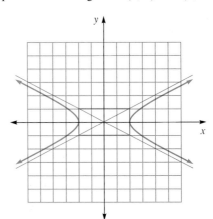

REVIEW EXERCISES—CHAPTER 7

▼ Write the following in standard form:

1. $3x^2 = 9 - 2x$ **2.** $-7 = 5x(2 - x)$

3. $\dfrac{x^2}{3} - 4 = \dfrac{3x}{2}$

▼ Solve by factoring:

4. $x^2 - 49 = 0$ **5.** $x^2 + 2x - 15 = 0$

6. $6x^2 + 7x - 3 = 0$ **7.** $8x = 5 - 4x^2$

8. $x^2 - 2x = 9x + 3x^2 - 21$

▼ Solve using the method described in Section 7.1.

9. $a^2 = 81$ **10.** $x^2 = 12$ **11.** $3y^2 = 60$

12. $2x^2 - \dfrac{8}{9} = 0$ **13.** $\dfrac{t^2}{4} - 7 = 0$

▼ Fill in the blanks.

14. $x^2 + 6x +$ ____ $= (x +$ ____$)^2$

15. $y^2 - 5y +$ ____ $= (y -$ ____$)^2$

16. $t^2 - \dfrac{1}{2}t +$ ____ $= (t -$ ____$)^2$

▼ Solve by completing the square.

17. $x^2 - 4x - 5 = 0$ **18.** $x^2 - 6x + 5 = 0$

19. $y^2 - y - 1 = 0$ **20.** $3t^2 - 3t - 2 = 0$

21. $2m^2 - 3m - 2 = 0$ **22.** $3x^2 + 6x - 9 = 0$

▼ Solve using the quadratic formula.

23. $x^2 - x - 12 = 0$ **24.** $m^2 - 3m + 5 = 0$

25. $x^2 + 4x + 9 = 0$ **26.** $y^2 + 9y + 1 = 0$ **33.** $x^2 - 6x + 1 = 0$ **34.** $2x^2 - 3x = -2$

27. $x^2 - 10 = 0$ **28.** $y^2 = 5y - 4$

▼ Solve for x:

35. $\dfrac{3}{x^2} - 1 = \dfrac{2}{x}$ **36.** $\dfrac{4}{x-1} - \dfrac{1}{2x-5} = 1$

29. $x(x - 1) + 2(x + 5) = 12$

37. $\dfrac{1}{x+1} - 1 = \dfrac{1}{x-1}$ **38.** $2x^4 - 7x^2 - 4 = 0$

30. $(x + 3)(x - 1) = 2(x + 1)$

39. $x^4 - 7x^2 - 18 = 0$

▼ Use the discriminant to determine the nature of the solutions for each of the following equations.

40. $(x - 2)^2 + 3(x - 2) + 2 = 0$

31. $x^2 - 6x + 3 = 0$ **32.** $x^2 + 2x + 6 = 0$

41. $\sqrt{2x^2 + 4} = x + 2$

42. $\sqrt{2x - 1} + 1 = \sqrt{3x + 1}$

43. $\sqrt{5x + 1} - \sqrt{x + 1} = -2$

▼ Solve the following application problems.

44. Five times the square of a whole number minus that number is equal to 18. Find the number.

45. The length of a rectangle is 3 in. more than its width. If the diagonal of the rectangle is 15 in., find the length and width of the rectangle.

46. A boat travels 30 miles upstream and 30 miles back in a total of 8 hours. If the speed of the current is 2 mph, what is the speed of the boat in still water?

47. How many sides does a polygon have if it has 35 diagonals?

48. Graph the parabola $y = -2(x + 1)^2 + 1$. Identify the vertex and any x-intercepts.

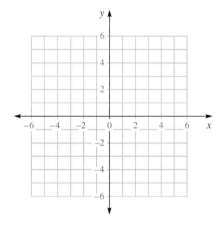

49. Graph the parabola $y = x^2 + 3x - 4$. Identify the vertex and any x-intercepts.

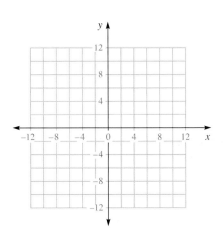

50. Samantha Wright wants to fence in a rectangular flower bed using her house as one of the sides. If she has 40 ft of fencing available, what should the dimensions of the flower bed be so that the area will be maximum? What is this maximum area?

51. Graph the circle $x^2 + 6x + y^2 - 2y + 6 = 0$.

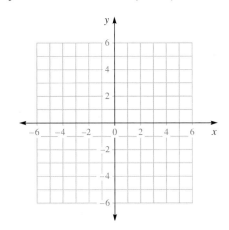

52. Graph the ellipse $\dfrac{x^2}{25} + \dfrac{y^2}{9} = 1$.

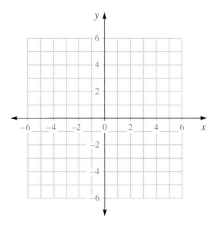

53. Graph the hyperbola $\dfrac{y^2}{16} - \dfrac{x^2}{9} = 1$.

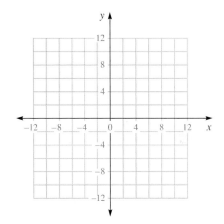

NAME   ▮ ▮ ▮ ▮ ▮ CLASS

ACHIEVEMENT TEST—CHAPTER 7

• • • • • • • ▼

▼ Write in standard form:

1. $2 + 3x^2 = -9x$ 1. _____

2. $\dfrac{x^2}{3} - \dfrac{5}{2} = \dfrac{x}{4}$ 2. _____

▼ Solve by factoring:

3. $x^2 + x - 12 = 0$ 3. _____

4. $6x^2 + 13x = 5$ 4. _____

▼ Solve by the method described in Section 7.1:

5. $4x^2 = 80$ 5. _____

▼ Fill in the blanks:

6. $x^2 - 4x + \underline{\quad} = (x - \underline{\quad})^2$ 6. _____

▼ Solve by completing the square:

7. $x^2 - 8x + 1 = 0$ 7. _____

▼ Solve using the quadratic formula:

8. $2x^2 - 3x - 1 = 0$ 8.

                                                               _____

9. $x^2 - 4x + 6 = 0$ 9. _____

▼ Use the discriminant to determine the nature of the solutions.

10. $9x^2 + 12x + 4 = 0$ 10. _____

11. $2x^2 - 2x + 5 = 0$ 11. _____

▼ Solve by any method.

12. $\dfrac{5}{x + 3} + 3 = \dfrac{8}{x}$ 12. _____

                                          13. _____

13. $x^4 - 2x^2 - 8 = 0$

14. $\sqrt{2x^2 - 1} = 2x - 3$ 14. _____

15. $\sqrt{2x - 1} - \sqrt{3x + 1} = 1$ 15. _____

▼ Solve the following application problems.

16. The length of a rectangle is 2 in. more than twice its width. What are the   16. _____
    dimensions of the rectangle if the area is 60 in.²?

17. The sum of a number and its reciprocal os $4\frac{13}{18}$. Find the number.   17. _____

18. A boat travels upstream 8 miles and downstream 20 miles. The total trip   18. _____
    takes 4 hours. If the current has a speed of 3 mph, find the speed of the boat
    in still water.

19. An object is thrown into the air with an initial velocity of 48 ft/sec from the top of a 64-ft-high cliff. The height of the object at any time t is given by

$$h = -16t^2 + 48t + 64$$

Find the maximum height attained by the object.

19. _____

20. Graph the parabola $y = (x + 2)^2 - 2$. Identify the vertex and any x-intercepts.

20. _____

_____

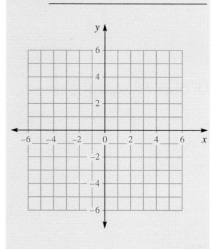

21. Graph the parabola $y = -x^2 + x + 6$. Identify the vertex and any x-intercepts.

21. _____

_____

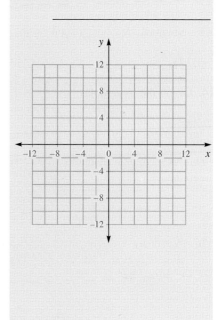

22. Graph the circle $(x + 3)^2 + (y - 2)^2 = 9$.

22.

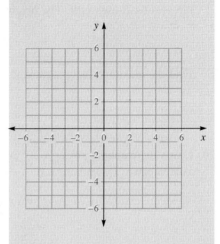

23. Graph the ellipse $x^2 + 4y^2 = 4$.

23.

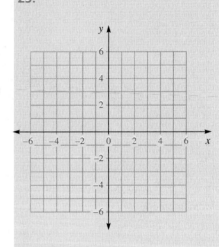

24. Graph the hyperbola $\dfrac{x^2}{3} - \dfrac{y^2}{4} = 1$.

24.

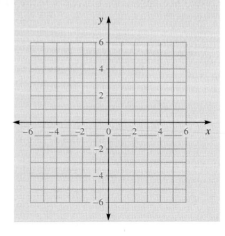

Systems of Linear Equations and Inequalities

INTRODUCTION

Previously, we have studied extensively about functions and their graphs. We have also learned about linear equations and their solutions.

Now, in this chapter we will combine what we have learned from each of these areas to investigate groups of linear equations called **systems of linear equations.** We will also look at applications of systems of equations and then take things a step further and solve **systems of linear inequalities.**

CHAPTER 8—NUMBER KNOWLEDGE

Mathematics in Africa

What is known about mathematics on the continent of Africa? If you read most history of mathematics books, you will find very little reference to the subject at all. The one exception will be a rather detailed account of ancient Egyptian mathematics. The Moscow Papyrus (1850 B.C.) and the Rhind Papyrus (1650 B.C.) are the main sources of information about Egyptian mathematics. Together they contain 110 problems, mostly of a practical nature.

Information about mathematics in other parts of Africa comes mainly from anthropologists. Written records are scarce and much of what we know has come from art and oral traditions. Claudia Zaslavsky, in her book *Africa Counts,*[1] tells us that most of the mathematics in Africa involves calculation, measurement, and shape.

A question frequently asked is, "Do Africans count like we do?" The answer is yes. Students in Africa and most of the rest of the world use the numbers 0, 1, 2, 3, 4, and so on (the Hindu–Arabic numerals), like we do in the United States. One major difference in African counting is called *gesture-counting,* where a standardized system of hand gestures either accompanies or replaces the spoken names for numbers. Gesture-counting is particularly useful in the marketplace because of the large number of languages spoken in Africa (estimated at over 1,000). See Figure 8.1 for photographs of these hand gestures.

[1]Zaslavsky, Claudia, *Africa Counts:* Prindle, Weber, and Schmidt (Boston, 1973).

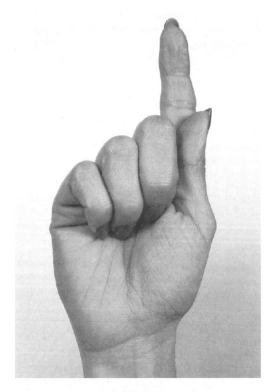

N a b o
(o n e)

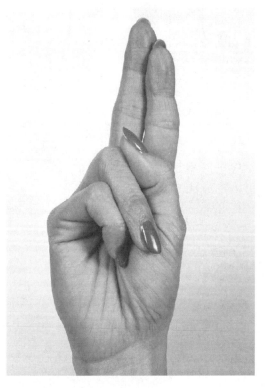

A r e
(t w o)

FIGURE 8.1.

Finger gestures of the Arusha Maasai. (Source: Gulliver)

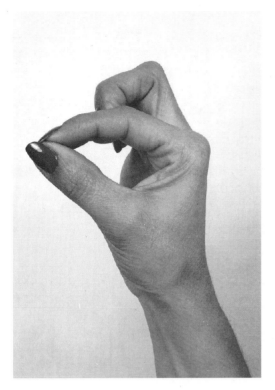

Uni
(three)

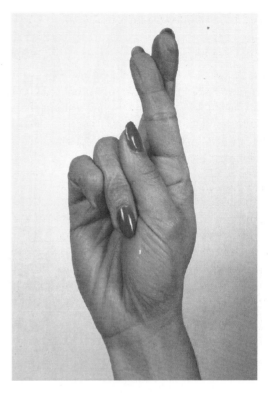

Onguan
(four)

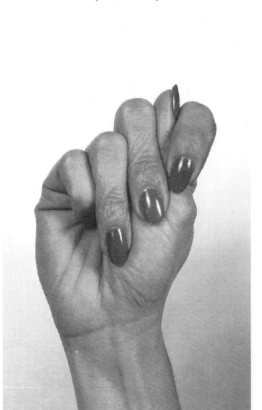

Imiet
(five)

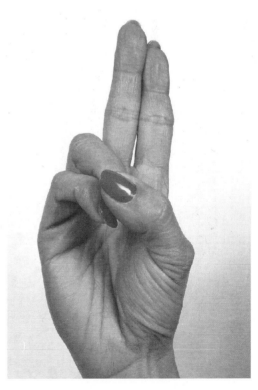

Ile
(six)

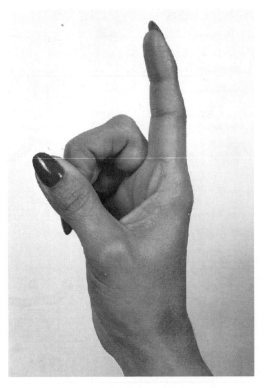

Naapishana
(seven)

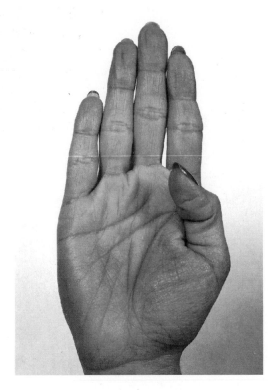

Isiet
(eight)

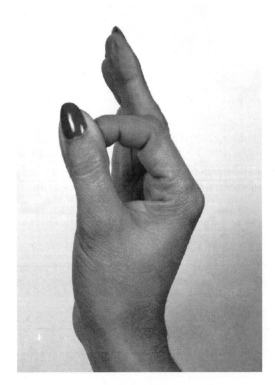

Enderuj
(nine)

8.1 Graphing, the Addition Method and the Substitution Method

If we consider two linear equations together, they form what we call a **system of linear equations.** For example, the two equations

$$y = -\frac{3}{2}x + 4$$
$$y = x - 1$$

form a linear system. The solution to the system is an ordered pair (x, y) that satisfies both of the equations. If we graph each equation on the same set of axes, this solution becomes apparent; see Figure 8.2.

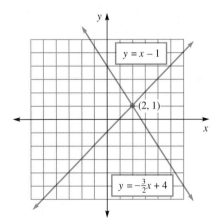

FIGURE 8.2.

The point of intersection is the ordered pair $(2, 1)$, and since it lies on both lines, it satisfies both of the equations. The ordered pair $(2, 1)$ is the solution for the system of equations.

If we graph two lines on the same set of axes, there are three possible things that can happen, as pictured in Figure 8.3:

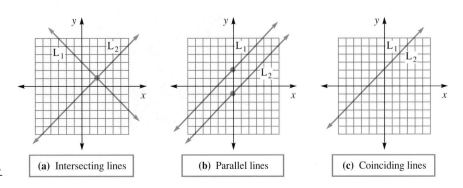

FIGURE 8.3.   (a) Intersecting lines   (b) Parallel lines   (c) Coinciding lines

a. The lines **intersect** each other in exactly one point. This point of intersection is the *only* solution to the system. This system is called **consistent.**

b. The lines are **parallel** and never intersect. Therefore, there is *no* solution to the system. This system of equations is called **inconsistent.**

c. The graphs of the two equations are the same line. We say the lines **coincide** and every ordered pair satisfying one equation will also satisfy the other equation. There are an infinite number of solutions and the system of equations is called **dependent.**

There are major problems in using graphing to solve systems of linear equations. Graphing can be very time-consuming and if the solution is not a pair of integers, an exact solution may be difficult or impossible to find.

Fortunately, there are algebraic methods for solving systems of linear equations. The first of these to be discussed is called **the addition method,** (also called **the addition–subtraction method** or **the elimination method**).

The Addition Method

· · · · ·

EXAMPLE 1   Solve the system

$$2x - y = 1$$
$$x + y = 8$$

SOLUTION   Notice that the coefficients of the *y*-terms are opposites. Therefore, if we add the two equations together, the *y*-terms will be eliminated:

$$\begin{array}{ll} 2x - y = 1 & \\ \underline{x + y = 8} & \text{Add.} \\ 3x + 0 = 9 & \text{The variable } y \text{ has been eliminated.} \\ 3x = 9 & \text{Solve for } x. \\ \boxed{x = 3} & \text{This is } part \text{ of our solution.} \end{array}$$

Now we substitute 3 for *x* in either one of our original equations to obtain the value for *y*. Substituting $x = 3$ in the second equation gives us:

$$\begin{array}{ll} x + y = 8 & \text{Substitute 3 for } x. \\ \boxed{3} + y = 8 & \text{Solve for } y. \\ \boxed{y = 5} & \end{array}$$

The solution to the system is $(3, 5)$, and we check it in both equations as follows:

Check

$$\begin{array}{c|c} 2x - y = 1 & x + y = 8 \\ 2(3) - 5 \overset{?}{=} 1 & 3 + 5 \overset{?}{=} 8 \\ 6 - 5 \overset{?}{=} 1 & 8 = 8 \quad ✔ \\ 1 = 1 \quad ✔ & \end{array}$$

▮ ▮

Once you have found a value for one of the variables in the system, be sure not to forget to solve for the other variable. The problem is not finished until you have found a value for *both* variables in the system.

EXAMPLE 2   Solve the system

$$5x - 2y = 14$$
$$3x + 6y = -6$$

SOLUTION   Neither the *x*-terms nor the *y*-terms are opposites. However, if we multiply both sides of the first equation by 3, the coefficients of the *y*-terms will then be opposites. Adding this new equation to the second equation yields the desired result.

$$
\begin{array}{rl}
15x - 6y = \ \ 42 & \text{Multiply the first equation by 3.} \\
\underline{3x + 6y = -6} & \text{Add the equations.} \\
18x \ \ \ \ \ \ = \ \ 36 & \text{The } y\text{-term is eliminated.} \\
x = 2 & \text{Solve for } x.
\end{array}
$$

To find the value for *y*, substitute $x = 2$ in either of the original equations.

$$
\begin{array}{rl}
3x + 6y = -6 & \text{Use equation two.} \\
3 \cdot \boxed{2} + 6y = -6 & \text{Substitute } x = 2. \\
6y = -12 & \text{Solve for } y. \\
y = -2
\end{array}
$$

The solution to the system is $(2, -2)$, which we must now check in *both* of the original equations.

Check
$$
\begin{array}{l}
x = 2, \ y = -2 \\
5x - 2y = 14 \\
5(2) - 2(-2) \overset{?}{=} 14 \\
10 + 4 \overset{?}{=} 14 \\
14 = 14 \quad \text{✓}
\end{array}
\qquad
\begin{array}{l}
x = 2, \ y = -2 \\
3x + 6y = -6 \\
3(2) + 6(-2) \overset{?}{=} -6 \\
6 - 12 \overset{?}{=} -6 \\
-6 = -6 \quad \text{✓} \qquad \blacksquare\blacksquare
\end{array}
$$

EXAMPLE 3   Solve the system of Example 2 by eliminating the *x*-term.

SOLUTION
$$5x - 2y = 14$$
$$3x + 6y = -6$$

To make the coefficients of the *x*-terms opposites, we need to multiply *each* equation by some number. If we mutliply the first equation by 3 and the second equation by -5, the coefficients of the *x*-terms will be opposites.

$$
\begin{array}{rl}
15x - \ \ 6y = \ \ 42 & \text{Multiply the first equation by 3.} \\
\underline{-15x - 30y = \ \ 30} & \text{Multiply the second equation by } -5. \\
-36y = \ \ 72 & \text{Add to eliminate the } x\text{-term.} \\
y = -2 & \text{Solve for } y.
\end{array}
$$

Substituting $y = -2$ into the first equation yields

$$5x - 2(-2) = 14$$
$$5x = 10$$
$$x = 2$$

As before, the solution is $(2, -2)$.   ▮▮

The Substitution Method

· · · · ·

Another algebraic method for solving systems of linear equations is called **the substitution method.** We solve one equation for one of the variables and substitute this value into the other equation.

EXAMPLE 4   Solve the system

$$x + y = 4$$
$$x = y + 2$$

SOLUTION   The second equation says that x and $y + 2$ are equal to each other. Therefore we can substitute $y + 2$ for x in the first equation.

$$\boxed{x} + y = 4$$
$$\boxed{(y + 2)} + y = 4 \qquad \text{Substitute } y + 2 \text{ for } x \text{ in the first equation.}$$

This new equation contains only y and we can solve it.

$$2y + 2 = 4$$
$$2y = 2$$
$$y = 1$$

Now we substitute $y = 1$ in either of the original equations and solve for x:

$$x + y = 4 \qquad \text{First equation.}$$
$$x + \boxed{1} = 4 \qquad \text{Substitute } y = 1.$$
$$x = 3$$

The solution is $(3, 1)$.

Check

$x + y = 4$	$x = y + 2$
$3 + 1 \overset{?}{=} 4$	$3 \overset{?}{=} 1 + 2$
$4 = 4$ ☑	$3 = 3$ ☑

▮▮

EXAMPLE 5   Solve the following system by the substitution method:

$$2x + y = 10$$
$$4x + 3y = 8$$

SOLUTION   In this case we must solve one of the equations for one of the variables. If one of the variables in either equation has a coefficient of 1, it will be easiest to solve for that variable. We solve the first equation for y and substitute the expression obtained into the second equation:

$$2x + y = 10 \qquad \text{Solve for } y.$$
$$y = 10 - 2x$$

Now we substitute $y = 10 - 2x$ in the second equation and solve for x:

$$4x + 3\,y\, = 8$$
$$4x + 3\,(10 - 2x)\, = 8$$
$$4x + 30 - 6x = 8$$
$$-2x = -22$$
$$x = 11$$

Next we substitute $x = 11$ in $2x + y = 10$. This gives us

$$2 \cdot 11 + y = 10$$
$$22 + y = 10$$
$$y = -12$$

Our solution is $(11, -12)$.

Check
$$2x + y = 10 \qquad\qquad 4x + 3y = 8$$
$$2 \cdot 11 + (-12) \overset{?}{=} 10 \qquad\qquad 4 \cdot 11 + 3(-12) \overset{?}{=} 8$$
$$22 - 12 \overset{?}{=} 10 \qquad\qquad 44 - 36 \overset{?}{=} 8$$
$$10 = 10 \quad \text{✔} \qquad\qquad 8 = 8 \quad \text{✔} \quad \blacksquare$$

STOP   **Which method should you use?**   The substitution method lends itself to situations in which one of the variables has a coefficient of 1. Otherwise the addition method is generally easier. However, either method can always be used and both should be learned.

E X A M P L E   6   Solve the system

$$\frac{1}{2}x + \frac{2}{3}y = 2$$

$$\frac{3}{4}x - \frac{1}{3}y = 2$$

SOLUTION   We clear both equations of fractions:

EQUATION 1

$$6 \cdot \frac{1}{2}x + 6 \cdot \frac{2}{3}y = 6 \cdot 2 \qquad \text{Multiply by 6, the LCD.}$$

$$3x + 4y = 12 \qquad \text{The fractions are gone.}$$

EQUATION 2

$$12 \cdot \frac{3}{4}x - 12 \cdot \frac{1}{3}y = 12 \cdot 2 \qquad \text{Multiply by 12, the LCD.}$$

$$9x - 4y = 24 \qquad \text{The fractions are gone.}$$

Now we solve the new equations as before, using the addition method.

$$
\begin{aligned}
3x + 4y &= 12 \\
9x - 4y &= 24 \qquad \text{Add the equations.} \\
\hline
12x &= 36 \qquad \text{Solve for } x. \\
x &= 3
\end{aligned}
$$

Next we substitute 3 for x in Equation 1:

$$3 \cdot 3 + 4y = 12 \qquad \text{Substitute 3 for } x.$$

$$9 + 4y = 12$$

$$4y = 3 \qquad \text{Solve for } y.$$

$$y = \frac{3}{4}$$

The solution is $\left(3, \frac{3}{4}\right)$.   ▮▮

Parallel Lines (Inconsistent Equations)

The following example illustrates how to recognize a system with a graph consisting of parallel lines.

EXAMPLE 7   Solve:

$$
\begin{aligned}
x + y &= 4 \\
3x + 3y &= -2
\end{aligned}
$$

SOLUTION   We solve the first equation for x and substitute into the second equation.

$$x + y = 4$$

$$x = 4 - y \qquad \text{Solved for } x.$$

Next we substitute $x = 4 - y$ in $3x + 3y = -2$:

$$3\,(4 - y) + 3y = -2$$
$$12 - 3y + 3y = -2$$
$$14 = 0 \qquad \text{Both variables are eliminated} \\ \text{and a contradiction results.}$$

Since there is no ordered pair (x, y) for which $14 = 0$, we conclude that there is *no solution*, the lines are *parallel*, and we call the system of equations *inconsistent*. ▮▮

Coinciding Lines (Dependent Equations)

• • • • •

A system with graphs that are coinciding lines (the same line) has an infinite number of solutions, since every solution to one equation will be a solution to the other equation. The next example shows how we identify coinciding lines.

E X A M P L E 8   Solve:
$$6x - 9y = -6$$
$$-4x + 6y = 4$$

S O L U T I O N   We will use the addition method. We multiply the first equation by 2 and the second equation by 3.

$$\begin{array}{ll} 12x - 18y = -12 & \text{Multiply both sides of Equation one by 2.} \\ \underline{-12x + 18y = 12} & \text{Multiply both sides of Equation two by 3.} \\ 0 = 0 & \text{Both variables are eliminated and an identity results.} \end{array}$$

This means that each of the given equations is equivalent to the other. Therefore their graphs are the same, and every solution to the first equation is also a solution to the second equation. We call the system of equations *dependent,* and every point on the line is a solution to the system, so there is *an infinite number of solutions*. ▮▮

To solve a system of equations by the addition method:

1. Multiply one or both of the equations by numbers that will make the coefficients of either of the variables opposites.

2. Add the equations.

*3. Solve the resulting equation. (See Notes 1 and 2.)

4. Substitute the value obtained in Step 3 back in either of the original equations and solve for the remaining variable.

To solve a system of equations by the substitution method:

1. Solve either equation for one variable in terms of the other variable.

*2. Substitute this result in the other equation and solve. (See Notes 1 and 2.)

3. Substitute the results from Step 2 in either of the original equations and solve for the remaining variable.

---

*Note 1:   If both variables are eliminated and the resulting statement is a contradiction, then the lines are **parallel** and there is **no solution.**
*Note 2:   If both variables are eliminated and the resulting statement is an identity, then the lines **coincide** and any solution to either equation is a solution to the system, so there is an **infinite number of solutions.**

8.1   Exercises

▮ ▮ ▮ ▮ ▮

▼ Solve the following systems by the addition method. Some systems may be parallel (no solution) or coinciding (an infinite number of solutions).

7. $3x - 2y = 4$
  $3x - 2y = 5$

8. $3x + 7y = 6$
  $-6x - 5y = -12$

1. $x + y = 3$
  $x - y = 7$

2. $7x - 3y = -6$
  $-4x + 3y = -3$

9. $5x + 2y = -1$
  $4x + 3y = -12$

10. $3x - 6y = 5$
  $2x - 4y = 2$

3. $-x + 6y = 10$
  $x + \ y = -3$

4. $6x - 4y = 1$
  $x + 5y = 3$

11. $x + 2y = 4$
  $-3x - 6y = -12$

12. $2x + 2y = -4$
  $5x + 7y = 2$

5. $2x - 3y = 4$
  $-4x + 6y = 1$

6. $2x + 7y = 12$
  $-4x + 3y = -7$

13. $\frac{1}{2}x + \frac{1}{3}y = 4$
$\frac{1}{4}x + \frac{1}{3}y = 3$

14. $\frac{2}{3}x + y = 10$
$5x - \frac{1}{4}y = 13$

21. $2x - y = 4$
$-4x + 2y = -8$

22. $4x - y = 3$
$-4x - 3y = 1$

▼ Solve the following systems by the substitution method. Some systems may be parallel (no solution) or coinciding (an infinite number of solutions).

23. $2x - 3y = -6$
$3x - y = 5$

24. $4x - 6y = 0$
$2x + y = 4$

15. $y = 2x - 3$
$x - y = 1$

16. $x - y = 4$
$y = 2 - x$

25. $2x - y = 1$
$-4x + 2y = -2$

26. $\frac{1}{2}x + y = 2$
$-x + y = 2$

17. $x + 2y = 7$
$x - y = -5$

18. $y = 3 - 4x$
$4x - y = 3$

 27. Without graphing or solving, how can you tell whether a system of equations is consistent, inconsistent, or dependent?

19. $x + y = 10$
$x + y = 6$

20. $2x + y = -1$
$3x - 2y = -12$

8.2 Systems of Linear Equations in Three Variables

▮ ▮ ▮ ▮ ▮

An equation containing two variables, x and y, has an ordered pair (x, y) as a solution. Similarly, an equation containing three variables, x, y, and z, will have an ordered triple (x, y, z) as a solution.

For example, the triple $(1, -1, -2)$ is a solution to the equation

$$2x + 4y - 3z = 4$$

If we replace x, y, and z by 1, -1, and -2, respectively, we obtain a true statement.

$$2(1) + 4(-1) - 3(-2) \overset{?}{=} 4$$

$$2 - 4 + 6 \overset{?}{=} 4$$

$$4 = 4 \quad \checkmark$$

A solution to a system of three linear equations in three variables is an **ordered triple** (x, y, z) that satisfies all three equations. For example, you can verify that $(-2, 3, -1)$ is the solution to the following system:

$$3x + y - z = -2$$

$$-x + 2y + 5z = 3$$

$$2x + y - 2z = 1$$

by substituting $x = -2$, $y = 3$, and $z = -1$ in each equation.

The method used to solve such systems is similar to the one we used to solve a system in two variables.

To solve a system of three equations in three variables:

1. Eliminate one of the variables from any two of the three equations by the addition method.

2. Eliminate the same variable from the remaining equation and either of the other two equations by the addition method.

3. Solve the system of equations (in two variables) that you obtained in Steps 1 and 2.

4. Substitute the values of the variables obtained in Step 3 in one of the given equations. This gives the value of the third variable.

5. Check by substituting the solution in each of the original equations.

6. If any of the above steps yields a contradiction (like $0 = 4$), the system has no solution. If any step yields an identity (like $0 = 0$), the system has no unique solution.

EXAMPLE 1   Solve the system

$$\begin{array}{ll} \text{(A)} & 3x + y - z = -2 \\ \text{(B)} & -x + 2y + 5z = 3 \\ \text{(C)} & 2x + y - 2z = 1 \end{array}$$

SOLUTION   We multiply Equation (A) by -2 and add it to Equation (B) to eliminate y:

$$\begin{array}{ll} -6x - 2y + 2z = 4 & -2\text{(A)} \\ \underline{-x + 2y + 5z = 3} & \text{(B)} \\ -7x \qquad + 7z = 7 & \text{Divide by 7 to simplify.} \\ \text{(D)} \qquad -x + z = 1 & \text{Call this Equation (D).} \end{array}$$

Now we multiply Equation (C) by -2 and add it to Equation (B) to eliminate y:

$$
\begin{array}{ll}
-4x - 2y + 4z = -2 & -2(C) \\
\underline{-x + 2y + 5z = 3} & (B) \\
\quad\ \ -5x \qquad + 9z = 1 & \text{Call this Equation (E).}
\end{array}
$$

(E)

Now we have two equations, (D) and (E), in two variables, x and z.

$$
\begin{array}{ll}
(D) & -x + \ z = 1 \\
(E) & -5x + 9z = 1
\end{array}
$$

We next multiply Equation (D) by -5 and add it to Equation (E) to eliminate x:

$$
\begin{array}{ll}
5x - 5z = -5 & -5(D) \\
\underline{-5x + 9z = 1} & (E) \\
\qquad\ \ 4z = -4 & \text{Solve for } z. \\
\boxed{z = -1}
\end{array}
$$

Now we substitute $z = -1$ in Equation (D) and solve for x.

$$
\begin{array}{ll}
-x + z = 1 & (D) \\
-x - 1 = 1 & \text{Solve for } x. \\
\boxed{x = -2}
\end{array}
$$

Finally, we substitute $x = -2$ and $z = -1$ in Equation (A) and solve for y.

$$
\begin{array}{ll}
3x + y - z = -2 & (A) \\
3(-2) + y - (-1) = -2 & \text{Substitute } x = -2 \text{ and } z = -1. \\
-6 + y + 1 = -2 & \text{Solve for } y. \\
\boxed{y = 3}
\end{array}
$$

The solution is $(-2, 3, -1)$.
Check $x = -2$, $y = 3$, $z = -1$

$$
\begin{array}{ll}
(A) & 3(-2) + 3 - (-1) \overset{?}{=} -2 \\
& \qquad\qquad\qquad -2 = -2 \quad ✔ \\
(B) & -(-2) + 2(3) + 5(-1) \overset{?}{=} 3 \\
& \qquad\qquad\qquad\qquad 3 = 3 \quad ✔ \\
(C) & 2(-2) + 3 - 2(-1) \overset{?}{=} 1 \\
& \qquad\qquad\qquad\quad 1 = 1 \quad ✔ \qquad\qquad ∎
\end{array}
$$

EXAMPLE 2 Solve the system

$$
\begin{array}{ll}
(A) & -x + 2y - 3z = -1 \\
(B) & 2x - 4y + 6z = 5 \\
(C) & 7x + 3y - 4z = 2
\end{array}
$$

SOLUTION   If we try to eliminate x by multiplying Equation (A) by 2 and adding it to Equation (B),

$$
\begin{array}{ll}
-2x + 4y - 6z = -2 & \text{2(A)} \\
\underline{2x - 4y + 6z = 5} & \text{(B)} \\
\quad\quad\quad\quad 0 = 3 &
\end{array}
$$

we immediately obtain the contradiction $0 = 3$. We conclude that the system has no solution.

EXAMPLE 3   Solve the system

$$
\begin{array}{lll}
\text{(A)} & \quad x + z = 3 \\
\text{(B)} & \quad x + y = 6 \\
\text{(C)} & \quad y + z = 5
\end{array}
$$

SOLUTION   Since there is no y-term in Equation (A), we will eliminate the y-term from Equations (B) and (C). We multiply (C) by -1 and add it to (B):

$$
\begin{array}{lll}
& x + y \quad\;\; = 6 & \text{(B)} \\
& \underline{-y - z = -5} & -1\text{(C)} \\
\text{(D)} \quad & x \quad\quad - z = 1 & \text{Call this equation (D).}
\end{array}
$$

Now Equations (A) and (D) are two equations in the two variables x and z. We can eliminate z by adding them together.

$$
\begin{array}{ll}
x + z = 3 & \text{(A)} \\
\underline{x - z = 1} & \text{(D)} \\
2x \quad\;\; = 4 & \text{Solve for } x. \\
\quad\; \boxed{x = 2} &
\end{array}
$$

Substituting $x = 2$ in Equation (A) yields

$$
\begin{array}{ll}
x + z = 3 & \text{(A)} \\
2 + z = 3 & \text{Substitute } x = 2. \\
\;\boxed{z = 1} &
\end{array}
$$

Substituting $z = 1$ in Equation (C) yields

$$
\begin{array}{ll}
y + z = 5 & \text{(C)} \\
y + 1 = 5 & \text{Substitute } z = 1. \\
\;\boxed{y = 4} &
\end{array}
$$

The solution is $(2, 4, 1)$.
   Check $x = 2$, $y = 4$, $z = 1$

$$
\begin{array}{lll}
\text{(A)} & \quad x + z = 3: & 2 + 1 = 3 \quad ✓ \\
\text{(B)} & \quad x + y = 6: & 2 + 4 = 6 \quad ✓ \\
\text{(C)} & \quad y + z = 5: & 4 + 1 = 5 \quad ✓
\end{array}
$$

▮▮

8.2 Exercises

▼ Solve the following systems of equations.

1. $3x + y - 6z = 7$
$5x - 2y - z = 4$
$-2x + y - z = -3$

2. $x + y + z = 2$
$x - y - z = 0$
$2x - y + z = -1$

3. $x - y + 2z = 7$
$-2x + y + z = 0$
$x + y + z = 6$

4. $x + y - 3z = -2$
$2x - y + z = 9$
$-3x - y + z = -6$

5. $2x - 3y = 1$
$x + y + z = 1$
$x + 4y - z = -8$

6. $4x - z = 0$
$2y + 2z = 6$
$2x + y = 2$

7. $2x + y + z = -3$
$x + 2y + 3z = 4$
$x - y + 2z = -3$

8. $x + 2y + 3z = 4$
$x + y + z = 1$
$x + 3y + 7z = 13$

9. $2x - y + z = 4$
$8x + y + z = 2$
$x + 2y - z = -8$

10. $x - 2y + 3z = 1$
$3x + y + 2z = 3$
$4x - 8y + 12z = 7$

11. $x + 3y + 8z = 1$
$3x - y + 2z = -1$
$2x - 3y + z = 8$

12. $5x + 4y - 6z = -5$
$4x - 7y + 8z = 14$
$2x - 3y + 5z = 11$

13. $2x + 4y + z = 0$
$5x + 3y - 2z = 1$
$4x - 7y - 7z = 6$

14. $-2x + 3y + 3z = 2$
$x + 2y - z = 1$
$2x - y - z = 0$

15. $-x + y + z = -12$
$x + z = 0$
$2x - 3y + z = 6$

16. $x + 8y + z = 1$
$2x + 4y - 3z = 9$
$3x - 4y + 4z = -6$

8.3   Solving Application Problems Using Systems of Equations

▌ ▌ ▌ ▌ ▌

Application problems involving more than one unknown can frequently be simplified by using systems of linear equations to solve them. In each of the following examples we represent each unknown quantity by a different variable.

EXAMPLE 1   The sum of two numbers is 64 and their difference is 14. Find the numbers.

SOLUTION   Let x = the first number
and y = the second number. Then

$$
\begin{array}{ll}
x + y = 64 & \text{Their sum is 64.} \\
\underline{x - y = 14} & \text{Their difference is 14.} \\
2x \quad\;\; = 78 & \text{Add the equations.} \\
\;\; x = 39 &
\end{array}
$$

We substitute $x = 39$ in the equation $x + y = 64$:

$$39 + y = 64$$
$$y = 25$$

The two numbers are 39 and 25.

Check   Sum is 64. | Difference is 14.

  $39 + 25 = 64$ ✔ | $29 - 25 = 14$ ✔   ▌▌

Coin problems and mixture problems are solved using a similar approach. In each we find a *quantity equation* and a *value equation*.

EXAMPLE 2   Elizabeth has a total of 20 coins, consisting only of dimes and quarters, with a total value of $3.95. How many does she have of each coin?

SOLUTION   Let d = the number of dimes
and q = the number of quarters.
There are two equations to be formulated, a quantity equation and a value equation.

Quantity equation

(no. of dimes)	+	(no. of quarters)	=	(total no. of coins)
d	+	q	=	20

Value equation

(value of dimes)	+	(value of quarters)	=	(total value of coins)	
(10)(no. of dimes)	+	(25)(no. of quarters)	=	395¢	($3.95 = 395¢)
$10d$	+	$25q$	=	395	

We solve $d + q = 20$ for d,

$$d = \boxed{20 - q}$$

and substitute it into $10d + 25q = 395$.

$$10 \underline{(20 - q)} + 25q = 395$$
$$200 - 10q + 25q = 395$$
$$15q = 195$$
$$q = 13 \text{ quarters}$$

We substitute $q = 13$ into $d + q = 20$ to find d:

$$d + \boxed{13} = 20$$
$$d = 7 \text{ dimes}$$

Check

$$13 \text{ quarters} = 13 \times \$.25 = \$3.25$$
$$+ \underline{\ 7 \text{ dimes}\ } = \ 7 \times \$.10 = \underline{\ \ .70}$$
$$20 \text{ coins} = \text{total value} = \$3.95$$

EXAMPLE 3   The Green-Grow Garden Center wants to make a mixture of grass seed from type X seed, worth \$4.05/lb, and type Y seed, worth \$3.00/lb. How much of each type is needed to make 100 lb of mixture worth \$3.21/lb?

SOLUTION   Let $x = $ the number of pounds of type X seed
and $y = $ the number of pounds of type Y seed.

Quantity equation

(no. of lb of X)	+	(no. of lb of Y)	=	(no. of lb of mixture)
x	+	y	=	100

Value equation

(value of type X)	+	(value of type Y)	=	(total value of mixture)
$(x \cdot \$4.05/\text{lb})$	+	$(y \cdot \$3.00/\text{lb})$	=	$(100 \text{ lb} \cdot \$3.21/\text{lb})$

$$405x + 300y = (100)(321) \qquad \text{Change all values to cents.}$$

Our system of equations is now

$$x + y = 100$$
$$405x + 300y = 32100$$

Solving for x in Equation 1 yields $x = 100 - y$. Substituting in Equation 2 gives us

$$405(100 - y) + 300y = 32100$$

$$40500 - 405y + 300y = 32100$$

$$-105y = -8400$$

$$y = 80 \text{ lb of type } Y \text{ seed}$$

$$x = 100 - y = 100 - 80 = 20 \text{ lb of type } X \text{ seed} \quad ▮▮$$

EXAMPLE 4   How much 50% sulfuric acid solution and 10% sulfuric acid solution should be mixed to make 20 gallons of 20% sulfuric acid solution?

SOLUTION   Let x = the number of gallons of 50% solution needed
and y = the number of gallons of 10% solution needed.
We need to end up with 20 gallons of solution, so one equation involves quantity:

$$x + y = 20$$

The amount of acid in the x gallons of 50% acid solution is $0.50x$, and the amount of acid in the y gallons of 10% acid solution is $0.10y$.
   The total amount of acid in these must add up to the amount of acid in the 20 gallons of 20% mixture. This gives us the second equation.

$$0.50x + 0.10y = 0.20(20)$$

We clear of decimals by multiplying both sides of the equation by 10:

$$5x + y = 40$$

Now we solve the system.

$$x + y = 20$$
$$5x + y = 40$$

We multiply the first equation by -1 and add.

$$-x - y = -20$$
$$\underline{5x + y = 40}$$
$$4x \quad\quad = 20$$
$$x = 5$$

Finally, we substitute $x = 5$ into $x + y = 20$.

$$5 + y = 20$$

$$y = 15$$

Our solution is 5 gallons of the 50% solution plus 15 gallons of the 10% solution yields 20 gallons of the 20% solution.   ▮▮

8.3   Exercises

▮ ▮ ▮ ▮ ▮

▼ Solve each of the following problems using a system of equations.

1. The sum of two numbers is 58 and their difference is 6. Find the numbers.

2. The sum of two numbers is 75 and their difference is 15. Find the numbers.

3. The sum of two numbers is 52. One number is three times the other. Find the numbers.

4. The difference of two numbers is 13. The larger number is 2 less than twice the smaller. Find the numbers.

5. The difference between two numbers is 12. The larger number is 8 less than twice the smaller. Find the numbers.

6. The sum of two angles in a triangle is 90°, while their difference is 34°. Find the angles.

7. The sum of three numbers is 13. The sum of the two smaller numbers is 3 less than the largest. The sum of the smallest and the largest is 10. Use a system of three equations with three variables to find the numbers.

8. The sum of two angles is 180° and their difference is 72°. Find the angles.

9. The sum of the angles of a triangle is 180°. The difference of the largest and smallest angles is 40°. The sum of the two smallest angles is 20° more than the largest. Find the angles.

10. A man has some $5 bills and $20 bills. If he has a total of 13 bills and the total value of the money is $155, how many of each denomination does he have?

11. A boy has $3.50 in nickels and dimes. How many does he have of each type of coin if he has 45 coins in all?

12. The total value of 30 coins is $4.50. If the coins consist of quarters and dimes, how many of each coin are there?

13. How many nickels and quarters would you have if you had 15 coins worth $1.95?

14. A theater charges $7.00 for adults and $3.50 for children. If there are 109 people attending and the total receipts amount to $563.50, how many of each type of ticket were sold?

15. The admission charge at a school play is $1.00 for students and $3.00 for nonstudents. If a total of 247 people attend the play and the total money taken in is $377, how many of each type of ticket are sold?

16. A father and his three children take a bus ride. An adult ticket costs $6 more than a child's. If the total cost of the tickets is $54, how much does each type of ticket cost?

17. Tickets to a high school football game are $1.00 for students and $2.50 for nonstudents. If 610 people pay to see the game and the total gate receipts are $892, how many students and how many nonstudents pay to see the game?

18. The sum of the digits of a two-digit number is 14. The units digit is 2 greater than the tens digit. Find the number.

19. The sum of the digits of a three-digit number is 15. The units digit is 4 greater than the tens digit and the hundreds digit is 4 less than the tens digit. Find the number.

20. If chocolates cost $9.50/lb and caramels cost $5.50/lb, how many pounds of each are there in a 6-lb mixture of chocolates and caramels costing a total of $47.00?

21. A mixture of 80% alcohol solution is to be made from mixing 70% and 82% alcohol solutions. How many gallons of each solution are needed to make 12 gallons of the 80% solution?

22. How many ounces of a 30% acid solution and a 90% acid solution must be mixed to get 30 ounces of 50% acid solution?

23. Boxes *A*, *B*, and *C* each weigh a different amount. Three boxes of *A*, 1 of *B*, and 3 of *C* weigh 33 ounces. Four boxes of *A*, 2 of *B*, and 1 of *C* weigh 28 ounces. Four boxes of *A*, 1 of *B*, and 2 of *C* weigh 31 ounces. What is the weight of each box?

24. Carol spent a total of $4.96 for 24 stamps; she purchased 4¢, 19¢, and 29¢ denominations. If she bought the same number of 19¢ stamps as she did 29¢ stamps and six fewer 4¢ stamps than 19¢ stamps, how many of each kind did she buy?

25. Jack Shanley operates a hot dog stand in New York City. He sells hot dogs for $1.50 each and soda for $1.00 per can. If he sells a total of 255 items and his total receipts amount to $335.00, how many hot dogs and cans of soda does he sell?

26. The Finger Lakes Performing Arts Center in Canandaigua, N.Y., seats 3,000 people. Tickets for a Sting concert held there sold for $16.00 and $25.00. How many of each price ticket were sold if total sales were $58,800?

27. Tickets to a Crosby, Stills, and Nash concert sold for $11.50 and $14.50. If the concert hall holds 1500 people and total sales amounted to $19,350, how many of each price ticket were sold?

28. The radiator in Frank's car holds 12 quarts. How many quarts of pure antifreeze must he add to a mixture of antifreeze and water that is 40% antifreeze to make a 50% antifreeze mixture that will fill his radiator?

29. The sum of the three angles of any triangle is 180°. Find each of the angles if the second angle is three times as large as the first and the third angle is 100° greater than the sum of the other two angles.

31. A flock of birds is sitting on three telephone wires. There are three more birds on the first wire than there are on the second one; and there are three times as many birds on the third wire as there are on the first. If all of the birds on the first wire were to fly over to the second wire, there would then be twice as many birds on the third wire as there are on the second. How many birds are sitting on each wire?

30. In a triangle, the second angle is 10° greater than the first and the third angle is 10° greater than the second. If the sum of the three angles is 180°, find the three angles.

8.4   Determinants

▮ ▮ ▮ ▮ ▮

Determinants will enable us to solve systems of equations in an easy, somewhat mechanical manner. However, we first need to see what determinants are and how we find their value.

A **2-by-2,** (written 2×2) **determinant** is a real number associated with a square array of numbers enclosed by vertical lines:

$$\begin{vmatrix} a_1 & b_1 \\ a_2 & b_2 \end{vmatrix}$$

Its value is given by $a_1b_2 - a_2b_1$, which can be remembered by the diagram

$$\begin{vmatrix} a_1 & b_1 \\ a_2 & b_2 \end{vmatrix} = a_1b_2 - a_2b_1$$

EXAMPLE 1   Find the value of each determinant.

(a) $\begin{vmatrix} 4 & 5 \\ 2 & 3 \end{vmatrix} = (4)(3) - (2)(5) = 12 - 10 = 2$

(b) $\begin{vmatrix} -2 & -4 \\ 3 & -1 \end{vmatrix} = (-2)(-1) - (3)(-4) = 2 + 12 = 14$ ▮▮

EXAMPLE 2   Find x if

$$\begin{vmatrix} 3 & x \\ 2 & 6 \end{vmatrix} = 4$$

SOLUTION

$$\begin{vmatrix} 3 & x \\ 2 & 6 \end{vmatrix} = 14$$

$$18 - 2x = 14$$

$$-2x = -4$$

$$x = 2$$

■ ■

Evaluating 3-by-3 Determinants

• • • • •

A **3 × 3 determinant** has a value given by

$$\begin{vmatrix} a_1 & b_1 & c_1 \\ a_2 & b_2 & c_2 \\ a_3 & b_3 & c_3 \end{vmatrix} = a_1 b_2 c_3 + b_1 c_2 a_3 + c_1 a_2 b_3 - (c_1 b_2 a_3 + a_1 c_2 b_3 + b_1 a_2 c_3)$$

If this looks unbelievably complicated, don't despair; there are two methods that we can use to simplify the process considerably.

❚ **Method I—Subtraction of Diagonals** • • If we write the determinant with the first two columns repeated at the right, the value of the determinant is **the sum of the products of the main diagonals minus the sum of the products of the secondary diagonals** (see the diagram):

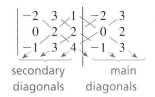

main diagonals   minus   secondary diagonals

$$(a_1 b_2 c_3 + b_1 c_2 a_3 + c_1 a_2 b_3) - (c_1 b_2 a_3 + a_1 c_2 b_3 + b_1 a_2 c_3)$$

EXAMPLE 3   Find the value of $\begin{vmatrix} -2 & 3 & 1 \\ 0 & 2 & 2 \\ -1 & 3 & 4 \end{vmatrix}$.

SOLUTION   We rewrite the first two columns at right:

$$\begin{vmatrix} -2 & 3 & 1 \\ 0 & 2 & 2 \\ -1 & 3 & 4 \end{vmatrix} \begin{matrix} -2 & 3 \\ 0 & 2 \\ -1 & 3 \end{matrix}$$

secondary   main
diagonals   diagonals

main diagonals   secondary diagonals

$$[(-2)(2)(4) + (3)(2)(-1) + (1)(0)(3)] - [(1)(2)(-1) + (-2)(2)(3) + (3)(0)(4)]$$

$$= \qquad (-16 - 6 + 0) \qquad - \qquad (-2 - 12 + 0)$$

$$= -22 + 14$$

$$= -8$$

■ ■

▮ **Method II—Expansion by Minors** • • The **minor of an element of a determinant** is the determinant that results from deleting the row and column in which that entry occurs. Concentrating on the first column, for example,

$$\begin{vmatrix} a_1 & b_1 & c_1 \\ a_2 & b_2 & c_2 \\ a_3 & b_3 & c_3 \end{vmatrix} \text{ yields } \textbf{the minor of } a_1; \text{ which is } \begin{vmatrix} b_2 & c_2 \\ b_3 & c_3 \end{vmatrix}$$

$$\begin{vmatrix} a_1 & b_1 & c_1 \\ a_2 & b_2 & c_2 \\ a_3 & b_3 & c_3 \end{vmatrix} \text{ yields } \textbf{the minor of } a_2; \text{ which is } \begin{vmatrix} b_1 & c_1 \\ b_3 & c_3 \end{vmatrix}$$

$$\begin{vmatrix} a_1 & b_1 & c_1 \\ a_2 & b_2 & c_2 \\ a_3 & b_3 & c_3 \end{vmatrix} \text{ yields } \textbf{the minor of } a_3; \text{ which is } \begin{vmatrix} b_1 & c_1 \\ b_2 & c_2 \end{vmatrix}$$

When we evaluate a determinant using expansion by minors, the signs connecting the minors follow the pattern

$$\begin{vmatrix} + & - & + \\ - & + & - \\ + & - & + \end{vmatrix}$$

The pattern starts with $+$ in the upper left corner and alternates $+$ and $-$ as we move either across or down.

To evaluate a 3 × 3 determinant using expansion by minors:

1. Select any row or column to expand along.

2. Write the products of each element and its minor for the row or column selected.

3. Connect these products with the signs from the sign pattern for the row or column selected.

4. Evaluate the 2 × 2 determinants and simplify.

Follow the next example carefully. You will find that the process is not as complicated as it appears at first glance. We will evaluate the same determinant as in Example 3 to show that the result is the same for either method.

EXAMPLE 4   Evaluate the following 3 × 3 determinant by expansion of minors about the first column.

$$\begin{vmatrix} -2 & 3 & 1 \\ 0 & 2 & 2 \\ -1 & 3 & 4 \end{vmatrix}$$

SOLUTION   The products of the elements of the first column and their minors are:

$$-2 \cdot \begin{vmatrix} 2 & 2 \\ 3 & 4 \end{vmatrix} \qquad 0 \cdot \begin{vmatrix} 3 & 1 \\ 3 & 4 \end{vmatrix} \qquad -1 \cdot \begin{vmatrix} 3 & 1 \\ 2 & 2 \end{vmatrix}$$

The signs from the sign pattern for the first column are $+, -, +$. Connecting the products with these signs gives us

$$\boxed{+} \quad -2 \cdot \begin{vmatrix} 2 & 2 \\ 3 & 4 \end{vmatrix} \quad \boxed{-} \quad 0 \cdot \begin{vmatrix} 3 & 1 \\ 3 & 4 \end{vmatrix} \quad \boxed{+} \quad -1 \cdot \begin{vmatrix} 3 & 1 \\ 2 & 2 \end{vmatrix}$$

$$= -2[(2)(4) - (3)(2)] - 0[(3)(4) - (3)(1)] - 1[(3)(2) - (2)(1)]$$

$$= -2(2) - 0(9) - 1(4)$$

$$= -4 - 0 - 4$$

$$= -8$$

We could have expanded about any row or column. In the next example we will expand the same determinant about the third row. You'll see that the result is the same.

E X A M P L E   5   Evaluate by expansion of minors about the third row:

$$\begin{vmatrix} -2 & 3 & 1 \\ 0 & 2 & 2 \\ -1 & 3 & 4 \end{vmatrix}$$

S O L U T I O N   We write the products of the elements of the third row and their minors, connected by the signs for the third row: $+, -, +$:

$$\begin{vmatrix} -2 & 3 & 1 \\ 0 & 2 & 2 \\ -1 & 3 & 4 \end{vmatrix} = -1 \cdot \begin{vmatrix} 3 & 1 \\ 2 & 2 \end{vmatrix} - 3 \cdot \begin{vmatrix} -2 & 1 \\ 0 & 2 \end{vmatrix} + 4 \cdot \begin{vmatrix} -2 & 3 \\ 0 & 2 \end{vmatrix}$$

$$= -1[(3)(2) - (2)(1)] - 3[(-2)(2) - (0)(1)] + 4[(-2)(2) - (0)(3)]$$

$$= -1(4) - 3(-4) + 4(-4)$$

$$= -4 + 12 - 16$$

$$= -8$$

It is generally a good idea to expand about a row or column that contains zeros, since each zero eliminates the need to evaluate a minor determinant.

8.4   Exercises

▼   Evaluate the following 2 × 2 determinants.

1. $\begin{vmatrix} 2 & 3 \\ 4 & 3 \end{vmatrix}$

2. $\begin{vmatrix} 2 & -1 \\ 3 & 4 \end{vmatrix}$

3. $\begin{vmatrix} -8 & 2 \\ 1 & 0 \end{vmatrix}$

4. $\begin{vmatrix} -1 & -1 \\ -3 & -4 \end{vmatrix}$

5. $\begin{vmatrix} -2 & 3 \\ -1 & 0 \end{vmatrix}$

6. $\begin{vmatrix} -3 & 1 \\ 4 & -4 \end{vmatrix}$

7. $\begin{vmatrix} 0 & 0 \\ 4 & -2 \end{vmatrix}$

8. $\begin{vmatrix} -2 & -3 \\ -2 & -3 \end{vmatrix}$

9. $\begin{vmatrix} 4 & -2 \\ -8 & 4 \end{vmatrix}$

▼ Solve for *x*:

10. $\begin{vmatrix} 2 & 3 \\ 1 & x \end{vmatrix} = 7$

11. $\begin{vmatrix} 3 & x \\ 4 & -5 \end{vmatrix} = 1$

12. $\begin{vmatrix} 3 & x \\ 4 & -2x \end{vmatrix} = -16$

13. $\begin{vmatrix} x^2 & x \\ 4 & 1 \end{vmatrix} = -3$

▼ Find the value of each determinant using Method I.

14. $\begin{vmatrix} 2 & 0 & -1 \\ 3 & 2 & 1 \\ 1 & -4 & 2 \end{vmatrix}$

15. $\begin{vmatrix} 1 & 2 & 1 \\ 2 & 1 & 1 \\ 0 & 2 & -3 \end{vmatrix}$

16. $\begin{vmatrix} 3 & 4 & 5 \\ 4 & 5 & 6 \\ 5 & 6 & 7 \end{vmatrix}$

17. $\begin{vmatrix} 1 & 3 & 5 \\ 0 & 1 & 0 \\ 2 & 4 & 3 \end{vmatrix}$

18. $\begin{vmatrix} 2 & 0 & 2 \\ 3 & 4 & 5 \\ 2 & -1 & 0 \end{vmatrix}$

19. $\begin{vmatrix} 6 & 1 & 1 \\ 3 & -1 & 1 \\ 2 & 2 & -1 \end{vmatrix}$

▼ Evaluate by Method II (expansion by minors):

20. $\begin{vmatrix} 0 & 0 & 4 \\ 2 & -1 & 3 \\ 5 & 1 & -6 \end{vmatrix}$

21. $\begin{vmatrix} 2 & -1 & 3 \\ 7 & 1 & 2 \\ 5 & 2 & -1 \end{vmatrix}$

22. $\begin{vmatrix} 3 & 1 & 2 \\ -2 & 3 & 1 \\ 3 & 4 & -6 \end{vmatrix}$

23. $\begin{vmatrix} 3 & -1 & 3 \\ -1 & 2 & -1 \\ 5 & 0 & 5 \end{vmatrix}$

24. $\begin{vmatrix} 4 & 1 & 1 \\ 3 & -1 & -2 \\ 2 & -3 & -3 \end{vmatrix}$

25. $\begin{vmatrix} 1 & 4 & 1 \\ 2 & 3 & -2 \\ 4 & 2 & -3 \end{vmatrix}$

8.5   Cramer's Rule

▮ ▮ ▮ ▮ ▮

Determinants are useful in solving systems of equations in which the number of variables is the same as the number of equations. Let's solve the following system of equations using the addition method.

(A) $a_1x + b_1y = c_1$

(B) $a_2x + b_2y = c_2$

$a_1b_2x + b_1b_2y = b_2c_1$	Multiply (A) by b_2.
$-a_2b_1x - b_1b_2y = -b_1c_2$	Multiply (B) by $-b_1$.
$a_1b_2x - a_2b_1x = c_1b_2 - c_2b_1$	Add.
$(a_1b_2 - a_2b_1)x = c_1b_2 - c_2b_1$	Factor out x.
$x = \dfrac{c_1b_2 - c_2b_1}{a_1b_2 - a_2b_1}$	Solve for x.

The expression $c_1b_2 - c_2b_1$ is the value of the determinant

$$D_x = \begin{vmatrix} c_1 & b_1 \\ c_2 & b_2 \end{vmatrix}$$

and $a_1b_2 - a_2b_1$ is the value of the determinant

$$D = \begin{vmatrix} a_1 & b_1 \\ a_2 & b_2 \end{vmatrix}$$

So now we have

$$x = \frac{D_x}{D} = \frac{\begin{vmatrix} c_1 & b_1 \\ c_2 & b_2 \end{vmatrix}}{\begin{vmatrix} a_1 & b_1 \\ a_2 & b_2 \end{vmatrix}}$$

We can also solve for y in an analogous manner; this gives us

$$y = \frac{D_y}{D} = \frac{\begin{vmatrix} a_1 & c_1 \\ a_2 & c_2 \end{vmatrix}}{\begin{vmatrix} a_1 & b_1 \\ a_2 & b_2 \end{vmatrix}}$$

The denominator D, which is the same in both cases, is made up of the coefficients of x and y in the original system. The determinants D_x and D_y are formed by replacing in D the coefficients of x and y, respectively, by the constant terms. This leads us to the following rule.

Cramer's rule

The solution to a system of linear equations is given by

$$x = \frac{D_x}{D} \qquad \text{and} \qquad y = \frac{D_y}{D}$$

EXAMPLE 1   Use Cramer's rule to solve the system

$$3x - 4y = 2$$
$$2x + 5y = 1$$

SOLUTION

$$x = \frac{D_x}{D} = \frac{\begin{vmatrix} 2 & -4 \\ 1 & 5 \end{vmatrix}}{\begin{vmatrix} 3 & -4 \\ 2 & 5 \end{vmatrix}} = \frac{14}{23}$$

$$y = \frac{D_y}{D} = \frac{\begin{vmatrix} 3 & 2 \\ 2 & 1 \end{vmatrix}}{23} = \frac{-1}{23}$$

The solution is $\left(\frac{14}{23}, -\frac{1}{23}\right)$.   ▮▮

Cramer's rule also can be used to solve a system in three variables. Consider the system

$$a_1x + b_1y + c_1z = d_1$$
$$a_2x + b_2y + c_2z = d_2$$
$$a_3x + b_3y + c_3z = d_3$$

The solution, (x, y, z), is given by

$$x = \frac{D_x}{D}, \qquad y = \frac{D_y}{D}, \qquad z = \frac{D_z}{D}$$

where

$$D = \begin{vmatrix} a_1 & b_1 & c_1 \\ a_2 & b_2 & c_2 \\ a_3 & b_3 & c_3 \end{vmatrix}$$

$$D_x = \begin{vmatrix} d_1 & b_1 & c_1 \\ d_2 & b_2 & c_2 \\ d_3 & b_3 & c_3 \end{vmatrix}, \qquad D_y = \begin{vmatrix} a_1 & d_1 & c_1 \\ a_2 & d_2 & c_2 \\ a_3 & d_3 & c_3 \end{vmatrix}, \qquad D_z = \begin{vmatrix} a_1 & b_1 & d_1 \\ a_2 & b_2 & d_2 \\ a_3 & b_3 & d_3 \end{vmatrix}$$

EXAMPLE 2   Use Cramer's rule to solve the system

$$2x - 2y - z = -2$$
$$3x - 4y - 3z = -3$$
$$x + y + 4z = -4$$

SOLUTION   We evaluate D, D_x, D_y, and D_z by expanding down column 1:

$$D = \begin{vmatrix} 2 & -2 & -1 \\ 3 & -4 & -3 \\ 1 & 1 & 4 \end{vmatrix} = 2\begin{vmatrix} -4 & -3 \\ 1 & 4 \end{vmatrix} - 3\begin{vmatrix} -2 & -1 \\ 1 & 4 \end{vmatrix} + 1\begin{vmatrix} -2 & -1 \\ -4 & -3 \end{vmatrix}$$

$$= 2(-13) - 3(-7) + 1(2) = -3$$

$$D_x = \begin{vmatrix} -2 & -2 & -1 \\ -3 & -4 & -3 \\ -4 & 1 & 4 \end{vmatrix} = -2\begin{vmatrix} -4 & -3 \\ 1 & 4 \end{vmatrix} - (-3)\begin{vmatrix} -2 & -1 \\ 1 & 4 \end{vmatrix} - 4\begin{vmatrix} -2 & -1 \\ -4 & -3 \end{vmatrix}$$

$$= -2(-13) + 3(-7) - 4(2) = -3$$

$$D_y = \begin{vmatrix} 2 & -2 & -1 \\ 3 & -3 & -3 \\ 1 & -4 & 4 \end{vmatrix} = 2\begin{vmatrix} -3 & -3 \\ -4 & 4 \end{vmatrix} - 3\begin{vmatrix} -2 & -1 \\ -4 & 4 \end{vmatrix} + 1\begin{vmatrix} -2 & -1 \\ -3 & -3 \end{vmatrix}$$

$$= 2(-24) - 3(-12) + 1(3) = -9$$

$$D_z = \begin{vmatrix} 2 & -2 & -2 \\ 3 & -4 & -3 \\ 1 & 1 & -4 \end{vmatrix} = 2\begin{vmatrix} -4 & -3 \\ 1 & -4 \end{vmatrix} - 3\begin{vmatrix} -2 & -2 \\ 1 & -4 \end{vmatrix} + 1\begin{vmatrix} -2 & -2 \\ -4 & -3 \end{vmatrix}$$

$$= 2(19) - 3(10) + 1(-2) = 6$$

$$x = \frac{D_x}{D} = \frac{-3}{-3} = 1, \qquad y = \frac{D_y}{D} = \frac{-9}{-3} = 3, \qquad z = \frac{D_z}{D} = \frac{6}{-3} = -2$$

The solution is $(1, 3, -2)$.   ▮▮

For both 2×2 and 3×3 systems, if the value of D is 0, there is no unique solution to the system. For this reason, always evaluate D first. If $D = 0$, then D_x, D_y, and D_z need not be evaluated.

When solving a system of equations using Cramer's rule, the same three outcomes are possible as when we used the methods of sections 8.1 and 8.2.

1. **One solution:**   the determinant in the denominator does not equal zero; the system is **consistent.**

2. **No solution:**   the determinant in the denominator equals zero but any one of the determinants in the numerator does not equal zero; the system is **inconsistent.**

3. **Infinitely many solutions:**   the determinants in the denominator and all numerators equal zero; the system is **dependent.**

8.5   Exercises

▮▮▮▮▮

▼ Solve using Cramer's rule.

1. $2x + y = 5$
$3x - 5y = 1$

2. $2x + 3y = 2$
$3x + 7y = -2$

3. $3x - y = 5$
$2x - 3y = 6$

4. $4x - 3y = -9$
$5x - 2y = -13$

5. $5x + 3y = -1$
$7x - 4y = 2$

6. $7x - y = -2$
$3x + 5y = -2$

7. $5x + 2y = -4$
$4x - 6y = 12$

8. $3x + 2y = 4$
$6x + 2y = 7$

9. $x - 2y - 3z = 1$
$3x + 3y - z = 0$
$2x + 3y + z = 2$

10. $2x + 3y - 4z = 2$
$x + 2y - z = 4$
$5x - 2y - z = 4$

11. $x + 2y - z = 2$
$2x - 2y + z = 1$
$5x - y + 5z = 18$

12. $3x + 3y - z = 11$
$5x - 5y + z = 11$
$x - 4y - 4z = -5$

13. $-2x + 3y - z = 10$
$5x + 2y - 3z = -6$
$4x + 4y - 5z = 0$

14. $2x + 3y - z = 11$
$3x - 3y + 4z = 7$
$x - 2y - 3z = -4$

15. $2x + 2y - 5z = 10$
$4x - y = 0$
$6x + 2z = 1$

16. $2x + 4y - 3z = -3$
$4x + 4y - z = 2$
$6x + 2y + 2z = 8$

Additional systems of equations can be found in Exercises 8.1 and 8.2. All can be solved using Cramer's rule.

8.6   Linear Inequalities in Two Variables

┃ ┃ ┃ ┃ ┃

Solutions to Linear Inequalities in Two Variables

• • • • •

A **linear inequality in two variables** is an inequality containing two variables, each of which is raised to the first power. The inequality symbol can be any one of $<$, $\leq$, $>$, or $\geq$. For example,

$$x + 2y < 4$$

is an inequality in the two variables x and y. A **solution** to an inequality in two variables is an ordered pair of numbers that makes the inequality a true statement. For example, $(-3, 1)$ is a solution to $x + 2y < 4$ because if we replace x by -3 and y by 1, it yields a true statement.

$$x + 2y < 4$$

$$\boxed{-3} + 2 \cdot \boxed{1} \overset{?}{<} 4 \qquad \text{Substitute } x = -3 \text{ and } y = 1.$$

$$-1 < 4 \qquad \text{This is a true statement.}$$

But $(3, 2)$ is *not* a solution to $x + 2y < 4$, as we show here:

$$x + 2y < 4$$

$$\boxed{3} + 2 \cdot \boxed{2} \overset{?}{<} 4 \qquad \text{Substitute } x = 3 \text{ and } y = 2.$$

$$7 < 4 \qquad \text{This is a } \textit{false} \text{ statement.}$$

EXAMPLE 1   Are $(2, 1)$ and $(-3, -2)$ solutions to the inequality $4x - 3y \geq 5$?

SOLUTION

$$4x - 3y \geq 5$$

$$4 \cdot \boxed{2} - 3 \cdot \boxed{1} \overset{?}{\geq} 5 \qquad \text{Substitute } x = 2 \text{ and } y = 1.$$

$$8 - 3 \overset{?}{\geq} 5$$

$$5 \geq 5 \qquad \text{This is a true statement.}$$

Therefore $(2, 1)$ is a solution.

$$4x - 3y \geq 5$$

$$4\,\boxed{(-3)} - 3\,\boxed{(-2)} \overset{?}{\geq} 5 \qquad \text{Substitute } x = -3 \text{ and } y = -2.$$

$$-12 + 6 \overset{?}{\geq} 5$$

$$-6 \geq 5 \qquad \text{This is a false statement.}$$

Therefore $(-3, -2)$ is not a solution.

┃┃

Graphing Linear Inequalities in Two Variables

• • • • •

The graph of the solutions of a linear inequality is a region of the coordinate plane. The boundary line of this region is the graph of the *equation* that results from replacing the inequality symbol by an equal sign. If the inequality symbol is $\leq$ or $\geq$, then the boundary line is included in the solution and we draw a solid line. If the inequality symbol is $<$ or $>$, then the boundary line is not part of the solution and we indicate this by drawing a broken, or dashed, line.

EXAMPLE 2   Graph the inequality $y \leq x + 3$.

SOLUTION   We first find the boundary of the graph of the inequality by graphing the line $y = x + 3$ (replace $\leq$ by $=$). Since our inequality symbol is $\leq$, this line will be drawn as a solid line and will become part of the solution. See Figure 8.4.

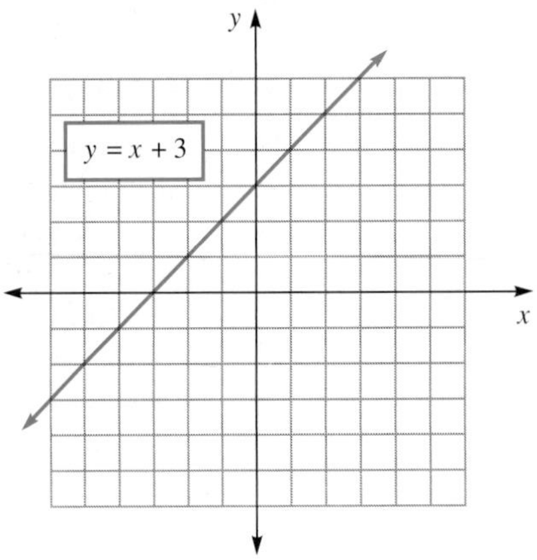

FIGURE 8.4.
The graph of the line
$y = x + 3$.

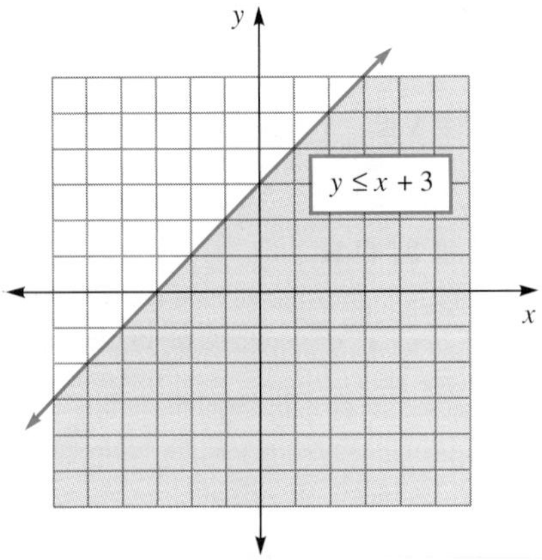

FIGURE 8.5.
The graph of the
inequality $y \leq x + 3$.

This boundary line divides the plane into two regions. One of these regions, along with the boundary line, will be the graph of the solution of $y \leq x + 3$. To determine which region is correct, we choose a test point not on the boundary line and substitute it coordinates in the original inequality $y \leq x + 3$. If the test point *satisifies* the original inequality, than *all* points on the *same* side of the boundary line will be in the solution set. If the test point does *not* satisfy the inequality, then *all* points on the *other* side of the boundary line will be in the solution set.

The point $(0, 0)$ is frequently used as a test point, provided it is not on the boundary line.

$$y \leq x + 3 \qquad \text{Check } (0, 0) \text{ in the inequality.}$$

$$0 \overset{?}{\leq} 0 + 3 \qquad \text{Substitute } x = 0 \text{ and } y = 0.$$

$$0 \leq 3 \qquad \text{This is a true statement.}$$

Since the point $(0, 0)$ is a solution to the original inequality $y \leq x + 3$, so is every other point on the same side of the boundary line. And because our inequality symbol is $\leq$, the boundary line is also part of the solution. The graph of $y \leq x + 3$ is shown in Figure 8.5. ▮▮

To graph a linear inequality in two variables:

1. Replace the inequality symbol with an equal sign and graph this equation. Use a solid line if the inequality symbol is $\leq$ or $\geq$. Use a broken line if the inequality symbol is $<$ or $>$.

2. Choose any point not on the boundary line and substitute its coordinates into the original inequality. If the point satisfies the inequality, then the graph is the region on the same side of the boundary line as the chosen point. If the point does not satisfy the inequality, then the graph is the region on the other side of the boundary line.

EXAMPLE 3   Graph the inequality $2x - y > 3$.

SOLUTION   We replace $>$ by $=$ and graph $2x - y = 3$ using a broken line. This line will *not* be part of the solution. We then choose $(0, 0)$ as a test point:

$$2x - y > 3$$

$$2 \cdot 0 - 0 \overset{?}{>} 3 \qquad \text{Substitute } x = 0 \text{ and } y = 0.$$

$$0 > 3 \qquad \text{A false statement}$$

The test point does not satisfy the original inequality, so the solution is the region on the other side of the boundary line (see Figure 8.6). ▮▮

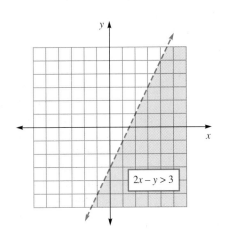

FIGURE 8.6.

E X A M P L E   4   Graph the inequality $y > -1$.

S O L U T I O N   The line $y = -1$, a horizontal line, is the boundary line. Since the original inequality symbol is $>$, we use a broken line. Any points above this line will have a y-coordinate greater than -1, so this is our solution (see Figure 8.7).

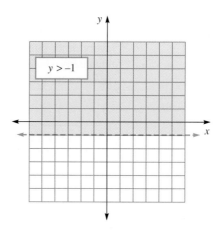

F I G U R E   8 . 7 .

8.6   Exercises

▮▮▮▮▮

▼ Graph the given inequalities.

1. $x + y > 3$

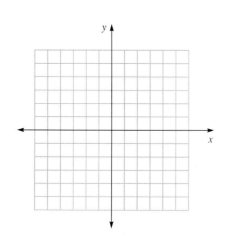

2. $x + y \geq 2$

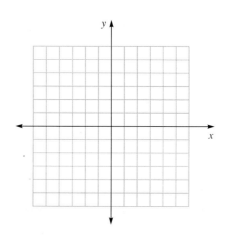

3. $x - y \leq 2$

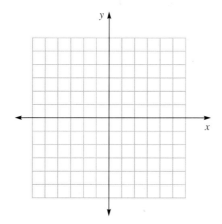

6. $2x + 3y \geq 6$

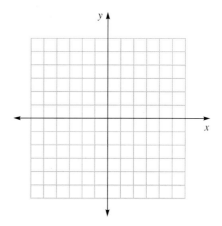

4. $x - y < -2$

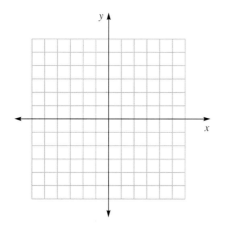

7. $3x + 4y < 0$

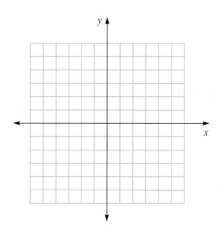

5. $2x + 3y \leq 6$

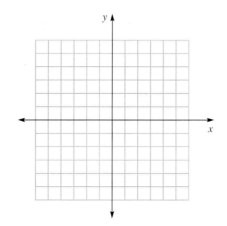

8. $3x + 4y \geq 0$

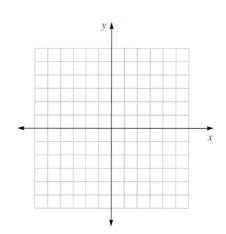

9. $x \leq 4$

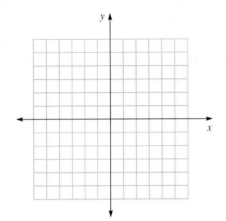

10. $x > -2$

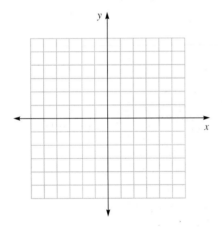

11. $y > 2$

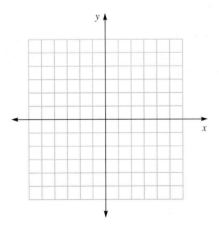

12. $y \geq -3$

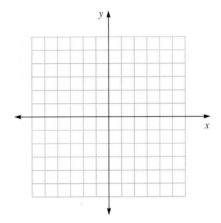

13. Find the inequality with a graph that lies above the line with slope 2 and y-intercept $(0, 3)$.

14. Find the inequality with a graph that lies on and below the line with slope -2 and y-intercept $(0, 1)$.

15. Find the inequality with a graph that lies on and below the line passing through the points $(2, 6)$ and $(-1, 3)$.

16. Find the inequality with a graph that lies below the line passing through the points $(4, 1)$ and $(2, 5)$.

8.7 Systems of Linear Inequalities in Two Variables

■ ■ ■ ■ ■

A system of linear inequalities in two variables consists of two or more linear inequalities in two variables. The graph of such a system contains all ordered pairs such that each one is a solution to *all* of the inequalities in the system.

EXAMPLE 1 Graph the system of inequalities

$$x + y > 4$$
$$2x - y < 3$$

SOLUTION We graph each inequality on the same set of axes using the methods of the preceding section. The solution to the system will be the *intersection* of the graphs of the two inequalities, which is the double-shaded portion on the graph (see Figure 8.8).

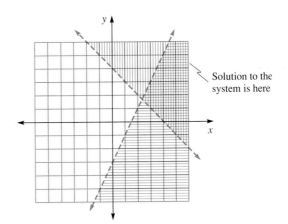

FIGURE 8.8.

Perhaps the easiest way to find this intersection is to shade the graph of one inequality using vertical lines and the other inequality using horizontal lines. The region that is shaded using *both* vertical and horizontal lines will be the solution to the system.   ■ ■

EXAMPLE 2 Graph the system

$$3x + y \leq 4$$
$$y > -2$$

SOLUTION We shade the graph of $3x + y \leq 4$ using vertical lines and the graph of $y > -2$ using horizontal lines (see Figure 8.9).   ■ ■

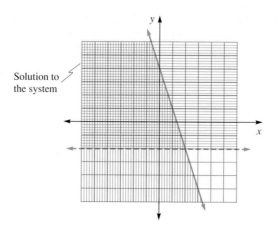

FIGURE 8.9.

EXAMPLE 3   Graph the system

$$y \leq x + 3$$
$$x \leq 2$$
$$x \geq 0$$
$$y \geq 0$$

SOLUTION   We have four separate inequalities, all of which must be graphed on the same set of coordinate axes. The solution to the system is the intersection of the solutions of all four inequalities; it is shown in Figure 8.10.

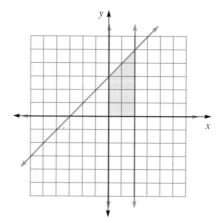

FIGURE 8.10.

Graphs of systems of multiple inequalities such as these are used in a problem-solving technique called **linear programming.**

8.7   Exercises

▼ Graph the given systems of inequalities.

1. $x + y < 3$
    $2x + y < 4$

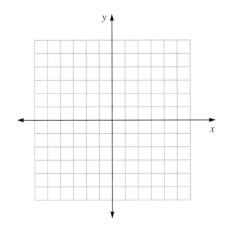

2. $x - y \leq 1$
   $x + 2y \geq 2$

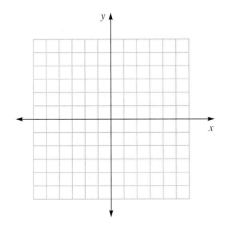

3. $2x + y \geq 4$
   $x - 2y \leq 2$

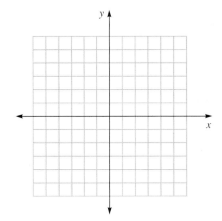

4. $3x - y < 1$
   $2x + y > 0$

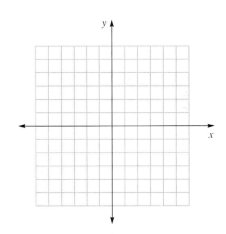

5. $x - y < 1$
   $x > 2$

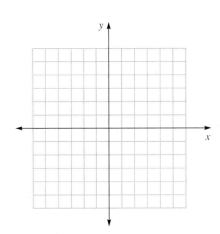

6. $x + y \leq 3$
   $\quad\quad y \leq 2$

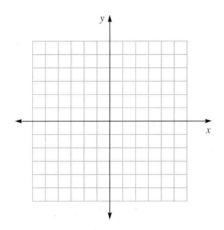

9. $2x + y \leq 1$
   $\quad\quad y \geq -2$
   $\quad\quad x \leq 2$

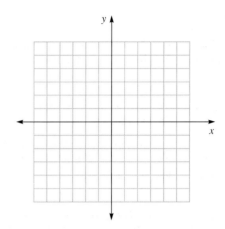

7. $x - 2 \leq 0$
   $3x + y \geq 6$
   $\quad\quad y \leq 6$

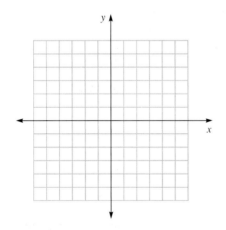

10. $y \geq x - 2$
    $\quad x \geq -1$
    $\quad y \leq 3$

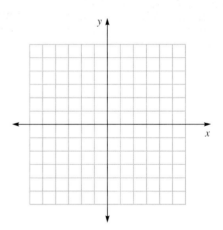

8. $\quad\quad y \leq 0$
   $\quad\quad x \geq -3$
   $2x - 3y \leq 6$

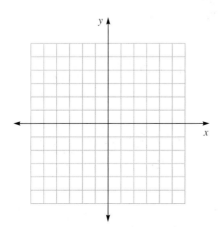

EXAMPLES

To determine whether a point (x, y) is a **solution to a system of linear equations,** check it in *each* equation.

Solve: $x - y = 5$
$x + 2y = -1$

When **solving a system of two linear equations in two variables** three different cases are possible:

1. One solution: the system is **consistent.** Graph: the lines intersect in one point. Addition or substitution method: yields one ordered pair for the solution.

2. No solution: the system is **inconsistent.** Graph: the lines are parallel. Addition or substitution method: both variables are eliminated and a false statement results.

3. An infinite number of solutions (every solution to one equation is also a solution to the other equation): the system is **dependent.** Graph: the lines coincide. Addition or substitution method: both variables are eliminated and a true statement results.

Addition Method:
Multiply the first equation by -1 and add it to the second equation:

$$-x + y = -5$$
$$\underline{x + 2y = -1}$$
$$3y = -6$$
$$y = -2$$

Substituting $y = -2$ in either of the original equations yields $x = 3$. The solution is $(3, -2)$

Substitution Method:
Solve the first equation for x:

$$x = y + 5$$

Substitute for x in the second equation:

$$x + 2y = -1$$
$$(y + 5) + 2y = -1$$
$$3y = -6$$
$$y = -2$$

Now substitute $y = -2$ in either of the original equations, yielding $x = 3$. The solution is $(3, -2)$.

To solve a system of three equations in three variables:

Solve: (A) $x + y - z = 1$
(B) $2x + y + z = 4$
(C) $x + 2y - z = 2$

1. Eliminate the same variable from two pairs of the given equations.

(A) + (B) $\begin{cases} x + y - z = 1 \\ 2x + y + z = 4 \end{cases}$
(D) $ 3x + 2y = 5$

(B) + (C) $\begin{cases} 2x + y + z = 4 \\ x + 2y - z = 2 \end{cases}$
(E) $ 3x + 3y = 6$

EXAMPLES

2. Solve the resulting system of two equations in two variables and substitute the solutions in an original equation to find the value for the third variable.

$$(-1)(D) + (E) \begin{cases} -3x - 2y = -5 \\ 3x + 3y = 6 \\ \hline \boxed{y = 1} \end{cases}$$

Substitute $y = 1$ in (D):

$$3x + 2(1) = 5$$

$$\boxed{x = 1}$$

Substitute $x = 1$ and $y = 1$ in (A):

$$1 + 1 - z = 1$$

$$\boxed{z = 1}$$

The solution is $(1, 1, 1)$.

The **value of a 2-by-2 (2×2) determinant** is given by

$$\begin{vmatrix} a_1 & b_1 \\ a_2 & b_2 \end{vmatrix} = a_1 b_2 - a_2 b_1$$

$$\begin{vmatrix} 3 & -4 \\ 1 & 2 \end{vmatrix} = 6 - (-4) = 10$$

A detailed explanation of two methods for finding the value of a 3×3 determinant is given in Section 8.4. The methods are called **subtraction of diagonals** and **expansion by minors**.

Cramer's rule: The solution to a system of linear equations is given by

$$x = \frac{D_x}{D} \quad \text{and} \quad y = \frac{D_y}{D}$$

Solve the system

$$3x - 2y = 4$$
$$5x - 3y = 2$$

where D is the determinant formed by the coefficients of x and y in the system, and the determinants D_x and D_y are formed by replacing in D the coefficients of x and y, respectively, by the constant terms.

$$x = \frac{D_x}{D} = \frac{\begin{vmatrix} 4 & -2 \\ 2 & -3 \end{vmatrix}}{\begin{vmatrix} 3 & -2 \\ 5 & -3 \end{vmatrix}}$$

$$= \frac{-12 - (-4)}{-9 - (-10)} = \frac{-8}{1} = -8$$

Cramer's rule can also be extended to solve three equations in three unknowns.

$$y = \frac{D_y}{D} = \frac{\begin{vmatrix} 3 & 4 \\ 5 & 2 \end{vmatrix}}{1} = \frac{6 - 20}{1} = -14$$

A **solution to a linear inequality in two variables** is an ordered pair that makes the inequality a true statement.

$(2, -1)$ is a solution to $2x - 3y < 10$ because

$$(2)(2) - 3(-1) < 10$$

$$7 < 10$$

is a true statement.

To graph a linear inequality in two variables:

1. Graph the boundary line (replace the inequality symbol by =). This line is broken if the inequality symbol is > or < and solid if the inequality symbol is ≥ or ≤.

2. Choose any point not on the boundary line. If the point satisfies the inequality, shade in the side of the line that contains the point. If the point does *not* satisfy the inequality, shade in the other side of the line.

Graph $y < 2x + 3$.

First graph the *broken* line $y = 2x + 3$ (since the inequality is <).

Next try $(0, 0)$ in $y < 2x + 3$.

$$0 \stackrel{?}{<} 2 \cdot 0 + 3$$

$$0 < 3$$

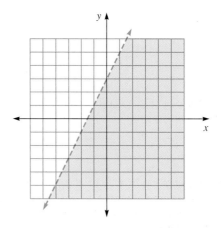

To graph a system of linear inequalities in two variables, graph all of the given inequalities on the same set of axes. The intersection of the graphs is the solution to the system.

Graph: $3x + y \le 3$
$$y \ge 1$$

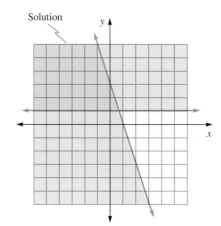

REVIEW EXERCISES—CHAPTER 8

▾ Solve by the addition method.

1. $3x - 2y = 1$
$2x + 2y = 4$

2. $3x + 2y = 14$
$2x - 3y = 7$

3. $4x - 6y = -2$
$-2x + 3y = 1$

4. $3x - y = -5$
$6x - 2y = 2$

▾ Solve by the substitution method.

5. $x + y = 5$
$x - y = 1$

6. $x + 2y = -3$
$-2x - 9y = 6$

7. $3y - 6x = 3$
$-y + 2x = 4$

8. $6x + 3y = 3$
$-2x - y = -1$

▾ Solve the given systems.

9. $2x + y + z = 3$
$2x - 3y - z = 3$
$x + 2y + 2z = 3$

10. $3x + 2y + z = 3$
$x + 2y - z = 7$
$x + 3y - 2z = 10$

▾ Solve using a system of linear equations.

11. The sum of two numbers is 44 and their difference is 18. Find the numbers.

12. How many nickels and how many dimes do you have if you have 26 coins with a total value of $1.90?

13. A hardware store mixes 200 pounds of grass seed costing $4.44/lb. The mixture consists of two kinds seed, one costing $3.60/lb and one costing $4.80/lb. How many pounds of each type are in the mixture?

14. Evaluate: $\begin{vmatrix} 3 & 4 \\ -1 & 2 \end{vmatrix}$

15. Solve for x: $\begin{vmatrix} 3 & 1 \\ -4 & x \end{vmatrix} = 10$

16. Evaluate: $\begin{vmatrix} 2 & 0 & 2 \\ -1 & 3 & -2 \\ 0 & 2 & 1 \end{vmatrix}$

17. Solve the following system using Cramer's rule:

$$3x + 2y = -4$$
$$5x - 3y = -1$$

18. Solve the following system using Cramer's rule:

$$3x + 4y - z = 5$$
$$2x + 3y + z = 3$$
$$x - 2y + z = 5$$

▼ Graph the given inequalities.

19. $y < x + 2$

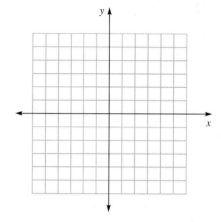

20. $2x + 3y \geq -4$

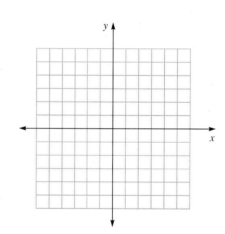

▼ Graph the systems of inequalities.

21. $2x + y > 4$
      $x - 2y \leq 4$

22. $x \geq 1$
             $y \leq 2$
      $x - 2y \geq 4$

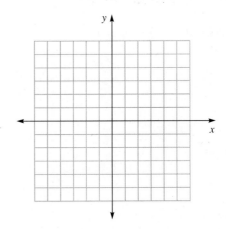

ACHIEVEMENT TEST—CHAPTER 8

▼ Solve by the addition method:

1. $3x - 2y = 4$
$2x + 5y = 9$

1. _____

2. $2x - 6y = 6$
$-3x + 9y = -9$

2. _____

_____

▼ Solve by the substitution method:

3. $x + 2y = 4$
$3x + 6y = -1$

3. _____

4. $-2x + y = 3$
$3x - 4y = -2$

4. _____

5. Solve the following system:

$$3x - y + z = 5$$
$$2x + 2y + 3z = 8$$
$$x + y - 3z = -5$$

5. _____

▼ Solve using a system of equations.

6. The sum of two angles is 90°, while their difference is 38°. Find the angles.

6. _____

7. Tickets to a dance are $3.00 for students and $5.00 for nonstudents. If there were a total of 120 people at the dance and ticket sales amounted to $432, how many students and how many nonstudents were at the dance?

7. _____

8. Christine spent a total of $10.32 for 38 stamps, purchasing 19¢, 29¢, and 30¢ denominations. She bought twice as many 29¢ stamps as she did 30¢ stamps and she bought two more 30¢ stamps than she did 19¢ stamps. How many of each stamp did she buy?

8. _____

_____

9. Evaluate $\begin{vmatrix} 3 & 1 \\ -2 & 4 \end{vmatrix}$.

9. _____

10. Evaluate $\begin{vmatrix} 2 & 3 & -1 \\ 0 & 4 & 1 \\ 1 & 3 & -2 \end{vmatrix}$.

10. _____

11. Solve the following system of equations using Cramer's rule:

$$3x + 5y = -1$$
$$4x - 2y = 3$$

11. _____

12. Solve the system using Cramer's rule.

$$3x + 4y - z = 7$$
$$x - 4y + 2z = -9$$
$$3x + 5y + z = 14$$

12. _____

13. Graph the inequality $2x - y < 2$.

13.

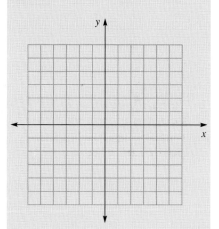

14. Graph the following system of inequalities:

$$x - 2y \geq 6$$
$$x + y > 3$$

14.

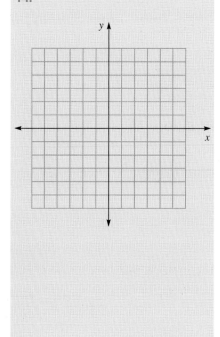

15. Graph the following system of inequalities:

$$x \leq 4$$
$$y \leq 2$$
$$2x + y \geq 4$$

15.

Exponential and Logarithmic Functions

INTRODUCTION

In this chapter we turn our attention to two new types of functions: exponential functions and logarithmic functions. As you will see, both exponential functions and logarithmic functions have many applications to situations in which something's growth is dependent upon the quantity present at any given time—like investments or a bacterial culture.

CHAPTER 9—NUMBER KNOWLEDGE

Gender and Mathematics

Mathematicians as a group are not particularly well known. However, if male mathematicians have had little recognition, women in mathematics have had even less.

For a variety of reasons, some still unclear, the role of the mathematician has, in the past, generally been reserved for men. As a result, mathematically talented women have found it difficult to use their talents to achieve success in the field of mathematics. In the United States, the first woman to receive a Ph.D. in mathematics was Winifred Merrill in 1886.

Fortunately, much has happened recently to improve the opportunities for women who want to study mathematics. Current research indicates that admission to college and university mathematics programs is no longer discriminatory against women.[1] In fact, today almost half of the students who graduate with a major in mathematics are women.

However, there is also evidence that many teachers and the public in general continue to look at mathematics as an unfeminine endeavor.[2] Lynn Osen in her book *Women in Mathematics* refers to what she calls the *feminine mathtique,* which ". . . encourages the notion

that to enjoy mathematics in its many forms is to be, in some obscure way, at variance with one's womanhood. It breeds tasteless jokes and stereotypes about the helpless, checkbook-bumbling female, the mindless housewife, the empty-headed husband-chasing coed, the intuitive (but illogical) woman who hates arithmetic."[3]

There is still not equality in either the availability of jobs or salary levels for mathematically trained women in both industry and in teaching. In 1971 the Society of American Women in Mathematics (AWM) was formed. This organization is dedicated to extending the choices that women have available to them in mathematics.

"Today, women mathematicians are making fundamental contributions to contemporary mathematics. Among them are many brilliant, interesting, vital women who find their work a fulfilling part of their lives."[4]

For more information about AWM activities and programs contact:

Association for Women in Mathematics
4114 Computer & Space Sciences Bldg.
University of Maryland
College Park, Maryland 20742-2461

9.1   Exponential Functions and Their Graphs

Functions such as $y = b^x$, where b is a constant greater than zero but not equal to 1 and x and y are variables, occur widely in mathematics, business, and science. Functions of this kind are called **exponential functions** because the variable x appears as an exponent.

The **domain** of an exponential function is the set of real numbers, but b is restricted as follows:

1. $b \neq 0$, because 0^0 is indeterminate and $0^x = 0$ if $x \neq 0$.

2. b cannot be negative, because expressions such as $(-9)^{1/2} = \sqrt{-9}$ are not real numbers.

3. $b \neq 1$, because $1^x = 1$ for all x.

[1]Perl, Teri, *Math Equals, Biographies of Women Mathematicians,* Addison Wesley (Reading, Mass., 1978) p. 81.
[2]Fennema and Leder, Eds., *Mathematics and Gender,* Teachers College Press, Columbia University (New York, 1990).
[3]Osen, Lynn, *Women in Mathematics,* The M.I.T. Press (Cambridge, Mass., 1974) p. 165.
[4]Perl, Teri, *Math Equals, Biographies of Women Mathematicians,* Addison Wesley (Reading, Mass., 1978) p. 203.

Let's now look at the graph of a typical exponential function.

EXAMPLE 1   Graph $y = 2^x$.

SOLUTION   We calculate a few ordered pairs and then plot them in Figure 9.1.

x	y
0	1
1	2
2	4
−1	$\frac{1}{2}$
−2	$\frac{1}{4}$

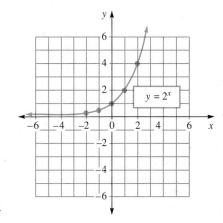

FIGURE 9.1.

Notice that in the preceding example, y was always positive no matter how small x got. The **range** of an exponential function is the set of positive real numbers.

EXAMPLE 2   Graph $y = \left(\dfrac{1}{2}\right)^x$.

SOLUTION   Again we calculate some ordered pairs, which we plot in Figure 9.2.

x	y
0	1
1	$\frac{1}{2}$
2	$\frac{1}{4}$
−1	2
−2	4

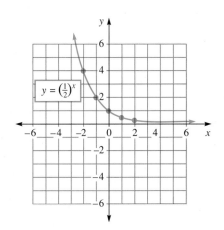

FIGURE 9.2.

In general, when the base b is greater than 1, the graph of $y = b^x$ will look similar to the graph in Figure 9.1. If the base is a fraction ($0 < b < 1$), the graph of $y = b^x$ will look similar to the graph in Figure 9.2.

EXAMPLE 3   Population growth provides a good example of an exponential function. The following chart represents the population of the United States for various years.

Year	1800	1830	1860	1890	1920	1950	1980
Population (in millions)	5.3	12.8	31.4	63.0	105.7	151.3	226.5

When we graph these numbers (see Figure 9.3), you can see that the curve has the basic shape of an exponential function. If we continue the graph to the right for future years, you can see why there is concern for the problems connected with population growth—energy usage, sewage disposal, and pollution of various kinds: air, water, noise, heat, etc.

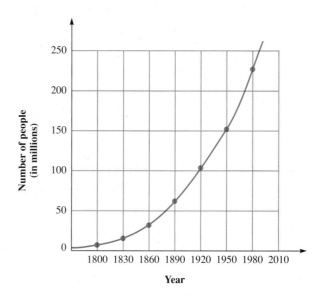

FIGURE 9.3.   ▮▮

9.1   Exercises

▮▮▮▮▮

▼ Graph the given exponential functions.

1. $y = 3^x$

2. $y = 4^x$

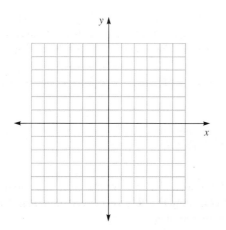

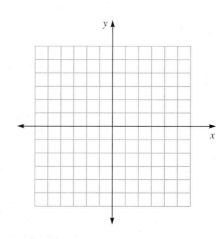

3. $y = \left(\frac{1}{3}\right)^x$

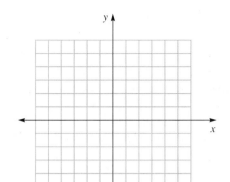

5. $y = 10^x$

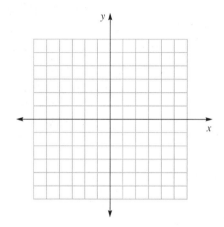

4. $y = \left(\frac{1}{4}\right)^x$

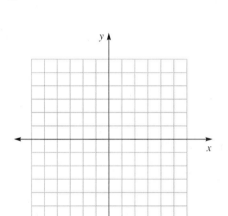

6. $y = \left(\frac{2}{3}\right)^x$

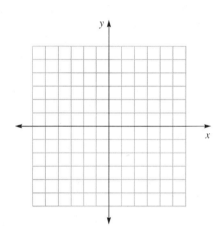

9.2   Logarithmic Functions

If we examine the graph of an exponential function (see Figures 9.1 and 9.2), we find that it is a one-to-one function, since any horizontal line will intersect the graph in at most one point. This means that the inverse of an exponential function like $y = 2^x$ must also be a function.

To find the inverse, we interchange x and y and solve for y in terms of x. Interchanging x and y yields

$$x = 2^y$$

But how do we solve for y? Unfortunately, none of the equation-solving techniques we have learned so far tells us how to solve for a variable when it is in an exponent.

We will now investigate the concept of logarithms, which will enable us to solve an equation of the form $x = 2^y$ for y.

Consider the equation

$$2^y = 8$$

By inspection, we can see that $y = 3$, since we know that $2^3 = 8$. We could have stated the equation $2^y = 8$ in words as

"y is the exponent on 2 that yields 8."

When we work with an exponent that is separated from its base, the word **logarithm** is used in place of *exponent*. Also, we know that the 2 in the expression 2^x is called the **base**. With this terminology, our statement becomes

"y is the logarithm to the base 2 of 8."

This statement is written in symbols as

$$y = \log_2 8$$

We have shown that the two equations

$$2^y = 8 \qquad \text{and} \qquad y = \log_2 8$$

exponential form   logarithmic form

are really just equivalent forms of the same equation.

We generalize this in the following definition:

Definition: If $b > 0$, $b \neq 1$, then the **exponential equation**

$$b^y = x$$

is equivalent to the **logarithmic equation**

$$y = \log_b x$$

For example,

exponent	power		logarithm	power
↓	↓		↓	↓
2^3	= 8	is equivalent to	3	= $\log_2 8$
↑				↑
base				base

The following table illustrates this relationship of equivalence between the exponential and logarithmic forms.

Exponential Form	Log Form	Notes
$8 = 2^3$	$\log_2 8 = 3$	3 is the exponent or logarithm.
$100 = 10^2$	$\log_{10} 100 = 2$	2 is the exponent or logarithm.
$0.1 = 10^{-1}$	$\log_{10} 0.1 = -1$	-1 is the exponent or logarithm.
$\frac{1}{9} = 3^{-2}$	$\log_3 \left(\frac{1}{9}\right) = -2$	-2 is the exponent or logarithm.

Recall that the graph of a one-to-one function and its inverse are mirror images of one another about the line $y = x$. We can see this property if we graph the log function $y = \log_2 x$ (or $x = 2^y$ in its equivalent exponential form) and its inverse, the exponential function $y = 2^x$ (see Figure 9.4). Note the property of symmetry about the line $y = x$. The domain of a logarithmic function is the set of positive real numbers, as you can see from the graph. You cannot find the logarithm of zero or of a negative number. For example, the $\log_2 (-4) = y$ when written in exponential form becomes $2^y = -4$. However, 2 raised to any exponent cannot equal -4.

EXPONENTIAL FUNCTION

x	$y = 2^x$
0	1
1	2
2	4
3	8
-1	$\frac{1}{2}$
-2	$\frac{1}{4}$
-3	$\frac{1}{8}$

LOGARITHMIC FUNCTION

x	$y = \log_2 x$ (or $x = 2^y$)
1	0
2	1
4	2
8	3
$\frac{1}{2}$	-1
$\frac{1}{4}$	-2
$\frac{1}{8}$	-3

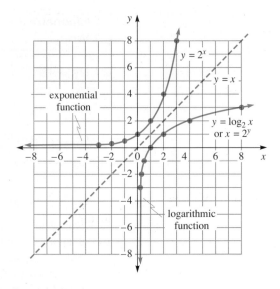

FIGURE 9.4.

EXAMPLE 1   Write $5^2 = 25$ in logarithmic form.

SOLUTION   2 is the exponent (or log), 5 is the base and 25 is the power, so

$$5^2 = 25 \Leftrightarrow \log_5 25 = 2 \qquad (\Leftrightarrow \text{ means "is equivalent to"}) \qquad \blacksquare$$

EXAMPLE 2   Write each of the following equations in logarithmic form.

(a) $3^4 = 81$

(b) $3^{-2} = \dfrac{1}{9}$

(c) $6^0 = 1$

(d) $5^1 = 5$

SOLUTION   (a) $3^4 = 81 \Leftrightarrow \log_3 81 = 4$

(b) $3^{-2} = \dfrac{1}{9} \Leftrightarrow \log_3 \dfrac{1}{9} = -2$

(c) $6^0 = 1 \Leftrightarrow \log_6 1 = 0$

(d) $5^1 = 5 \Leftrightarrow \log_5 5 = 1$ $\blacksquare$

EXAMPLE 3   Write $\log_4 16 = 2$ in exponential form.

SOLUTION   2 is the log or exponent, 4 is the base, and 16 is the power:

$$\log_4 16 = 2 \Leftrightarrow 4^2 = 16$$ $\blacksquare$

EXAMPLE 4   Write each of the following equations in exponential form.

(a) $\log_4 64 = 3$

(b) $\log_9 3 = \dfrac{1}{2}$

(c) $\log_4 1 = 0$

(d) $\log_3 3 = 1$

SOLUTION **(a)** $\log_4 64 = 3 \Leftrightarrow 4^3 = 64$

(b) $\log_9 3 = \dfrac{1}{2} \Leftrightarrow 9^{1/2} = 3$

(c) $\log_4 1 = 0 \Leftrightarrow 4^0 = 1$

(d) $\log_3 3 = 1 \Leftrightarrow 3^1 = 3$ ∎

Examples 2(c) and (d) and 4(c) and (d) suggest the following properties of logarithms.

---

1. $\log_b b = 1, \quad (b > 0,\ b \neq 1)$

2. $\log_b 1 = 0, \quad (b > 0,\ b \neq 1)$

Equation 1 is true because $b^1 = b$, and Equation 2 is true because $b^0 = 1$.

EXAMPLE 5 Find the value of $\log_4 16$.

SOLUTION If we let $\log_4 16 = y$ and write it in exponential form, we obtain

$$\log_4 16 = y \Leftrightarrow 4^y = 16$$

By inspection, we know that $y = 2$, since $4^2 = 16$. Therefore $\log_4 16 = 2$. ∎

EXAMPLE 6 Solve for x: $\log_2 x = 5$.

SOLUTION We write $\log_2 x = 5$ in its equivalent exponential form:

$$\log_2 x = 5 \Leftrightarrow x = 2^5 \text{ or } x = 32$$ ∎

EXAMPLE 7 Solve for x: $\log_x 64 = 3$.

SOLUTION We write in the equation exponential form:

$$\log_x 64 = 3 \Leftrightarrow x^3 = 64$$

Now we take the cube root of both sides of the equation:

$$x = \sqrt[3]{64} = 4$$ ∎

EXAMPLE 8 Solve for x: $\log_5 x = -2$.

SOLUTION $\log_5 x = -2 \Leftrightarrow x = 5^{-2} \text{ or } x = \dfrac{1}{25}$ ∎

E X A M P L E 9   Solve for x: $\log_x \dfrac{1}{8} = -3$.

S O L U T I O N $\log_x \dfrac{1}{8} = -3 \Leftrightarrow x^{-3} = \dfrac{1}{8}$

Therefore,

$$\dfrac{1}{x^3} = \dfrac{1}{8} \qquad \text{Write } x^{-3} \text{ as } \dfrac{1}{x^3}.$$

$$x^3 = 8 \qquad \text{Cross-multiply.}$$

$$x = 2 \qquad \text{Take the cube root of each side.} \qquad \blacksquare$$

9.2   Exercises

▮▮▮▮▮

▼ Write each equation in logarithmic form.

1. $2^3 = 8$

2. $3^2 = 9$

3. $5^3 = 125$

4. $10^3 = 1000$

5. $2^{-5} = \frac{1}{32}$

6. $\left(\frac{1}{2}\right)^{-3} = 8$

7. $6^0 = 1$

8. $10^{-2} = 0.01$

9. $\left(\frac{1}{2}\right)^{-5} = 32$

10. $4^{-3} = \frac{1}{64}$

11. $a^r = y$

12. $t^{-3} = 5$

▼ Write each equation in exponential form.

13. $\log_2 8 = 3$

14. $\log_3 81 = 4$

15. $\log_{10} 1000 = 3$

16. $\log_2 32 = 5$

17. $\log_b 1 = 0$

18. $\log_6 6 = 1$

19. $\log_8 4 = \frac{2}{3}$

20. $\log_{10} 0.1 = -1$

21. $\log_a \dfrac{1}{a} = -1$

22. $\log_{10} 0.0001 = -4$

23. $\log_3 \left(\frac{1}{27}\right) = -3$

24. $a = \log_b x$

▼ Solve for x in each of the following.

25. $\log_4 x = 2$

26. $\log_4 x = 3$

27. $\log_4 x = -2$

28. $\log_{10} x = -2$

29. $\log_x 9 = 2$

30. $\log_x 16 = 4$

31. $\log_x 12 = 1$

32. $\log_x \left(\frac{1}{4}\right) = -1$

33. $\log_3 9 = x$

34. $\log_2 16 = x$

35. $\log_4 64 = x$

36. $\log_2 \left(\frac{1}{4}\right) = x$

9.3   Properties of Logarithms

Since logarithms are exponents, we use properties of exponents to develop three properties of logarithms. In all of the following we assume that u, v, and b are positive real numbers, $b \neq 1$, and that r is any real number.

Property in Symbols	Property in Words
1. $\log_b (u \cdot v) = \log_b u + \log_b v$ The **product rule**	The log of a product is the sum of the logs.
2. $\log_b \left(\dfrac{u}{v} \right) = \log_b u - \log_b v$ The **quotient rule**	The log of a quotient is the difference of the logs.
3. $\log_b u^r = r \cdot \log_b u$ The **power rule**	The log of a number raised to an exponent is the exponent times the log of the number.

Proof of Property 1   Let $u = b^x$ and $v = b^y$. Then in log form, $x = \log_b u$ and $y = \log_b v$. However, $u \cdot v = b^x \cdot b^y = b^{x+y}$, so writing $u \cdot v = b^{x+y}$ in log form gives us

$$\log_b (u \cdot v) = x + y$$

But $x = \log_b u$ and $y = \log_b v$, so

$$\log_b (u \cdot v) = \log_b u + \log_b v \qquad \blacksquare$$

Properties 2 and 3 are proved in a similar manner, and the proofs are omitted here.

Students frequently use the following incorrect rules:

$$\left.\begin{array}{c} \log_b (u + v) = \log_b u + \log_b v. \\[4pt] \log_b (u - v) = \log_b u - \log_b v \\[4pt] \log_b (u \cdot v) = \log_b u \cdot \log_b v \\[4pt] \dfrac{\log_b u}{\log_b v} = \log_b \dfrac{u}{v} \\[8pt] \dfrac{\log_b u}{\log_b v} = \log_b u - \log_b v \end{array}\right\}$$

These are all **INCORRECT** rules.

Remember that logarithms are exponents and that, when we multiply, we *add* exponents; when we divide, we *subtract* exponents.

The correct rules are:

$$\left.\begin{array}{l} \log_b (u \cdot v) = \log_b u + \log_b v \\[2mm] \log_b \left(\dfrac{u}{v}\right) = \log_b u - \log_b v \end{array}\right\} \quad \text{These are the only \textbf{correct} rules.}$$

EXAMPLE 1   Using the properties of logs, expand $\log_2 \dfrac{6}{5}$.

SOLUTION $\log_2 \dfrac{6}{5} = \log_2 6 - \log_2 5$ Quotient rule   ▮▮

EXAMPLE 2   Express $\log_b 7 + \log_b x$ as a single logarithm.

SOLUTION $\log_b 7 + \log_b x = \log_b(7x)$ Use the product rule in reverse.   ▮▮

EXAMPLE 3   Expand $\log_3 (7x)^3$.

SOLUTION

$$\begin{aligned} \log_3 (7x)^3 &= 3 \log_3 7x && \text{Power rule} \\ &= 3(\log_3 7 + \log_3 x) && \text{Product rule} \\ &= 3 \log_3 7 + 3 \log_3 x \end{aligned}$$ ▮▮

EXAMPLE 4   Expand $\log_b \dfrac{x^2 \sqrt{y}}{z^2}$.

SOLUTION

$$\begin{aligned} \log_b \dfrac{x^2 \sqrt{y}}{z^2} &= \log_b \dfrac{x^2 \cdot y^{1/2}}{z^2} && \text{Write radical as an exponent.} \\ &= \log_b (x^2 \cdot y^{1/2}) - \log_b z^2 && \text{Quotient rule} \\ &= \log_b x^2 + \log_b y^{1/2} - \log_b z^2 && \text{Product rule} \\ &= 2 \log_b x + \tfrac{1}{2} \log_b y - 2 \log_b z && \text{Power rule} \end{aligned}$$ ▮▮

EXAMPLE 5   Write $\log_b x + 3 \log_b y - \tfrac{1}{3} \log_b z$ as a single logarithm.

SOLUTION   In general, we use the power rule first.

$$\begin{aligned} &\log_b x + 3 \log_b y - \tfrac{1}{3} \log_b z \\ &= \log_b x + \log_b y^3 - \log_b z^{1/3} && \text{Power rule} \\ &= \log_b (xy^3) - \log_b z^{1/3} && \text{Product rule} \\ &= \log_b \dfrac{xy^3}{z^{1/3}} && \text{Quotient rule} \\ &= \log_b \dfrac{xy^3}{\sqrt[3]{z}} && z^{1/3} = \sqrt[3]{z} \end{aligned}$$ ▮▮

E X A M P L E   6   Solve for x: $\log_3 5 + \log_3 x = 2$.

S O L U T I O N   We first observe that

$$\log_3 5 + \log_3 x = \log_3 5x = 2 \qquad \text{Product rule}$$

Now, writing $\log_3 5x = 2$ in exponential form gives us

$$5x = 3^2 = 9$$

$$x = \frac{9}{5}$$

9.3   Exercises

▼ Expand each expression using the properties of logarithms.

1. $\log_3 5x$

2. $\log_2 xy$

3. $\log_3 \dfrac{5}{x}$

4. $\log_2 \dfrac{x}{y}$

5. $\log_3 x^5$

6. $\log_2 x^y$

7. $\log_b x^3 y^2$

8. $\log_b \sqrt{x}$

9. $\log_{10} \sqrt[5]{y}$

10. $\log_b \dfrac{4x}{y}$

11. $\log_2 \dfrac{2z}{y}$

12. $\log_a \dfrac{x}{y}$

13. $\log_3 \sqrt[3]{\dfrac{x}{y}}$

14. $\log_a x^2 \sqrt{y}$

15. $\log_a \dfrac{x\sqrt[3]{y}}{\sqrt{y}}$

▼ Express each of the following as a single logarithm.

16. $\log_3 4 + \log_3 5$

17. $\log_3 5 - \log_3 x$

18. $\log_2 x + 2 \log_2 y$

19. $3 \log_b c - 2 \log_b c$

20. $\frac{1}{2} \log_{10} x + \log_{10} y$

21. $\frac{1}{2} \log_a x - 2 \log_a y$

22. $\frac{1}{2}\log_a x - \log_a y + 2\log_a z$

23. $\frac{1}{2}(\log_a x - \log_a y + 2\log_a z)$

▼ Solve each of the following for x.

24. $\log_3 x + \log_3 3 = 1$

25. $\log_3 x - \log_3 3 = 1$

26. $\log_2 x - \log_2 2 = 3$

27. $\log_2 x + 2\log_2 2 = 4$

▼ Say whether each of the following equations is true or false.

28. $\log_b \dfrac{u}{v} = \log_b u - \log_b v$

29. $\dfrac{\log_b u}{\log_b v} = \log_b u - \log_b v$

30. $\log_b u \cdot \log_b v = \log_b (u \cdot v)$

31. $\log_b u \cdot \log_b v = \log_b u + \log_b v$

32. $\log_b u + \log_b v = \log_b (u \cdot v)$

33. $\log_b u^r = r \log_b u$

9.4　Common Logarithms

▮ ▮ ▮ ▮ ▮

Any positive number except 1 can be used as a base for a logarithm ($y = \log_1 x \Leftrightarrow 1^y = x$ is true only for $x = 1$). Since ours is a base 10 number system, base 10 logarithms have traditionally been used to do calculations. Base 10 logarithms are called **common logarithms** and are used so often that the base is not written.

$$\log x \qquad \text{means} \qquad \log_{10} x$$

Common logarithms of powers of 10 are reasonably easy to find. Look at the following table.

Exponential Form	Log Form (base 10)
$10^4 = 10{,}000$	$\log 10{,}000 = 4$
$10^3 = 1000$	$\log 1000 = 3$
$10^2 = 100$	$\log 100 = 2$
$10^1 = 10$	$\log 10 = 1$
$10^0 = 1$	$\log 1 = 0$
$10^{-1} = 0.1$	$\log 0.1 = -1$
$10^{-2} = 0.01$	$\log 0.01 = -2$
$10^{-3} = 0.001$	$\log 0.001 = -3$
$10^{-4} = 0.0001$	$\log 0.0001 = -4$

To find the common logarithm of a number other than an integral power of 10, we can use a scientific calculator. Such a calculator has a logarithm key, **LOG**, and is the easiest way to find logarithms. Simply enter the number that you want the logarithm of and press the **LOG** key.

If a calculator is not available, you can use the table of common logarithms in Appendix A, which gives the logarithm of numbers between 1 and 10. To use the table, the column under n represents the first two digits of the number. The other numbers across the top represent the third digit. For numbers with more than three digits, use the nearest rounded value. Check the following values in the table to be certain you have the idea.

$$\log 6.21 = 0.7931$$

$$\log 2.00 = 0.3010$$

$$\log 1.16 = 0.0645$$

To find the logarithm of a number that is not between 1 and 10, we write the number in scientific notation and apply the product rule.

EXAMPLE 1   Find log 348.

SOLUTION

$\log 348 = \log (3.48 \times 10^2)$	Scientific notation
$= \log 3.48 + \log 10^2$	Product rule
$= 0.5416 + \log 10^2$	From the table
$= 0.5416 + 2$	$\log 10^2 = 2$
$\log 348 = 2.5416$	

In the preceding example, 2 is called the **characteristic** and 0.5416 is called the **mantissa**.

$$\log 348 = 2.5416$$
$$\uparrow \quad \uparrow$$
characteristic   mantissa

To find the common logarithm of a number using a table of logarithms:

1. Write the given number in scientific notation;

2. Find the log of that part of the number between 1 and 10 from the table (mantissa).

3. The log of the given number is the sum of the characteristic (the exponent of 10) and the mantissa.

EXAMPLE 2   Find log 3.19.

SOLUTION

$$\log 3.19 = \log (3.19 \times 10^0) \qquad \text{Scientific notation}$$

$$= \log 3.19 + \log 10^0 \qquad \text{Product rule}$$

$$= 0.5038 + 0 \qquad \text{From the table}$$

$$= 0.5038 \qquad\qquad\qquad ▮▮$$

EXAMPLE 3   Find log 0.00676.

SOLUTION

$$\log 0.00676 = \log (6.76 \times 10^{-3})$$

$$= \log 6.76 + \log 10^{-3}$$

$$= 0.8299 + (-3)$$

$$= 0.8299 - 3$$

Notice that we leave the mantissa and characteristic separate from each other. If we subtract, we obtain $0.8299 - 3 = -2.1701$, which is technically correct. However, it is usually preferable to indicate the negative characteristic and the positive mantissa. The table contains only positive mantissas. It is common practice to write

$$0.8299 - 3$$

$$\text{as} \qquad 7.8299 - 10$$

If we use a calculator with a log key to find the log of 0.00676, the display will show -2.1700533, which rounds to -2.1701. To convert this form to the form obtained from a table, we add and subtract the smallest whole number larger than 2.1701, which is 3:

$$\begin{array}{r} 3.0000 - 3 \\ -2.1701 \\ \hline 0.8299 - 3 \end{array} \qquad ▮▮$$

EXAMPLE 4   Find log 0.0588 by **(a)** using a calculator; **(b)** using a table.

SOLUTION   **(a)** We enter 0.0588 and press the **LOG** button. The display indicates -1.2306227, which rounds to -1.2306. We add and subtract 2, the smallest whole number larger than 1.2306.

$$\begin{array}{r} 2.0000 - 2 \\ -1.2306 \\ \hline 0.7694 - 2 \end{array}$$

or, equivalently, $8.7694 - 10$.

(b)

$$\log 0.0588 = \log (5.88 \times 10^{-2})$$

$$= \log 5.88 + (-2)$$

$$= 0.7694 - 2 \qquad \text{From the table·}$$

We can also write this as $8.7694 - 10$. ▮▮

Now that we have learned how to find the logarithm of a number, the next goal is to learn how to find the number, given that we know its logarithm. This process is referred to as finding the **antilogarithm** or **inverse logarithm.**

As in the case of finding logarithms, the easiest way to find antilogarithms is by using a scientific calculator. We use a combination of the inverse key, **INV**, and the logarithm key, **LOG**. First we enter the logarithm whose antilogarithm we wish to find. We press the inverse key, **INV**, followed by the logarithm key, **LOG**. The antilogarithm will appear in the display.

EXAMPLE 5   If log $x = 3.6503$, find the antilogarithm, x, by (**a**) using a calculator; (**b**) using a table.

SOLUTION   (**a**) We enter 3.6503 and press **INV**, followed by **LOG**. The number 4469.9226 will appear in the display; this is the antilogarithm. Rounded to the nearest unit, $x = 4470$.

(**b**) The mantissa is 0.6503 and the characteristic is 3. In the body of the table, we find .6503 in row 4.4 and under column 7. This gives us:

$$\log x = 3.6503$$

$= 0.6503 + 3$	The characteristic is 3.
$x = 4.47 \times 10^3$	3 is the power of 10.
$= 4470$	Move the decimal point 3 places to the right. ▮▮

EXAMPLE 6   If log $x = 0.1933 - 2$, find the antilogarithm, x, by (**a**) using a calculator; (**b**) using a table.

SOLUTION   (**a**) We key in 0.1933, **–**, 2, **=**. The display will read -1.8067. Next we press **INV** followed by **LOG** and the antilogarithm, $x = 0.0156063$, will appear in the display. Rounded to the nearest ten-thousandth, $x = 0.0156$.

(**b**) A mantissa of 0.1933 gives us 1.56 from the table. (Actually, 0.1931 is the closest mantissa in the table.) The characteristic is -2, which is the power of 10.

$$\log x = 0.1933 - 2$$
$$= 0.1933 + (-2)$$
$$x = 1.56 \times 10^{-2}$$
$$= 0.0156 \qquad ▮▮$$

EXAMPLE 7   Calculate $(1.08)^{18}$ using logarithms.

SOLUTION   We let $n = (1.08)^{18}$. Then

$\log n = \log (1.08)^{18}$	Take the log of both sides.
$= 18 \cdot \log 1.08$	Use the power rule.
$= (18)(0.0334)$	Find the log of 1.08
$= 0.6012$	

Now we find the antilog of 0.6012:

$$n \approx 3.99$$ ▮▮

EXAMPLE 8   Calculate $\left(\sqrt[3]{1160}\right)(24.3)^2$.

SOLUTION   If we let $n = (1160)^{1/3}(24.3)^2$, then

$$
\begin{aligned}
\log n &= \log\left[(1160)^{1/3}(24.3)^2\right] && \text{Take the log of both sides.}\\
&= \log\left[(1160)^{1/3}\right] + \log\left[(24.3)^2\right] && \text{Use the product rule.}\\
&= \tfrac{1}{3}\log 1160 + 2\log 24.3 && \text{Use the power rule.}\\
&= \tfrac{1}{3}(3.0645) + 2(1.3856) && \text{Find the logs of 1160 and 24.3.}\\
&= 1.0215 + 2.7712 && \\
&= 3.7927 && \text{Combine.}
\end{aligned}
$$

Taking the antilog, we get

$$
\begin{aligned}
n &= \text{antilog } 3.7927\\
&\approx 6204
\end{aligned}
$$ ▮▮

9.4   Exercises

▮▮▮▮▮

▼ Find the following logarithms using a calculator or the table in Appendix A.

1. $\log 1560$

2. $\log 325$

3. $\log 21.6$

4. $\log 621000$

5. $\log 71300$

6. $\log 45000$

7. $\log 0.00631$

8. $\log 0.0426$

9. $\log 1.68$

▼ Solve for x in each of the following equations.

10. $\log x = 0.3892$

11. $\log x = 0.5988$

12. $\log x = 2.6117$

13. $\log x = 2.9818$

14. $\log x = 3.8751$

15. $\log x = 0.6160 - 2$

16. $\log x = 0.6284 - 3$

17. $\log x = 8.6599 - 10$

18. $\log x = 7.9020 - 10$

▼ Change each of the following numbers to a form that gives a characteristic and a *positive* mantissa.

19. -2.9031

20. -3.4698

21. -3.1752

22. -1.2557

23. -2.4112

24. -5.2388

▼ Evaluate each of the following using logarithms.

25. $(255)(12600)$

26. $\dfrac{78500}{342}$

27. $(1.09)^{25}$

28. $(3.62)^5$

29. $(4.6)(3.81)^4$

30. $(2.06)^4(3.01)^7$

31. $\sqrt[3]{406,000}$

32. $\sqrt[4]{20,600}$

33. $\sqrt[3]{(4.68)^5}$

34. $\dfrac{(2340)(2.61)}{425}$

9.5 Solutions to Exponential Equations and Change of Base

▮ ▮ ▮ ▮ ▮

Solving Exponential Equations

• • • • •

An **exponential equation** is one in which the variable appears in an exponent. One technique for solving such equations is to take the logarithm of both sides of the equation and apply the power rule. This removes the variable from the exponent.

EXAMPLE 1   Solve for x: $7^x = 25$.

SOLUTION

$$7^x = 25$$

$$\log 7^x = \log 25 \qquad \text{Take the log of both sides.}$$

$$x \cdot \log 7 = \log 25 \qquad \text{Apply the power rule.}$$

$$x(0.8451) = 1.3979 \qquad \text{Find the logs.}$$

$$x = \frac{1.3979}{0.8451} \qquad \text{Solve for } x.$$

$$= 1.6541$$

EXAMPLE 2   Solve for x: $10^{3x-1} = 25$.

SOLUTION

$$10^{3x-1} = 25$$

$$\log 10^{3x-1} = \log 25 \qquad \text{Take the log of both sides.}$$

$$(3x-1)\log 10 = \log 25 \qquad \text{Apply the power rule.}$$

$$(3x-1)(1) = 1.3979 \qquad \text{Find the logs.}$$

$$3x = 2.3979 \qquad \text{Solve for } x.$$

$$x = 0.7993$$

Changing the Base of a Logarithm

• • • • •

Bases other than 10 occur frequently in science, engineering, and mathematics. The number e is an irrational number that has an approximate value of 2.71828. The base e is frequently used in calculus. Logarithms to the base e are called **natural logarithms** (usually denoted **ln**). Although 10 and e are the most frequently used bases, any positive number except 1 can be used as a base. Consider the following:

$$\text{Let } u = \log_a x$$

$$\text{then} \quad a^u = x$$

If we take the log to the base b of both sides of the second equation, watch what happens:

$$\log_b a^u = \log_b x$$

$$u \log_b a = \log_b x \qquad \text{Apply the power rule.}$$

$$u = \frac{\log_b x}{\log_b a}$$

$$\log_a x = \frac{\log_b x}{\log_b a} \qquad \text{Substitute } u = \log_a x.$$

This formula allows us to change from base a to base b.

> If a and b are positive numbers not equal to 1 and $x > 0$, then
>
> $$\log_a x = \frac{\log_b x}{\log_b a}$$

EXAMPLE 3   Find $\log_5 25$.

SOLUTION   Since we cannot find base 5 logarithms from a table, we will write the base 5 log in terms of a base 10 log, which we *can* find in the table.

$$\log_5 25 = \frac{\log_{10} 25}{\log_{10} 5}$$

$$= \frac{1.3979}{0.6990}$$

$$\log_5 25 = 2$$

This particular solution can be checked by writing $\log_5 25 = 2$ in exponential form: $5^2 = 25$.   ▮▮

EXAMPLE 4   Find $\log_8 625$.

SOLUTION

$$\log_8 625 = \frac{\log_{10} 625}{\log_{10} 8}$$

$$= \frac{2.7956}{0.9031}$$

$$\log_8 625 = 3.0956$$

9.5   Exercises

▼ Solve for x:

1. $4^x = 25$

2. $6^x = 30$

3. $3^x = 40$

4. $10^x = 200$

5. $10^{2x} = 100$

6. $5^{2x} = 100$

7. $8^{x-1} = 20$

8. $9^{x-1} = 3$

9. $2^{2x-2} = 8$

10. $3^{2x-1} = 27$

11. $5^{-x} = 14$

12. $10^{-x} = 25$

▼ Use the change-of-base formula to find each of the given logarithms.

13. $\log_5 12$

14. $\log_6 15$

15. $\log_2 10$

16. $\log_3 20$

17. $\log_{27} 9$

18. $\log_{16} 4$

19. $\log_8 9$

20. $\log_9 8$

21. $\log_{12} 30$

22. $\log_5 225$

23. $\log_6 225$

24. $\log_7 225$

9.6   Applications of Logarithms

▮ ▮ ▮ ▮ ▮

Population Growth

• • • • •

EXAMPLE 1   A city with a present population of 25,000 is growing at the rate of 5% per year. In t years, the population p will be given by

$$p = 25,000(1.05)^t$$

How many years will it take for the town to double its present population?

SOLUTION   Here p equals 50,000 in the formula and we must solve for t.

$$50,000 = 25,000(1.05)^t \qquad \text{Divide both sides by 25,000.}$$
$$2 = (1.05)^t \qquad \text{Take the log of both sides.}$$
$$\log 2 = \log (1.05)^t \qquad \text{Use the power rule.}$$
$$\log 2 = t \cdot \log 1.05 \qquad \text{Find the logs of 2 and 1.05.}$$
$$0.3010 = t \cdot (0.0212) \qquad \text{Solve for } t.$$
$$t = \frac{0.3010}{0.0212}$$
$$t \approx 14.2 \text{ years} \qquad\qquad \blacksquare\blacksquare$$

The pH of a Solution

• • • • •

The acidity or alkalinity of a solution is determined by the concentration of hydrogen ions (H^+) in the solution. The pH of a solution is measured by the formula

$$\text{pH} = -\log (H^+)$$

where a neutral solution such as distilled water has a pH of 7, acids have a pH less than 7, and alkaline solutions have a pH greater than 7.

EXAMPLE 2   Find the pH of normal rainfall in which the concentration of hydrogen ions in the rainwater is $H^+ = 2.5 \times 10^{-6}$.

SOLUTION
$$\text{pH} = -\log (H^+)$$
$$= -\log (2.5 \times 10^{-6})$$
$$= -[0.3979 + (-6)]$$
$$= -0.3979 + 6$$
$$\text{pH} \approx 5.6$$

Therefore normal rainfall has a pH of about 5.6, which is slightly acidic.   ▮▮

EXAMPLE 3   Samples of acid rain have been collected with an ion concentration of $H^+ = 1.25 \times 10^{-3}$. Find the pH of this rainwater.

SOLUTION

$$pH = -\log(H^+)$$
$$= -\log(1.25 \times 10^{-3})$$
$$= -[0.0969 + (-3)]$$
$$= -0.0969 + 3$$
$$pH \approx 2.9$$

(Vinegar has a pH of about 2.9.)   ▌▌

Compound Interest

• • • • •

The equation $A = P\left(1 + \dfrac{r}{n}\right)^{nt}$ is called **the compound interest formula,** where A is the amount present after t years when a principal of P dollars is invested at an interest rate r, compounded n times per year.

EXAMPLE 4   If you invest \$1,000 at a rate of 5% per year compounded quarterly, how much will you have after 25 years?

SOLUTION   Here $P = 1000$, $r = 0.05$, $n = 4$, and $t = 25$. So we have

$$A = P\left(1 + \frac{r}{n}\right)^{nt}$$
$$= 1000\left(1 + \frac{0.05}{4}\right)^{4 \cdot 25}$$
$$= 1000(1.0125)^{100}$$

Taking the log of both sides,

$\log A = \log[(1000)(1.0125)^{100}]$	Apply the product rule.
$= \log 1000 + \log(1.0125^{100})$	Apply the power rule.
$= \log 1000 + 100 \log 1.0125$	Find the logs of 1000 and 1.0125
$= 3 + 100(0.005395)$	Combine.
$= 3.5395$	

Now we take the antilog of both sides:

$$A \approx \$3,463.38$$ ▌▌

EXAMPLE 5   How long will it take \$10,000 to double if it is invested at 6% compounded quarterly?

SOLUTION   We substitute $A = 20{,}000$, $P = 10{,}000$, $r = 0.06$, and $n = 4$ into the formula:

$$A = P\left(1 + \frac{r}{n}\right)^{nt}$$

$$20{,}000 = 10{,}000\left(1 + \frac{0.06}{4}\right)^{4t}$$

$$2 = (1.015)^{4t}$$

$$\log 2 = \log (1.015)^{4t}$$

$$= 4t \log 1.015$$

$$0.3010 = 4t(0.00647)$$

$$t = \frac{0.3010}{4(0.00647)}$$

$$= 11.6 \text{ years}$$

It will take almost 12 years for the money to double.   ▮ ▮

9.6   Exercises

▼ Solve the following.

1. A city of 40,000 people is growing at the rate of 6% per year. The present water system can supply a population of at most 60,000. How many years will it take before this maximum is reached if the population will be $p = 40{,}000(1.06)^t$ in t years?

2. A city in the southwest is growing at the rate of 4% per year. If the current population p is 150,000, what will it be in 25 years if the present rate of growth continues? In this case $p = 150{,}000(1.04)^t$, where t is the time in years.

3. The present world population is approximately 5.5 billion and is growing at the rate of about 2% per year. At the present growth rate, how long will it take for the population p to double? Here p (in billions) $= 5.5(1.02)^t$, where t is the time in years.

4. Using the formula $\mathrm{pH} = -\log H^+$, find the pH of a solution with an ion concentration of $H^+ = 6 \times 10^{-4}$.

5. What is the pH of milk if the ion concentration is $H^+ = 4 \times 10^{-7}$?

6. Find the pH of hydrochloric acid, which contains an ion concentration of $H^+ = 5 \times 10^{-2}$.

7. Students who take a mathematics exam and then are retested each week on the same material forget at the rate of $f(t) = 82 - 12 \log(1 + t)$, where t is the number of weeks elapsed since the first exam.

   a. What will the average score be after 4 weeks have elapsed?

   b. What was the average score on the first exam?

   c. How many weeks will elapse before the average score is 70?

8. If $5000 is invested at an annual rate of 6% compounded quarterly, how much will be in the account after 4 years?

9. How much money will be in an account at the end of 10 years if you start with $12,000 and the annual interest rate is 5% compounded quarterly?

10. How much must be invested at 4% compounded semiannually in order to earn $25,000 at the end of 20 years?

11. How long will it take a $2000 investment to double at an interest rate of 4% compounded annually?

12. How long will it take a $2000 investment to double at an interest rate of 8% compounded annually? Compare this result with that of Exercise 11. Does doubling the interest rate cut the time in half?

SUMMARY—CHAPTER 9

The **exponential equation** $b^y = x$ is equivalent to the **logarithmic equation** $y = \log_b x$.

$6^2 = 36 \Leftrightarrow \log_6 36 = 2$

Properties of logarithms:

1. $\log_b b = 1$

$\log_4 4 = 1$

2. $\log_b 1 = 0$

$\log 1 = 0$

3. $\log_b(u \cdot v) = \log_b u + \log_b v$

$\log_2(3 \cdot 5) = \log_2 3 + \log_2 5$

4. $\log_b\left(\dfrac{u}{v}\right) = \log_b u - \log_b v$

$\log_6\left(\dfrac{3}{5}\right) = \log_6 3 - \log_6 5$

5. $\log_b(u^r) = r \cdot \log_b u$

$\log_7(4^3) = 3 \cdot \log_7 4$

log x means $\log_{10} x$ and is called a **common logarithm.**

To find the common logarithm of a number.

1. Using a scientific calculator, enter the number and press the **LOG** key.

2. Using the table from Appendix A:

   a. Write the number in scientific notation.

   b. Find the log of the number between 1 and 10 in the table (the **mantissa**).

   c. The log of the given number is the sum of the **characteristic** (the power of 10) and the mantissa.

$$\begin{aligned}\log 425 &= \log(4.25 \times 10^2) \\ &= \log 4.25 + \log(10^2) \\ &= 0.6284 + 2 \\ &= 2.6284\end{aligned}$$

To change a logarithm from base a to base b, use the formula

$$\log_a x = \frac{\log_b x}{\log_b a}$$

$$\begin{aligned}\log_5 30 &= \frac{\log_{10} 30}{\log_{10} 5} \\ &= \frac{1.4771}{0.6990} \\ &= 2.1132\end{aligned}$$

REVIEW EXERCISES—CHAPTER 9

▼ Write each equation in logarithmic form.

1. $10^2 = 100$

2. $10^{-3} = \frac{1}{1000}$

3. $\left(\frac{1}{2}\right)^2 = \frac{1}{4}$

4. $\left(\frac{1}{2}\right)^{-2} = 4$

▼ Write each equation in exponential form.

5. $\log_{10} 10{,}000 = 4$

6. $\log_4 64 = 3$

7. $\log_{10} 0.01 = -2$

8. $\log_{27} 9 = \frac{2}{3}$

▼ Solve for x in each of the following:

9. $\log_3 x = 3$

10. $\log_x 32 = 5$

11. $\log_2 8 = x$

12. $\log_x \frac{1}{8} = -2$

▼ Expand using the properties of logarithms.

13. $\log_5 4x$

14. $\log_3 xy^4$

15. $\log_5 \left(\dfrac{3x}{y}\right)$

16. $\log_3 \left(\dfrac{4\sqrt{x}}{y^2}\right)$

▼ Express as a single logarithm.

17. $\log_3 6 + \log_3 4$

18. $\log_2 7 + 2\log_2 3$

19. $\frac{1}{2}\log_{10} 9 - \log_{10} 10$

20. $3\log_4 2 - 2\log_4 2$

▼ Find the following common logarithms.

21. $\log 18.7$

22. $\log 4250$

23. $\log 0.0946$

24. $\log 12$

▼ Solve for x in each of the following equations:

25. $\log x = 1.8751$

26. $\log x = 2.9330$

27. $\log x = 0.5988 - 2$

28. $\log x = 8.4133 - 10$

▼ Evaluate using logarithms.

29. $(4.31)^6$

30. $\sqrt[5]{2197}$

▼ Solve for x:

31. $3^x = 20$

32. $3^{(3x+7)} = 81$

▼ Use the change-of-base formula to find each of the following:

33. $\log_6 18$

34. $\log_5 31$

35. How long will it take a city with a present population of 20,000 to reach a population of 50,000 if it is growing at a rate of 5% per year? [$50{,}000 = 20{,}000(1.05)^t$]

36. Find the pH of a solution with an ion concentration of $H^+ = 3 \times 10^{-3}$ ($pH = -\log H^+$).

ACHIEVEMENT TEST—CHAPTER 9

• • • • • • • ▼

1. Write $5^3 = 125$ in logarithmic form.

   1. _____

2. Write $\log_3 81 = 4$ in exponential form.

   2. _____

3. Solve for x: $\log_4 64 = x$.

   3. _____

4. Expand using the properties of logarithms: $\log_5\left(\dfrac{x^2}{y}\right)$.

   4. _____

5. Express $\log_4 6 + 2\log_4 3$ as a single logarithm.

   5. _____

6. Find $\log 186$.

   6. _____

7. Find $\log 0.113$.

   7. _____

8. Solve for x if $\log x = 1.3979$.

   8. _____

9. Evaluate using logarithms: $\sqrt[3]{496}$.

   9. _____

10. Solve for x: $5^x = 72$.

   10. _____

11. Use a change of base to find $\log_3 12$.

   11. _____

12. How long will it take for \$3000 to triple if it is invested at 5%, compounded quarterly? $\left[A = P\left(1 + \dfrac{r}{n}\right)^{nt}\right]$

   12. _____

CUMULATIVE REVIEW—CHAPTERS 7, 8, 9

• • • • • • • ▼

▼ Solve for x:

1. $2x^2 + 9x - 5 = 0$ 1. _____

2. $\frac{1}{2}x^2 = 6$ 2. _____

3. $x^2 - 6x + 10 = 0$ 3. _____

4. $x^4 - 2x^2 - 8 = 0$ 4. _____

5. $\dfrac{6}{x} - \dfrac{6}{x+1} = 1$ 5. _____

6. $x - 3 = \sqrt{2x - 3}$ 6. _____

7. A fully loaded truck travels a distance of 300 miles from St. Louis, Mo, to Memphis, Tenn. On its return trip the truck is empty and it travels 10 mph faster. If the total time for the round trip is 11 hours, find the speed both going and returning.

8. Graph the parabola $y = -x^2 + 4x - 5$. Identify the vertex and any x-intercepts.

9. Graph the hyperbola $\dfrac{y^2}{9} - \dfrac{x^2}{16} = 1$.

7. _____

_____

8. _____

_____

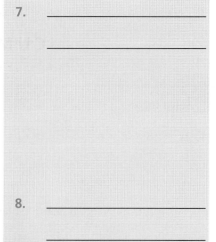

9.

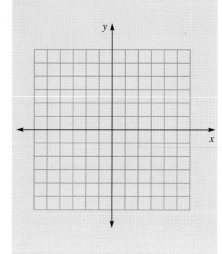

10. Solve the following system of equations using the addition method:

$$5x - 4y = -8$$
$$2x + 3y = 6$$

10. _____

11. Solve the following system of equations using the substitution method:

$$2x + y = 6$$
$$3x - 2y = 16$$

11. _____

12. Solve the given system of equations using Cramer's rule:

$$x - 2y - z = 1$$
$$2x - y - 2z = -1$$
$$x + y + z = 4$$

12. _____

13. An adult movie ticket costs $6.50 and a child's ticket costs $4.00. If a total of 175 tickets are sold, netting a total value of $1050.00, how many of each type of ticket are sold?

13. _____

14. Write $2^{-3} = \frac{1}{8}$ in logarithmic form.

14. _____

15. Write $\log_4 16 = 2$ in exponential form.

15. _____

16. Write $3 \log_{10} x - 2 \log_{10}(x - 3)$ as a single logarithm.

16. _____

17. Solve for x: $5^x = 20$.

17. _____

18. If you invest \$5,000 at a rate of 4% per year compounded quarterly, how much will you have at the end of 10 years?
$$\left[\text{Use } A = p\left(1 + \frac{r}{n}\right)^{nt}. \right]$$

18. _____

Table 1

TABLE 1. COMMON FOUR-PLACE LOGARITHMS OF NUMBERS

n	0	1	2	3	4	5	6	7	8	9
10	0000	0043	0086	0128	0170	0212	0253	0294	0334	0374
11	0414	0453	0492	0531	0569	0607	0645	0682	0719	0755
12	0792	0828	0864	0899	0934	0969	1004	1038	1072	1106
13	1139	1173	1206	1239	1271	1303	1335	1367	1399	1430
14	1461	1492	1523	1553	1584	1614	1644	1673	1703	1732
15	1761	1790	1818	1847	1875	1903	1931	1959	1987	2014
16	2041	2068	2095	2122	2148	2175	2201	2227	2253	2279
17	2304	2330	2355	2380	2405	2430	2455	2480	2504	2529
18	2553	2577	2601	2625	2648	2672	2695	2718	2742	2765
19	2788	2810	2833	2856	2878	2900	2923	2945	2967	2989
20	3010	3032	3054	3075	3096	3118	3139	3160	3181	3201
21	3222	3243	3263	3284	3304	3324	3345	3365	3385	3404
22	3424	3444	3464	3483	3502	3522	3541	3560	3579	3598
23	3617	3636	3655	3674	3692	3711	3729	3747	3766	3784
24	3802	3820	3838	3856	3874	3892	3909	3927	3945	3962
25	3979	3997	4014	4031	4048	4065	4082	4099	4116	4133
26	4150	4166	4183	4200	4216	4232	4249	4265	4281	4298
27	4314	4330	4346	4362	4378	4393	4409	4425	4440	4456
28	4472	4487	4502	4518	4533	4548	4564	4579	4594	4609
29	4624	4639	4654	4669	4683	4698	4713	4728	4742	4757
30	4771	4786	4800	4814	4829	4843	4857	4871	4886	4900
31	4914	4928	4942	4955	4969	4983	4997	5011	5024	5038
32	5051	5065	5079	5092	5105	5119	5132	5145	5159	5172
33	5185	5198	5211	5224	5237	5250	5263	5276	5289	5302
34	5315	5328	5340	5353	5366	5378	5391	5403	5416	5428
35	5441	5453	5465	5478	5490	5502	5514	5527	5539	5551
36	5563	5575	5587	5599	5611	5623	5635	5647	5658	5670
37	5682	5694	5705	5717	5729	5740	5752	5763	5775	5786
38	5798	5809	5821	5832	5843	5855	5866	5877	5888	5899
39	5911	5922	5933	5944	5955	5966	5977	5988	5999	6010
40	6021	6031	6042	6053	6064	6075	6085	6096	6107	6117
41	6128	6138	6149	6160	6170	6180	6191	6201	6212	6222
42	6232	6243	6253	6263	6274	6284	6294	6304	6314	6325
43	6335	6345	6355	6365	6375	6385	6395	6405	6415	6425
44	6435	6444	6454	6464	6474	6484	6493	6503	6513	6522
45	6532	6542	6551	6561	6571	6580	6590	6599	6609	6618
46	6628	6637	6646	6656	6665	6675	6684	6693	6702	6712
47	6721	6730	6739	6749	6758	6767	6776	6785	6794	6803
48	6812	6821	6830	6839	6848	6857	6866	6875	6884	6893
49	6902	6911	6920	6928	6937	6946	6955	6964	6972	6981
50	6990	6998	7007	7016	7024	7033	7042	7050	7059	7067
51	7076	7084	7093	7101	7110	7118	7126	7135	7143	7152
52	7160	7168	7177	7185	7193	7202	7210	7218	7226	7235
53	7243	7251	7259	7267	7275	7284	7292	7300	7308	7316
54	7324	7332	7340	7348	7356	7364	7372	7380	7388	7396

n	0	1	2	3	4	5	6	7	8	9
55	7404	7412	7419	7427	7435	7443	7451	7459	7466	7474
56	7482	7490	7497	7505	7513	7520	7528	7536	7543	7551
57	7559	7566	7574	7582	7589	7597	7604	7612	7619	7627
58	7634	7642	7649	7657	7664	7672	7679	7686	7694	7701
59	7709	7716	7723	7731	7738	7745	7752	7760	7767	7774
60	7782	7789	7796	7803	7810	7818	7825	7832	7839	7846
61	7853	7860	7868	7875	7882	7889	7896	7903	7910	7917
62	7924	7931	7938	7945	7952	7959	7966	7973	7980	7987
63	7993	8000	8007	8014	8021	8028	8035	8041	8048	8055
64	8062	8069	8075	8082	8089	8096	8102	8109	8116	8122
65	8129	8136	8142	8149	8156	8162	8169	8176	8182	8189
66	8195	8202	8209	8215	8222	8228	8235	8241	8248	8254
67	8261	8267	8274	8280	8287	8293	8299	8306	8312	8319
68	8325	8331	8338	8344	8351	8357	8363	8370	8376	8382
69	8388	8395	8401	8407	8414	8420	8426	8432	8439	8445
70	8451	8457	8463	8470	8476	8482	8488	8494	8500	8506
71	8513	8519	8525	8531	8537	8543	8549	8555	8561	8567
72	8573	8579	8585	8591	8597	8603	8609	8615	8621	8627
73	8633	8639	8645	8651	8657	8663	8669	8675	8681	8686
74	8692	8698	8704	8710	8716	8722	8727	8733	8739	8745
75	8751	8756	8762	8768	8774	8779	8785	8791	8797	8802
76	8808	8814	8820	8825	8831	8837	8842	8848	8854	8859
77	8865	8871	8876	8882	8887	8893	8899	8904	8910	8915
78	8921	8927	8932	8938	3943	8949	8954	8960	8965	8971
79	8976	8982	8987	8993	8998	9004	9009	9015	9020	9025
80	9031	9036	9042	9047	9053	9058	9063	9069	9074	9079
81	9085	9090	9096	9101	9106	9112	9117	9122	9128	9133
82	9138	9143	9149	9154	9159	9165	9170	9175	9180	9186
83	9191	9196	9201	9206	9212	9217	9222	9227	9232	9238
84	9243	9248	9253	9258	9263	9269	9274	9279	9284	9289
85	9294	9299	9304	9309	9315	9320	9325	9330	9335	9340
86	9345	9350	9355	9360	9365	9370	9375	9380	9385	9390
87	9395	9400	9405	9410	9415	9420	9425	9430	9435	9440
88	9445	9450	9455	9460	9465	9469	9474	9479	9484	9489
89	9494	9499	9504	9509	9513	9518	9523	9528	9533	9538
90	9542	9547	9552	9557	9562	9566	9571	9576	9581	9586
91	9590	9595	9600	9605	9609	9614	9619	9624	9628	9633
92	9638	9643	9647	9652	9657	9661	9666	9671	9675	9680
93	9685	9689	9694	9699	9703	9708	9713	9717	9722	9727
94	9731	9736	9741	9745	9750	9754	9759	9763	9768	9773
95	9777	9782	9786	9791	9795	9800	9805	9809	9814	9818
96	9823	9827	9832	9836	9841	9845	9850	9854	9859	9863
97	9868	9872	9877	9881	9886	9890	9894	9899	9903	9908
98	9912	9917	9921	9926	9930	9934	9939	9943	9948	9952
99	9956	9961	9965	9969	9974	9978	9983	9987	9991	9996

APPENDIX B

Right Triangle Trigonometry

Basic Definitions

Many problems encountered in mathematics and science require the ability to "solve" right triangles.

Recall from your work in proportion that ratios of corresponding parts of similar triangles are equal. Consider the two similar triangles in Figures B.1 and B.2 (same shape but different size):

FIGURE B.1.

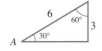

FIGURE B.2.

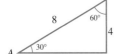

The ratios $\frac{3}{6}$ and $\frac{4}{8}$ are equal since they are both equal to $\frac{1}{2}$. The ratio of the corresponding sides in any other triangle similar to the given triangles will also equal $\frac{1}{2}$. Ratios are tabulated for angles of any right triangle (see Figure B.3) and are given in Table 2 in Appendix C. The ratios are named as follows:

$$\begin{array}{c} \text{sine } A \\ (\sin A) \end{array} = \frac{\text{opposite side}}{\text{hypotenuse}} = \frac{a}{c}$$

$$\begin{array}{c} \text{cosine } A \\ (\cos A) \end{array} = \frac{\text{adjacent side}}{\text{hypotenuse}} = \frac{b}{c}$$

$$\begin{array}{c} \text{tangent } A \\ (\tan A) \end{array} = \frac{\text{opposite side}}{\text{adjacent side}} = \frac{a}{b}$$

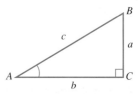

FIGURE B.3.

a = side opposite A
b = side adjacent to A
c = hypotenuse (longest side)

EXAMPLE 1

Find the sin, cos, and tan of angle A in Figure B.4.

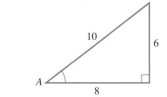

FIGURE B.4.

SOLUTION

$$\sin A = \frac{6}{10} = \frac{3}{5} = 0.6$$

$$\cos A = \frac{8}{10} = \frac{4}{5} = 0.8$$

$$\tan A = \frac{6}{8} = \frac{3}{4} = 0.75$$

The measure of angle A can be found with the aid of a scientific calculator. To start, make sure that the calculator is set in the *degree* mode. Key in 0.6 and press the **SIN^{-1}** key or, on some calculators, the sequence **INV** **SIN**. The measure of the angle, 36.869898, should appear in the display. Rounding to the nearest degree gives us 37°.

This result can also be obtained by using 0.8 with

COS⁻¹ or 0.75 with **TAN⁻¹** . Try it: if your calculator is slightly different, or if you have trouble, consult the instruction booklet that comes with it.

We can also find the measure of angle A by looking up any of the three ratios in Table 2, which is found in Appendix C. For example, the sine of what angle equals 0.6? The closest value in the table is $\sin 37° = 0.6018$, so $A \approx 37°$. Check $\cos 37°$ and $\tan 37°$.

Solving a Triangle.

To **solve a triangle,** we find the measures of the sides and the angles that are *not* known by using the sides and angles that *are* known.

EXAMPLE 2

Find side a in the triangle in Figure B.5.

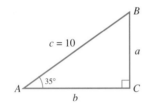

FIGURE B.5.

SOLUTION

The length of the *hypotenuse* is given ($c = 10$) and we want to find the length of the side *opposite* (side a) the known angle *(A)*. The trigonometric ratio involving the hypotenuse and the opposite side is the sine.

$$\sin A = \frac{\text{opposite}}{\text{hypotenuse}} = \frac{a}{c}$$

$$\sin 35° = \frac{a}{10} \qquad \text{Substitute } A = 35°, c = 10.$$

$$0.5736 = \frac{a}{10} \qquad \text{Find sin 35° using your calculator or Table 2.}$$

$$a = 5.736 \qquad \text{Multiply by 10.} \qquad \blacksquare\blacksquare$$

EXAMPLE 3

Find side a in the triangle pictured in Figure B.6.

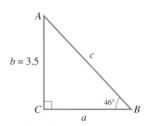

FIGURE B.6.

SOLUTION

The length of the opposite side is given ($b = 3.5$) and we are to find the length of the adjacent side *(a)*. The trigonometric ratio involving these two sides is the tangent.

$$\tan B = \frac{\text{opposite}}{\text{adjacent}} = \frac{b}{a}$$

$$\tan 46° = \frac{3.5}{a} \qquad \text{Substitute } B = 46°, b = 3.5.$$

$$1.0355 = \frac{3.5}{a} \qquad \text{Find tan 46°.}$$

$$a = \frac{3.5}{1.0355} = 3.38 \qquad\qquad \blacksquare\blacksquare$$

EXAMPLE 4

A ladder 20 ft long touches a wall at a point 16 ft above the ground (see Figure B.7). What angle does the ladder make with the ground?

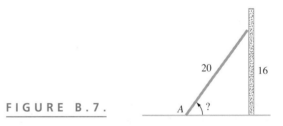

FIGURE B.7.

SOLUTION

The opposite side and the hypotenuse are known, so we will use the sine.

$$\sin A = \frac{\text{opposite}}{\text{hypotenuse}}$$

$$= \frac{16}{20} = 0.8$$

Using a calculator to find A, we key in 0.8 and press **SIN⁻¹** , yielding 53.130102 in the display. Rounding to the nearest degree gives us $A = 53°$.

Alternatively, using Table 2 we find that the sine of the angle closest to 0.8 is sin 53° = 0.7986. Therefore $A \approx 53°$. ▮▮

EXAMPLE 5

How high is a building that casts a shadow 130 ft long if the sun makes an angle of 28° with the ground?

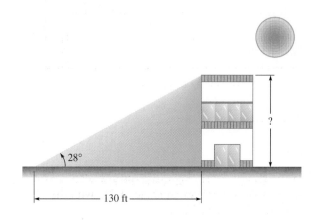

SOLUTION

The tangent is the trigonometric function that relates the opposite side (the height of the building) with the adjacent side (the length of the shadow). This gives us the equation

$$\tan 28° = \frac{h}{130} \qquad \tan = \frac{\text{opposite}}{\text{adjacent}}$$

$$0.5317 = \frac{h}{130} \qquad \text{Find } \tan 28°.$$

$$(0.5317)(130) = h \qquad \text{Multiply both sides by 130.}$$

$$69.12 = h$$

The building is approximately 69 ft high. ▮▮

FIGURE B.8.

EXERCISES: APPENDIX B

▼ In Exercises 1–6, sketch the right triangle *ABC* and find the unknown part.

1. If side $a = 12$ and angle $A = 35°$, find side b.

2. If side $b = 40$ and angle $A = 36°$, find side c.

3. If side $a = 20$ and angle $B = 30°$, find side b.

4. If side $c = 13$ and side $b = 5$, find angles A and B.

5. If angle $B = 30°$ and side $a = 10$, find side b.

6. If angle $A = 30°$ and side $a = 20$, find sides b and c.

▼ In Exercises 7–12, solve each of the given right triangles. (This means to find *all* of the unknown sides and angles.) In every case, angle C is the right angle.

7. angle $B = 74°$, side $b = 125$

8. angle $A = 60°$, side $b = 40$

9. side $a = 20$, side $c = 50$

10. side $b = 2.5$, side $c = 8$

11. angle $B = 12.1°$, side $b = 7$

12. angle $A = 57.2°$, side $c = 175$

13. Maria Hernandez stands at a point 800 ft from a point directly beneath a hot-air balloon. If she measures an angle of elevation of 35.2°, how high is the balloon? See Figure B.9.

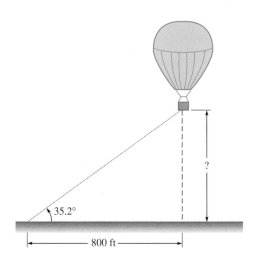

FIGURE B.9.

14. A vertical tower 130 ft high casts a shadow 200 ft long. What angle does the sun make with the ground? See Figure B.10.

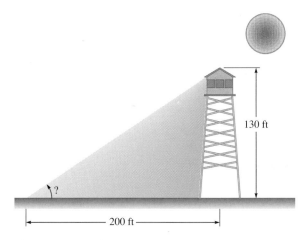

FIGURE B.10.

15. A TV camera is to be mounted in a store such that it is aimed directly at the cashier. Find the angle that it makes with the wall if it is mounted 7 ft above the floor and the wall is 20 ft from the cashier, as pictured in Figure B.11.

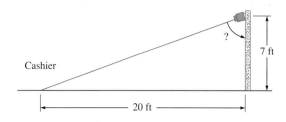

FIGURE B.11.

16. How high is a building that casts a shadow 100 ft long if the sun makes an angle of 31° with the ground? See Figure B.12.

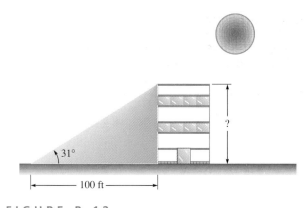

FIGURE B.12.

17. The cloud ceiling is the height of the lowest point of the cloud. To find it, a spotlight is located a specific distance from an airport and is pointed directly upward, and the angle of elevation from the airport to the illumination on the bottom of the cloud is measured (see Figure B.13.). From this information the cloud ceiling can be determined. Find the cloud ceiling if the spotlight is 1.5 miles from the airport and the angle of elevation is 37°.

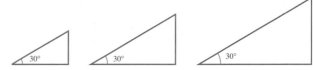

18. Why does sin 30° = $\frac{1}{2}$ regardless of the size of the right triangle? See Figure B.14.

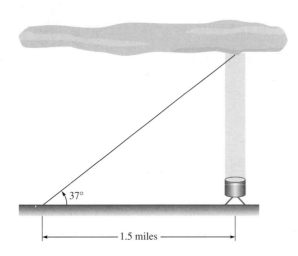

FIGURE B.13.

FIGURE B.14.

Table 2

TABLE 2.   TRIGONOMETRIC FUNCTIONS

Angle	Sine	Cosine	Tangent
0°	0.0000	1.0000	0.0000
1	.0175	.9998	.0175
2	.0349	.9994	.0349
3	.0523	.9986	.0524
4	.0698	.9976	.0699
5	.0872	.9962	.0875
6	.1045	.9945	.1051
7	.1219	.9925	.1228
8	.1392	.9903	.1405
9	.1564	.9877	.1584
10	.1736	.9848	.1763
11	.1908	.9816	.1944
12	.2079	.9781	.2126
13	.2250	.9744	.2309
14	.2419	.9703	.2493
15	.2588	.9659	.2679
16	.2756	.9613	.2867
17	.2924	.9563	.3057
18	.3090	.9511	.3249
19	.3256	.9455	.3443
20	.3420	.9397	.3640
21	.3584	.9336	.3839
22	.3746	.9272	.4040
23	.3907	.9205	.4245
24	.4067	.9135	.4452
25	.4226	.9063	.4663
26	.4384	.8988	.4877
27	.4540	.8910	.5095
28	.4695	.8829	.5317
29	.4848	.8746	.5543
30	.5000	.8660	.5774
31	.5150	.8571	.6009
32	.5299	.8480	.6249
33°	.5446	.8387	.6494
34	.5592	.8290	.6745
35	.5736	.8192	.7002
36	.5878	.8090	.7265
37	.6018	.7986	.7536
38	.6157	.7880	.7813
39	.6293	.7771	.8098
40	.6428	.7660	.8391
41	.6561	.7547	.8693
42	.6691	.7431	.9004
43	.6820	.7314	.9325
44	.6947	.7193	.9657
45	.7071	.7071	1.0000

Angle	Sine	Cosine	Tangent
46°	.7193	.6947	1.0355
47	.7314	.6820	1.0724
48	.7431	.6691	1.1106
49	.7547	.6561	1.1504
50	.7660	.6428	1.1918
51	.7771	.6293	1.2349
52	.7880	.6157	1.2799
53	.7986	.6018	1.3270
54	.8090	.5878	1.3764
55	.8192	.5736	1.4281
56	.8290	.5592	1.4826
57	.8387	.5446	1.5399
58	.8480	.5299	1.6003
59	.8572	.5150	1.6643
60	.8660	.5000	1.7321
61	.8746	.4848	1.8040
62	.8829	.4695	1.8807
63	.8910	.4540	1.9626
64	.8988	.4384	2.0503
65	.9063	.4226	2.1445
66	.9135	.4067	2.2460
67	.9205	.3907	2.3559
68	.9272	.3746	2.4751
69	.9336	.3584	2.6051
70	.9397	.3420	2.7475
71	.9455	.3256	2.9042
72	.9511	.3090	3.0777
73	.9563	.2924	3.2709
74	.9613	.2756	3.4874
75	.9660	.2588	3.7321
76	.9703	.2419	4.0108
77	.9744	.2250	4.3315
78	.9781	.2079	4.7046
79	.9816	.1908	5.1446
80°	.9848	.1736	5.6713
81	.9877	.1564	6.3138
82	.9903	.1392	7.1154
83	.9925	.1219	8.1443
84	.9945	.1045	9.5144
85	.9962	.0872	11.4301
86	.9976	.0698	14.3007
87	.9986	.0523	19.0811
88	.9994	.0349	28.6363
89	.9998	.0175	57.2900
90°	1.0000	0.0000	—

Using a Calculator

Different Types of Calculators

There are two major types of calculators: those that use *algebraic logic* and those that use *reverse Polish notation (RPN)*. Calculators that use RPN circuitry are usually identified by the presence of an **ENTER** key and the absence of an **=** key. In our discussion we will assume that your calculator is an algebraic logic calculator. If your calculator is the RPN type, the best help will probably be the instruction book provided with the calculator.

Order of Operations

Some less expensive calculators do not obey the correct order of operations. An easy experiment will tell you whether or not your calculator has the correct order of operations built into it. Key in the following sequence on your calculator:

$$3 \quad + \quad 4 \quad \times \quad 5 \quad =$$

If the answer in the display is *23*, then your calculator will multiply before it adds, which is the correct order of operations. If the answer in the display is *35*, then the calculator added the 3 + 4 before multiplying by 5, which is not the proper order of operations. In this appendix we will assume that your calculator automatically uses the correct order of operations. If your calculator is not of this type, then you must be sure that you enter the operations in the correct order.

▍ **Number of Digits** Not all calculators show the same number of digits in the display. If your display shows fewer digits than are indicated in our work that follows, round

off our answer to the same number of places shown in your display. The answers should then be the same.

One-Step Operations on the Calculator

Because everyone sooner or later presses a wrong button on the calculator, it is important first to estimate your answer whenever possible.

EXAMPLE 1

Add: 184.7 + 409.8.

SOLUTION

Estimate   200 + 400 = 600

Calculator   184.7 **+** 409.8 **=**   594.5

As you can see, our estimate is close to the answer in the display on our calculator.   ▮▮

EXAMPLE 2

Multiply: 46.31 × 72.08.

SOLUTION

Estimate   50 × 70 = 3500

Calculator   46.31 **×** 72.08 **=**   3338.0248

Some answers are not so easy to estimate, so be extra careful not to press any wrong buttons.   ▮▮

EXAMPLE 3

Evaluate the following:

	Problem	Key-In					Display
(a)	$38.26 - 14.56$	38.26	−	14.56	=		23.7
(b)	$(15.96)^2$	15.96	X²				254.7216
(c)	$279.4 \div 36$	279.4	÷	36	=		7.7611111
(d)	$\dfrac{348.91}{6.53}$	348.91	÷	6.53	=		53.431853
(e)	$91.6 - 115.8$	91.6	−	115.8	=		−24.2
(f)	$(-14.8)(9.4)$	14.8	±	×	9.4	=	−139.12

The ± key changes the sign of any number, so we use it to obtain -14.8. It is pressed *after* the number is keyed in.   ▮▮

▌ **Clearing An Incorrect Entry** Most calculators allow you to clear the last entry without having to start the problem over from the beginning. You may need to read your instruction booklet or experiment with the keys.

If your calculator has keys	To clear the last entry	To clear the entire problem
CE/C	CE/C once	CE/C twice
C and AC	C	AC
C and CE	CE	C

Study the following examples carefully to see how to change incorrect entries. Try them on your own calculator.

D.1   EXERCISES

▼ Do the following problems on your calculator. Also estimate your answer where applicable.

1. $3,264 + 5,986$ **2.** $6,158 + 9,946$

3. $16.236 + 18.91$ **4.** $13.66 + 33.104$

5. $13,069 - 5,998$ **6.** $68,982 - 53,114$

7. $19,216 - 39,427$ **8.** $8,436 - 26,555$

EXAMPLE 5

To correct a number that has been incorrectly entered, use the clear-entry key.

Multiply: 224×18

SOLUTION

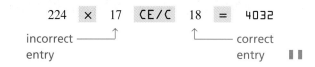

▮▮

EXAMPLE 6

To correct an operation that has been entered by mistake, simply re-enter the correct operation.

Multiply: 224×18

SOLUTION

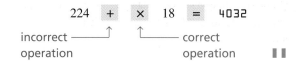

▮▮

EXAMPLE 7

To correct an error made early in the problem, you must clear the entire problem.

Multiply: 224×18

SOLUTION

▮▮

9. $3.201 - 14.823$ **10.** $7.426 - 13.009$

11. 68×43 **12.** 91×86

13. 9.63×21.6 **14.** 13.87×92.4

15. 268.5×199.1 **16.** 183.9×696.4

17. $7214 \div 13$ **18.** $1921 \div 46$

19. $23.6 \div 72.4$

20. $18.8 \div 38.4$

21. $\dfrac{248.9}{17.1}$

22. $\dfrac{311.6}{28.2}$

23. $\dfrac{19.3}{36.5}$

24. $\dfrac{11.1}{84.6}$

25. $(-62)(34.8)$

26. $(-26)(41.7)$

27. $(-32)(-447)$

28. $(-94)(-188)$

29. $\sqrt{1066}$

30. $\sqrt{342}$

31. $\sqrt{857}$

32. $\sqrt{921}$

33. $\sqrt{8850}$

34. $\sqrt{4127}$

35. $\sqrt{21.6}$

36. $\sqrt{9.86}$

37. 21^2

38. 16^2

39. 268^2

40. 139^2

41. $(31.42)^2$

42. $(63.78)^2$

43. $(0.4)^2$

44. $(0.6)^2$

Multi-step Operations

▮ **Memory** The **memory** in a calculator allows the calculator to store a number for use later on. Some common memory keys are listed below. If your calculator has a different key than those listed, check your instruction booklet.

Key	What it does
MIN or STO	Puts the displayed number into the memory.
MR	Recalls the number from memory for your use.
CM or AC	Erases the number from the memory.
M+	Adds the displayed number to the number in the memory.
M−	Subtracts the displayed number from the number in the memory.
RM or MRC CM	Pressed once, it recalls the number from the memory for your use. Pressed twice, it clears the number from the memory.

Experiment with the memory keys on your calculator and then use your calculator to follow along with the examples.

EXAMPLE 1

Find $\dfrac{62.4}{17 + 14.7}$.

SOLUTION

Since we are dividing 62.4 by the *entire* denominator, we evaluate the denominator first.

17 + 14.7 = `31.7` STO   Store 31.7 in the memory.

62.4 ÷ RCL = `1.9684543`   Divide 62.4 by 31.7, which is in the memory.

▮▮

EXAMPLE 2

Mrs. Serio went to the grocery store with $20.00 and bought two bottles of shampoo at $2.79 each, two dozen eggs at $1.23 per dozen, and six cans of tomatoes at $1.29 per can. Use your calculator to keep a running record of how much money she has left after each item.

SOLUTION

20 STO   Place $20 in storage.

2 × 2.79 M− `5.58`   The cost of the shampoo is $5.58.

MRC `14.42`   There is $14.42 left in memory.

2 × 1.23 M− `2.46`   The cost of the eggs is $2.46.

MCR `11.96`   There is $11.96 left of the original $20.

6 × 1.29 M− `7.74`   The tomatoes cost $7.74

MCR `4.42`   The change from the $20 will be $4.42. ▮▮

Order of Operations and Parentheses You must be especially careful when more than one operation is involved in a calculation. Regardless of the type of calculator you use, if you are careful about which operation is being done first, you will obtain the correct result. Always treat what appears inside parentheses as a single unit.

EXAMPLE 3

Find $(16 + 23) \times 14$.

SOLUTION

16 $\boxed{+}$ 23 $\boxed{=}$ 39   To add first, key in $\boxed{=}$.

$\boxed{\times}$ 14 $\boxed{=}$ 546   Now multiply.

EXAMPLE 4

Find 4^3.

SOLUTION

4 $\boxed{Y^X}$ 3 $\boxed{=}$ 64 $\boxed{Y^X}$ raises numbers to powers.

EXAMPLE 5

Find $\sqrt{79 - 46}$.

SOLUTION

79 $\boxed{-}$ 46 $\boxed{=}$ 33   Evaluate $79 - 46$ first.

$\boxed{\sqrt{}}$ 5.7445626   Find the square root.

Note that if you key in the sequence

79 $\boxed{-}$ 46 $\boxed{\sqrt{}}$ $\boxed{=}$

the result is 72.21767. Can you see why an incorrect result is obtained? Since the calculator performed the square root before it subtracted, we obtained $79 - \sqrt{46}$, which is not what we want to do.

EXAMPLE 6

$42 \div 11 - (156 - 118)$

SOLUTION

Work inside the parentheses first and store the result in the memory.

156 $\boxed{-}$ 118 $\boxed{=}$ 38 $\boxed{MIN}$ 38 is in the memory.

42 $\boxed{\div}$ 11 $\boxed{=}$ 3.18181818   Perform the division.

$\boxed{-}$ $\boxed{MR}$ $\boxed{=}$ -34.181818   Subtract what is in the memory.

The result is -34.181818.

D.2   EXERCISES

▼ Perform the following calculations.

1. $16 \times 24 + 56$

2. $17 \times 35 + 18$

3. $32 \times 18 - 72$

4. $46 \times 6 - 112$

5. $\sqrt{76 + 44}$

6. $\sqrt{36 + 58}$

7. $\sqrt{36 - 12}$

8. $\sqrt{114 - 75}$

9. 6^3

10. 11^3

11. 5^4

12. 6^4

13. 9^5

14. 8^6

15. $(3.2)^3$

16. $(6.1)^3$

17. $(9.46)^4$

18. $(3.86)^4$

19. $(14 - 9) \div (36 - 8)$

20. $(15 - 4) \div (56 - 19)$

21. $(36 + 4) \times (46 - 32)$

22. $(52 + 18) \times (76 - 52)$

23. $72.4 \div (13 + 52)$

24. $46.1 \div (75 - 8)$

25. $(26.4 - 7.6) \div 9.7$

26. $(58.1 - 13.7) \div 5.8$

27. $\dfrac{(14)(36)}{(75)(21)}$

28. $\dfrac{(24)(85)}{(13)(76)}$

29. $\dfrac{(23)(19)(52)}{(71)(43)}$

30. $\dfrac{(15)(77)(18)}{(52)(12)}$

31. $15(4^3)$

32. $17(5^4)$

33. $(-6.3)(2.38)$

34. $(-9.1)(5.61)$

35. $(-3.4)^3$

36. $(-8.7)^3$

37. $\dfrac{(14.7)(6.8)^2}{15.2}$

38. $\dfrac{(42.1)(3.7)^3}{42.8}$

39. $\dfrac{24\pi}{16}$

40. $\dfrac{32\pi}{18}$

NAME

▮ ▮ ▮ ▮ ▮ CLASS

ADDITIONAL CALCULATOR EXERCISES

▼ Estimate the answers. Then evaluate on your calculator.

1. $7,846 + 13,062$

      1. _____

           _____

2. $65.8 - 91.7$

      2. _____

           _____

3. $(189.1)(312.4)$

      3. _____

           _____

4. $7592 \div 811$

      4. _____

           _____

5. $(-624)(39)$

      5. _____

           _____

6. $\sqrt{3721}$

      6. _____

           _____

7. $\sqrt{15.25}$

      7. _____

           _____

8. $(21.4)^2$

      8. _____

           _____

▼ Evaluate the following on your calculator.

9. $46 \times 32 + 23$

      9. _____

10. $21 \times 76 - 225$

    10. _____

11. $\sqrt{785 - 521}$

    11. _____

12. 3^5

    12. _____

13. $(8.45)^4$

    13. _____

14. $75.8 \div (52.1 - 18.7)$ 14. _____

15. $(-5.7)^3$ 15. _____

16. $(-26.3)(73.6)$ 16. _____

17. $\dfrac{(2.01)(3.2)^2}{15.2}$ 17. _____

18. Karen Pawlitz has been offered a 5-year contract as a radiological technician   18. _____
    with a starting salary of $28,500. Under the terms of her contract, she is to
    receive an increase of $2,450 per year at the end of each year. What will her
    salary be at the end of the fifth year?

19. Gasoline costs $1.39/gal. It took 14.36 gal to fill Steve's tank. How much   19. _____
    did it cost for the gasoline?

20. After Derek left for college, his mother cleaned his room. She found 23   20. _____
    quarters, 41 dimes, 83 nickels, and 166 pennies. What is the total value of
    the money she found?

Graph Paper

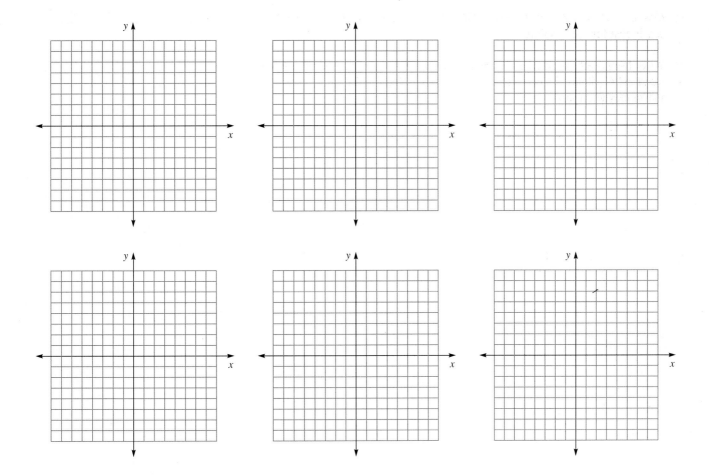

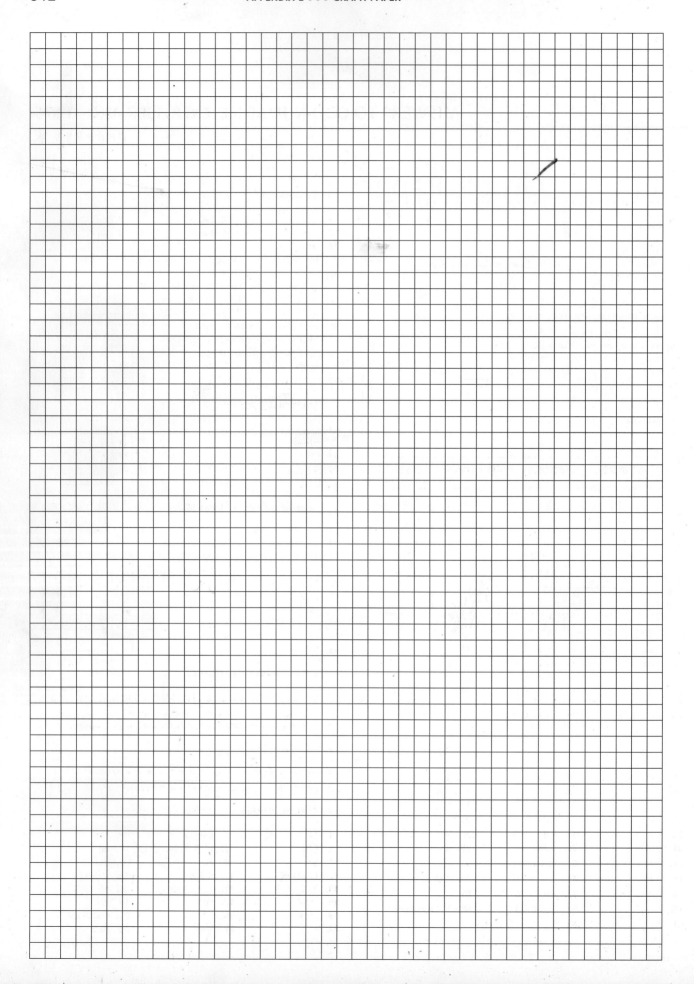

Chapter 1

■ ■ Exercise 1.1

1. $\{6, 7, 8, 9\}$
3. {New Hampshire, New Jersey, New Mexico, New York}
5. $\{6, 8, 10, 12, . . .\}$
7. $\varnothing$
9. {W, a, s, h, i, n, g, t, o, }
11. finite with cardinal number 70
13. finite with cardinal number 4
15. finite with cardinal number 26
17. true 　　19. true 　　21. true
23. false 　　25. false 　　27. false
29. $\{1, 2, 3, 4, 5, 7, 9\}$ 　31. $\{1, 2, 3, 4, 5, 6, 7\}$ 　33. $\varnothing$
35. $\{6, 7\}$ 　　37. $\varnothing$

■ ■ Exercise 1.2

	Number	Rational	Irrational	Real
1.	3	X		X
3.	π		X	X
5.	$\sqrt{9}$	X		X
7.	$-\sqrt{3}$		X	X
9.	0	X		X
11.	$0.\overline{62}$	X		X

13. false 　　15. false 　　17. true
19. false 　　21. true 　　23. true
25. false 　　27. false 　　29. true
31. $2 \cdot 7$ 　　33. $2 \cdot 2 \cdot 3$ 　　35. $2 \cdot 2 \cdot 2 \cdot 2 \cdot 2$
37. prime 　　39. $5 \cdot 17$ 　　41. $2 \cdot 2 \cdot 5 \cdot 5$
43. $3 \cdot 3 \cdot 5 \cdot 5$ 　45. $2 \cdot 3 \cdot 3 \cdot 7$ 　47. $13 \cdot 17$
49. $2 \cdot 2 \cdot 2 \cdot 2 \cdot 2 \cdot 2 \cdot 3$

■ ■ Exercise 1.3

1. -4 　　3. -10 　　5. 12
7. -5 　　9. 9 　　11. -8
13. 0 　　15. 6 　　17. $\frac{1}{8}$
19. $-\frac{17}{12}$ 　　21. -3 　　23. $-\frac{5}{8}$
25. -5.32 　　27. -8 　　29. -6
31. 0 　　33. -18 　　35. 44
37. -72 　　39. 9.086 　　41. $-\frac{1}{4}$
43. -16 　　45. 48 　　47. 192
49. $\frac{9}{8}$ 　　51. -4 　　53. 1
55. 0 　　57. undefined 　　59. indeterminate
61. 3 　　63. $\frac{3}{2}$

■ ■ Exercise 1.4

1. true 　　3. false 　　5. true
7. true 　　9. false 　　11. false
13. true
15. $6(3 + 5) = 6(8) = 48$ and $6 \cdot 3 + 6 \cdot 5 = 18 + 30 = 48$
17. $-8(4 + 7) = -8 \cdot 11 = -88$ and
　　$(-8)(4) + (-8)(7) = -32 + (-56) = -88$
19. $4.3(1.2 + 0.3) = (4.3)(1.5) = 6.45$ and
　　$(4.3)(1.2) + (4.3)(0.3) = 5.16 + 1.29 = 6.45$
21. multiplicative inverse property
23. multiplicative identity property
25. multiplicative inverse property
27. $(x)(6)$ 　　29. $3x + 12$ 　　31. 0
33. 1 　　35. $\frac{1}{8}$

■ ■ Exercise 1.5

1. 9 　　3. -4 　　5. 36
7. 15 　　9. 1 　　11. 7
13. 15 　　15. 11 　　17. 0
19. 7 　　21. 5 　　23. 0
25. 11 　　27. -1

■ ■ Chapter 1 Review Exercises

1. $\{10, 12, 14, 16\}$
3. 4
5. There are many answers; among them are $\varnothing$ and $\{a\}$.
7. $\{n, p\}$ 　　9. true 　　11. true
13. false 　　15. irrational 　　17. rational
19. $3 \cdot 3 \cdot 5$ 　　21. -13 　　23. -1
25. 12 　　27. -36 　　29. 24
31. 4 　　33. 0 　　35. indeterminate
37. false 　　39. true 　　41. 22
43. -5

■ ■ Critical Thinking Exercises

1. $\frac{22}{7}$ is only an approximation of π, which is irrational. Since $\frac{22}{7}$ is a fraction, it is, by definition, a rational number.
3. Zero is the only whole number that is not a natural number.
5. $\frac{-5}{0}$ is undefined, not 0.
7. 16 is a perfect square, so $\sqrt{16} = 4$, a rational number.
9. Zero is the only whole number that has no multiplicative inverse

■ ■ Chapter 1 Achievement Test

1. $\{4, 8, 12, 16, . . .\}$ 　2. infinite 　　3. 4
4. $\{10\}$ 　　5. $\{5, 8, 9, 10, 11, 12\}$ 　6. no
7. yes 　　8. no 　　9. yes
10. irrational 　　11. rational 　　12. rational
13. $2 \cdot 2 \cdot 3 \cdot 3$ 　14. $2 \cdot 3 \cdot 3 \cdot 5$ 　15. 56
16. -14 　　17. 10 　　18. 26
19. -14 　　20. 0 　　21. undefined
22. -32 　　23. true 　　24. false
25. true

Chapter 2

■ ■ Exercise 2.1

1. $x = 7$	**3.** $x = 8$	**5.** $a = -4$
7. $y = -6$	**9.** $t = -7$	**11.** $y = 2$
13. $x = 8$	**15.** $n = -7$	**17.** $p = 10$
19. $x = -32$	**21.** $x = -\frac{1}{2}$	**23.** $x = 2$
25. $x = 0$	**27.** $x = -2$	**29.** $x = \frac{2}{9}$
31. $x = \frac{4}{5}$		

■ ■ Exercise 2.2

1. $x = 4$	**3.** $x = 8$	**5.** $y = -7$
7. $x = 4$	**9.** $x = \frac{14}{5}$	**11.** $t = 0$
13. $x = 150$	**15.** $x = 0.2$	**17.** $x = 6$
19. $x = 6$	**21.** $y = 21$	**23.** $y = 27$
25. $t = -4$	**27.** $x = -\frac{11}{13}$	**29.** $w = \frac{8}{3}$
31. $x = \frac{29}{2}$		

■ ■ Exercise 2.3

1. contradiction	**3.** identity
5. contradiction	**7.** identity
9. conditional; $z = 1$	**11.** identity

■ ■ Exercise 2.4

1. $y = b - a$	**3.** $y = x - z$
5. $y = \dfrac{z - x}{a}$	**7.** $b = P - a - c$
9. $y = \dfrac{z}{5} - x$	**11.** $b = \dfrac{2A}{h}$
13. $b = \frac{1}{3} - 2a$	**15.** $x = -2y$
17. $P = \dfrac{CT}{V}$	**19.** $y = \frac{4}{3}z - \frac{2}{3}x$
21. $w = \dfrac{V}{lh}$	**23.** $m = \dfrac{y - y_1}{x - x_1}$
25. $r = \dfrac{I - P}{Pt}$	

■ ■ Exercise 2.5

1. 4	**3.** 36
5. 7	**7.** 16
9. 21 and 42	**11.** 12 ft and 28 ft
13. 4.8	**15.** 13 females and 17 males
17. 6 ft × 7 ft	**19.** 50 and 30
21. $180	**23.** 10 ft × 22 ft
25. 28, 30, and 32	**27.** 8 and 10
29. 29, 31, and 33	**31.** 8, 10, and 12
33. 27, 29, and 31	
35. any three consecutive even integers	
37. 28, 30, and 32	**39.** There are none.
41. 8, 10, 12, 14, and 16	**43.** −39, −37, and −35
45. 88	**47.** 87
49. 7	**51.** 6 miles
53. 596 miles/day; 630 miles/day	

■ ■ Exercise 2.6

1. $x > -2$

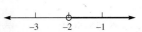

3. $x < 2$

5. $x > -3$

7. $x > -1$

9. $x < 5$

11. $x < 13$

13. $x > 3$

15. $b \leq 6$

17. $y < 40$

19. $x \leq -6$

21. $x > -16\frac{1}{2}$

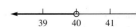

Exercise 2.7

1. $x = \pm 6$ **3.** no solution   **5.** $x = -8, 6$
7. $x = 11, -3$ **9.** $y = -1, 6$ **11.** no solution
13. $x = -7, 6$ **15.** no solution   **17.** $x = \frac{11}{3}, -1$
19. $y = \frac{1}{5}, 1$

Exercise 2.8

1. $x < -3$ or $x > 3$

3. $-3 < x < 3$

5. $-7 < x < 1$

7. $y \le 2$ or $y \ge 12$

9. $-2 < x < 2$

11. $-6 \le x \le 1$

13. no solution
15. The set of all real numbers

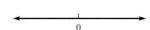

17. $x \le -2$ or $x \ge \frac{4}{3}$

19. no solution

21. $-8 < a < 12$

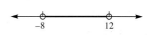

23. $x \le 2$ or $x \ge 3$

Chapter 2 Review Exercises

1. $x = 6$ **3.** $x = 6$ **5.** $x = -36$
7. $x = -30$ **9.** contradiction   **11.** $x = \frac{5z - 3y}{4}$
13. 7   **15.** 6   **17.** $x \le -4$
19. $x < -36$ **21.** $x = -13, 5$ **23.** $x = -1, 8$
25. $-6 \le x \le 6$

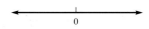

27. no solution
29. The set of all real numbers

Critical Thinking Exercises

1. $-7(x - 4) = -7x + 28$, not $-7x - 4$.
3. To isolate the x on the left side, we must *divide* by 8, not subtract 8.
5. The inequality symbol must be reversed when dividing by -6.
7. In fact, every real number is a solution.
9. $-3 < -x < 2$

Chapter 2 Achievement Test

1. $x = -6$ **2.** $x = 3$ **3.** $x = -\frac{5}{3}$
4. $x = 8$ **5.** $x = \frac{c - 2d - b}{a}$ **6.** $x \le -3$
7. $x < 3$ **8.** $x = 2, -5$ **9.** no solution
10. identity   **11.** contradiction
12. $1 < x < 7$

13. $x \le -\frac{1}{2}$ or $x \ge 1$

14. no solution
15. 18   **16.** 16, 18, 20   **17.** 98

Chapter 3

▮▮ Exercise 3.1

1. x^8
3. y^8
5. x^{12}
7. x^6
9. 2^5 or 32
11. x^3y^4
13. $2^3 \cdot 3^2$ or 72
15. y^{14}
17. y^{14}
19. 3^4 or 81
21. y^2
23. y^8
25. $\dfrac{y^3}{x}$
27. $16y^2$
29. $x^5y^5z^5$
31. $\dfrac{x^2}{9}$
33. 3
35. $16x^4y^4$
37. y^{4b}
39. x^{3a+5b}
41. x^ay^{4b}
43. x^{h+5}
45. x^{h-5}
47. 5^hx^h
49. 2^{a-b}
51. 5^9
53. y^{4a}
55. 6^yx^y
57. 3^{x^2}
59. x^9
61. x^6
63. 6^{15}
65. x
67. y^{14}
69. $9x^2$
71. x^3y^4
73. $\dfrac{x^3}{y}$
75. $\dfrac{4}{9}$

▮▮ Exercise 3.2

1. $\dfrac{1}{x^7}$
3. $\dfrac{1}{2^3}$ or $\dfrac{1}{8}$
5. x^4
7. 2^3 or 8
9. $\dfrac{1}{10}$
11. 10^2 or 100
13. $\dfrac{y^2}{x^2}$
15. $\dfrac{1}{x^2y^2}$
17. $\dfrac{1}{x^4y^4}$
19. $\dfrac{x^2}{y^2}$
21. $\dfrac{6}{x^3}$
23. 3
25. 1
27. $\dfrac{y^5}{x^5}$
29. 4^2 or 16
31. 5^3 or 125
33. 1
35. 1
37. 1
39. $\dfrac{z^4}{x^5y}$
41. $\dfrac{3}{2}$
43. b^2
45. 6^2 or 36
47. $\dfrac{3^2}{8^2}$ or $\dfrac{9}{64}$
49. 7^2 or 49
51. $\dfrac{1}{2^3} + \dfrac{1}{3^2}$ or $\dfrac{17}{72}$
53. $3^2 + \dfrac{1}{3^2}$ or $9\tfrac{1}{9}$
55. $\dfrac{1}{x^5}$
57. $\dfrac{1}{2^{11}}$
59. x^6
61. 7^8
63. $\dfrac{1}{y}$
65. x
67. $\dfrac{7}{x^6}$
69. $\dfrac{-2}{x^5}$
71. 1
73. 1
75. 1
77. $\dfrac{3}{5}$
79. 4
81. $\dfrac{1}{x}$
83. x^4

▮▮ Exercise 3.3

1. $\dfrac{y^2}{x^2}$
3. $125x^6$
5. $\dfrac{1}{125x^6}$
7. $6y^4$
9. x^3y^3
11. $\dfrac{x^2}{y^2}$
13. x^9y^6
15. $\dfrac{1}{xy^4}$
17. $\dfrac{x^4}{y^2}$
19. 1
21. $\dfrac{8x^6}{y^3}$
23. $\dfrac{y^4}{9x^2}$

25. $\dfrac{1}{x^6y^3}$
27. $\dfrac{1}{8x^9y^6}$
29. $\dfrac{y^{15}}{-8x^9}$
31. $\dfrac{x^8}{81}$
33. 1
35. $\dfrac{1}{x}$
37. $\dfrac{x^{12}}{y^{15}}$

▮▮ Exercise 3.4

1. 3
3. 4
5. 5
7. 1
9. 0
11. 8
13. 4
15. 8
17. 10
19. 3^4 has degree zero; 3 is not a variable. In fact, $3^4 = 81$, a constant.
21. $5x^2 - 4x$
23. $6x^3 - 2x^2 + 2x - 2$
25. $7n^5 + 14n^4 + 4n^3 - 5n^2 + 4n$
27. $6t^3 + 4t^2 + 32t + 2$
29. $3w^7 + 7w^4 - 21w^3 - 6w$
31. $-8y^2 + 8y - 7$
33. $12t^4 + 2t$
35. $-15a^2 - 4a - 3$
37. $-4x^3 + 14x^2 + 16x + 14$
39. $4y^9 + 6y^5 + 12$
41. $-2a^4 - 4a^3 + a^2 - 14a - 5$
43. $2x^5 - 5x^4 - 2x^3 + 3x^2 + 6x + 3$
45. $-4x^3 + 7x^2 - 5x + 14$
47. $11y^3 - 10y^2 + y - 1$
49. $-10x^4 + 6x - 3$
51. $-23x^3 + x^2 + x - 22$
53. $-a^3 - 9a^2 + 10a + 2$
55. $2a^4 - 2a^3 + 8a + 7$

▮▮ Exercise 3.5

1. $-30x^3$
3. $16y^4$
5. $6x^3y^2z^5$
7. $6x^2 + 10x$
9. $3x^4 + 9x^3 - 27x^2$
11. $24x^5 + 12x^4 + 18x^3 + 12x^2 - 42x$
13. $-60a^5 + 70a^4 + 30a^3 - 90a^2$
15. $x^3 + 7x^2 + 17x + 35$
17. $6x^3 - 49x^2 - 40x + 8$
19. $21x^3 + 55x^2 + 2x - 28$
21. $x^4 - 3x^3 - 3x^2 + 7x + 6$
23. $x^5 + 4x^4 - 2x^3 - 13x^2 - 29x - 3$
25. $4x^5 + 3x^4 + 6x^3 + 3x^2 - 18x - 18$
27. $21a^5 - 14a^4 + 25a^3 - 51a^2 - 10a - 25$

▮▮ Exercise 3.6

1. $x^2 + 6x + 8$
3. $a^2 - 7a + 10$
5. $x^2 - 2x - 15$
7. $n^2 - 9n + 14$
9. $x^2 - 4$
11. $6x^2 - x - 12$
13. $9x^2 - 4$
15. $x^2 + 6x + 9$
17. $x^2 - 14x + 49$
19. $y^2 - 18y + 81$
21. $12y^2 - 7y - 12$
23. $49x^2 - 4$
25. $9x^2 - 12x + 4$
27. $25n^2 - 9$
29. $49n^2 - 14n + 1$
31. $49x^2 + 28x + 4$
33. $x^2 + 4xy + 3y^2$
35. $9a^2 - b^2$
37. $50m^2 + 55mn - 21n^2$
39. $16a^2 - 40ab + 25b^2$

▮▮ Exercise 3.7

1. $9(4y - 3)$
3. $4a^2(a + 4)$
5. $3(a^2b + 2x^2y)$
7. $x^2(x^3 - x^2 + 1)$
9. $5x^2(5x^4 + 3x^2 + 2)$
11. $-4x(3x^2 + 4x + 6)$
13. $18m^2nq(3 - 2m^2n + nq)$
15. $8r^3(6t^2 - 4rt^4 + 3r^3)$
17. $(6 - x)(y + 2)$
19. $(t^2 + 1)(t + 5)$
21. $(x - 6)(x - y)$
23. $(x^4 + 1)(x + 1)$
25. $(2p - 9)(2p + 9)$
27. $3x(2x - 3)(2x + 3)$
29. $(3x - 2y)(3x + 2y)$
31. $y^2(y + 1)(y - 1)$
33. $(x^2 - 3)(x^2 + 3)$
35. $(9x^3 - y^6)(9x^3 + y^6)$
37. $(3x^3y - 7z^5)(3x^3y + 7z^5)$
39. $(x + 3y)(x^2 - 3xy + 9y^2)$
41. $(5x - 3)(25x^2 + 15x + 9)$
43. $7(a + b)(a^2 - ab + b^2)$
45. $7(s - 2t)(s^2 + 2st + 4t^2)$

Exercise 3.8

1. $(x - 3)(x - 1)$
3. $(y + 5)(y + 7)$
5. $(y + 10)(y + 1)$
7. $(a + 14)(a - 1)$
9. $(a - 2)(a - 7)$
11. $(a - 8)(a - 2)$
13. $(h - 3)(h - 9)$
15. $(h + 2)(h + 5)$
17. $(t + 15)(t - 2)$
19. $(m - 8)(m + 3)$
21. $(y + 7)(y + 8)$
23. $(3x + 4)(x - 1)$
25. $(3x + 2)(4x - 1)$
27. $(3y - 2)(y - 2)$
29. $(2h - 5)(2h - 1)$
31. prime
33. $(3y + 2)(y - 3)$
35. $(5x - 7)(x - 1)$
37. $(2y + 1)(y + 1)$
39. $(7w + 1)(w - 3)$
41. $(3x - 2)(2x - 3)$
43. $(5b - 4)(5b + 1)$
45. $(4a + 3)(2a - 7)$
47. prime
49. $(2y + 1)(4y - 9)$

Exercise 3.9

1. $2(x + 1)(x + 2)$
3. $3(2a - b)(2a + b)$
5. $4(2y + 1)(y - 2)$
7. $3(x + 2)(x - 4)$
9. $(3x - 5)(2x - 3)$
11. $(x + 6)(x - y)$
13. $(x^2 - 6)(x^2 + 2)$
15. $2(2t - 3)(4t - 5)$
17. $(2 - y)(4 + 2y + y^2)$
19. prime
21. $(a - 1)(b + 5)$
23. $(6a^2 - 1)(2a^2 + 1)$
25. $(s - t)(x - y)(x + y)$
27. $(2 - b)(a^2 + c^2)$
29. $(x + 2)(x - 2)(y + 2)$
31. $6(d^2 + 2c^2)$
33. $2h^2(3h + 1)(2h - 3)$
35. $(x^2 + 3)(x^2 - 3)(x^4 + 9)$
37. $5(x^{13} - 1)$
39. $2x(4x - 5y)(4x + 5y)$
41. $7(2x + 5)(x - 1)$
43. $(t - 2)(t + 2)(t^2 + 4)$
45. prime
47. $(9x + 2)(x - 6)$
49. $(2t - 9)(4t + 3)$

Exercise 3.10

1. $2x^3 + 3x^2 - 4x + 8$
3. $2x^3 + x^2 - 3 + \dfrac{x}{5}$
5. $-2 - \dfrac{3}{a} + \dfrac{4}{a^2} - \dfrac{6}{a^4}$
7. $x - 8$
9. $a - 3$
11. $2y - 3$
13. $2n + 3$
15. $2x + 5 + \dfrac{4}{3x + 5}$
17. $4x - 1 + \dfrac{6}{3x + 5}$
19. $x^2 - 7x + 28 + \dfrac{-12}{x + 4}$
21. $x + 7 + \dfrac{7}{6x - 1}$
23. $x^2 + 2x - 1$
25. $y^3 + 2y + 4$
27. $x^2 - 2x - 4$
29. $2x + 5$

Exercise 3.11

1. $x + 5$
3. $3x - 5$
5. $x^2 + 2x - 1 + \dfrac{3}{x + 2}$
7. $x^2 + 4x - 5 + \dfrac{4}{x + 2}$
9. $2x^2 - 3x + 1$
11. $3x^2 + 2x - 5 + \dfrac{2}{x - 2}$
13. $3x^2 - 6x + 7$
15. $2x^3 + 4x^2 - 3x - 6$
17. $4x^3 + 3x^2 - 5x + 2$
19. $x^3 - 5x^2 + 7x + 1$

Chapter 3 Review Exercises

1. a^{11}
3. x^{21}
5. x^{20}
7. y^6
9. $\dfrac{x^8}{y^8}$
11. $x^4 y^7$

13. $\dfrac{1}{x^5}$
15. x^4
17. 1
19. $\dfrac{n^5}{m^5}$
21. $\dfrac{y^5}{x^3}$
23. $\dfrac{9}{4}$
25. $\dfrac{27x^9}{y^6}$
27. $\dfrac{x^2}{y^6}$
29. $6x^3 + 3x^2 - 14x - 12$
31. $y^3 + 6y^2 + 10y - 20$
33. $12x^3 - 4x^2 + 3x + 4$
35. $4x^3 - 11x^2 + 7x + 4$
37. $16a^7 y^5$
39. $18x^2 - 33x - 40$
41. $2x^3 + 11x^2 + 8x - 16$
43. $9x^2 - 12x + 4$
45. $5x^2(7x^3 - 3x^2 + 2)$
47. $(x + 2)(x - 1)(x + 1)$
49. $(a + 1)(a^5 + 1)$
51. prime
53. $(y^2 - 2)(y^2 + 2)(y^4 + 4)$
55. $(4 - x)(16 + 4x + x^2)$
57. prime
59. $(6x + 1)(4x + 7)$
61. $(3h^2 + 4)(5h^2 - 2)$
63. $(x + 2)(x - 2)(x^2 - 2x + 4)(x^2 + 2x + 4)$
65. $-2y^3 - 3y^2 + 7 + \dfrac{2}{y}$
67. $x + 3$
69. $2y^2 - 3y + 5 + \dfrac{12}{2y + 3}$
71. $2x^2 - 5x + 3 + \dfrac{4}{x - 6}$
73. $2x^2 - 6x + 8$

Critical Thinking Exercises

1. $21^0 = 1$, not 0.
3. $5^2 \cdot 5^3 = 5^5$; we add the exponents but don't multiply the bases.
5. $4^{-2} = \dfrac{1}{4^2} = \dfrac{1}{16}$
7. $6^{-1} = \frac{1}{6}$
9. $x^a x^b = x^{a+b}$
11. $(4x)^3 = 4^3 x^3$; the exponent acts on everything in the parentheses.
13. $(2x - 5)^2 = 4x^2 - 20x - 25$
15. $x^2 - 5x + 6 = (x - 3)(x - 2)$
17. $-(2x^2 - 3x + 4) = -2x^2 + 3x - 4$; the negative sign affects all the terms in the parentheses.

Chapter 3 Achievement Test

1. a^{15}
2. $\dfrac{1}{x^6}$
3. $27x^{12}$
4. $\dfrac{16}{9}$
5. $\dfrac{x^4}{y^{10}}$
6. $\dfrac{x^2 y^6}{4}$
7. $-7x^4 + 6x^3 - 6x + 8$
8. $2y^3 - 3y^2 + 2$
9. $-24x^{11}$
10. $8x^2 + 14x - 15$
11. $36x^2 + 12x + 1$
12. $x^3 + 7x^2 + 7x - 15$
13. $6a^2 y(ax - 2y + 1)$
14. $(y + 4)(x - 5)$
15. $(y^2 + 1)(y + 1)$
16. $(6x - 7y)(6x + 7y)$
17. prime
18. $(2x - 3y)(4x^2 + 6xy + 9y^2)$
19. $(x - 7)(x - 6)$
20. $(3x + 7)(5x + 2)$
21. $4s(s + 3t)(s^2 - 3st + 9t^2)$
22. $(3x^2 + 1)(x^2 - 3)$
23. $(y - 1)(x + 2)(x - 2)$
24. $3y^4 + 5y^2 - y - \dfrac{9}{y^2}$
25. $2x + 3 + \dfrac{3}{x - 6}$
26. $2x + 3 + \dfrac{-4}{3x - 7}$
27. $5x^2 - 3x + 1$
28. $6x^2 - 12x + 7$

Cumulative Review Exercises—Chapters 1, 2, and 3

1. $\{-3, -1, 1, 3\}$
3. $\{e, x, a\}$
5. true
7. true
9. false
11. irrational
13. rational
15. $2 \cdot 3 \cdot 3 \cdot 7$
17. -48

19. indeterminate   **21.** 0   **23.** 96

25. $x = 13$ **27.** $x = \dfrac{4y + 3z}{2}$ **29.** $x = -\frac{3}{2}$ or $x = 2$

31. no solution   **33.** x^9 **35.** $16x^8$

37. 6   **39.** $x^4 - 3x^3 + 5x - 8$

41. $16x^2 - 40x + 25$ **43.** $5(x + 3y)(x - 3y)$

45. $(5x + 2)(x - 5)$ **47.** $(5x - 4)(2x - 1)$

49. $4x + 5 + \dfrac{-2}{2x - 3}$

Chapter 4

▮▮ Exercise 4.1

1. 3   **3.** −3   **5.** 4

7. −3, 1   **9.** 0   **11.** 2, −5

13. −3   **15.** 4   **17.** −5

19. $y - 2$ **21.** $x - 1$

▮▮ Exercise 4.2

1. $\dfrac{2}{3}$ **3.** $\dfrac{y^3}{2x^2}$ **5.** $\dfrac{1}{-3y}$

7. $\dfrac{1}{2}$ **9.** $\dfrac{x^2 y^4}{a^2 b^3}$ **11.** $\dfrac{2}{x + 3}$

13. $\dfrac{1}{x + 2}$ **15.** 3   **17.** 1

19. $\dfrac{x + 1}{x - 1}$ **21.** $\dfrac{1}{x + 6}$ **23.** $\dfrac{2x - 5}{x - 4}$

25. $\dfrac{t + 5}{t + 3}$ **27.** $\dfrac{x - 3}{x + 5}$ **29.** $-\dfrac{3a}{a + b}$

31. (a) no **(b)** yes **(c)** yes **(d)** no

▮▮ Exercise 4.3

1. $\dfrac{3}{8}$ **3.** $x^3 y$ **5.** 1

7. $\dfrac{b}{3}$ **9.** 2   **11.** $\dfrac{a}{a - 1}$

13. $\dfrac{2}{x - 2}$ **15.** $\dfrac{x + 7}{x - 4}$ **17.** $-\dfrac{1}{2x + 5}$

19. $\dfrac{x - 3}{x - 1}$ **21.** $\dfrac{1}{x - 3}$ **23.** −1

25. $\dfrac{x - 3}{x + 3}$

▮▮ Exercise 4.4

1. $\dfrac{1}{2}$ **3.** $\dfrac{2}{x}$ **5.** $\dfrac{2}{x}$

7. 4   **9.** $\dfrac{a + b + c}{x}$ **11.** 1

13. 120   **15.** $24x^2 y^3$ **17.** $x(x - 4)$

19. $(x + 1)(x + 1)$ **21.** $7x(7x + 1)(7x - 1)$

23. $a(a - 2)(a + 2)(a + 2)$ **27.** 6

29. $25x^2 y^3$ **31.** $49bc^2$

33. $4(x - 4)$ **35.** $(x - 2)(x + 2)$

37. $4(x - 4)(x + 1)$ **39.** $(x - 4)^2$

41. $\dfrac{35}{36}$ **43.** $\dfrac{2y - x}{x^2 y^2}$

45. $\dfrac{27y + 10}{12y^2}$ **47.** $\dfrac{a^2 + b^2 + 2c^2}{abc}$

49. $\dfrac{x^2 + 3x + 9}{x(x + 3)}$ **51.** $\dfrac{x^2}{y(x - y)}$

53. $\dfrac{a}{5(a + 2)}$ **55.** $\dfrac{-2x - 25}{(x + 2)(x - 5)}$

57. $\dfrac{2(y^2 - 5y + 10)}{5(y + 2)(y - 5)}$ **59.** $\dfrac{2}{x - 1}$

61. $-\dfrac{x^2 + 4x + 7}{(x - 3)(x + 3)}$ **63.** $\dfrac{3x^2 + 4x - 3}{(x + 1)(x - 1)(x + 2)}$

65. $\dfrac{2x^2 + 11x + 3}{2(x - 3)(x + 3)}$ **67.** $\dfrac{8}{a}$

69. $\dfrac{2}{x + 2}$ **71.** $\dfrac{x - 10}{(x - 2)(x + 2)}$

▮▮ Exercise 4.5

1. $\dfrac{b}{a^2}$ **3.** $\dfrac{1}{b}$ **5.** $\dfrac{3}{29}$

7. $\dfrac{xy^2}{2}$ **9.** $\dfrac{b + a}{a^2 b}$ **11.** $\dfrac{x^2 - 2xy}{2xy - y^2}$

13. $\dfrac{-3}{2x}$ **15.** $-2y$ **17.** $\dfrac{a + 2}{3 + 4a^2}$

19. $x^2 + y^2$ **21.** $\dfrac{2x - 3}{3x}$ **23.** $\dfrac{x - 1}{x + 2}$

25. x **27.** $\dfrac{-x + 5}{x^2 - 2x}$ **29.** $\dfrac{-x}{x^2 + x - 2}$

▮▮ Exercise 4.6

1. 20   **3.** $\frac{3}{2}$ **5.** no solution

7. −1   **9.** $-\frac{2}{3}$ **11.** 1

13. 14   **15.** no solution   **17.** no solution

19. $-\frac{1}{3}$ **21.** $\frac{1}{2}$ **23.** no solution

25. 1   **27.** no solution

▮▮ Exercise 4.7

1. 6, 14   **3.** 5   **5.** 3 and 5

7. $\frac{12}{13}$ hr   **9.** $1\frac{1}{3}$ hr   **11.** 6 in.

13. $R = 12$ ohms

15. $5\frac{1}{3}$ km/hr for the runner and $13\frac{1}{3}$ km/hr for the cyclist

17. approx. 136 miles

▮▮ Chapter 4 Review Exercises

1. $x = 1$ **3.** $x = \pm 2$ **5.** $-y - 4$

7. $\dfrac{3}{8}$ **9.** in lowest terms   **11.** $\dfrac{x + 3}{x - 3}$

13. $\dfrac{6}{x^2 y^2}$ **15.** $\dfrac{(x + 4)(x - 5)}{3x^2}$ **17.** −1

19. $\dfrac{x^2 - x + 8}{(x - 4)(x + 1)}$ **21.** $\dfrac{x}{x - 2}$ **23.** $\dfrac{17}{4}$

25. $\dfrac{5x^3 y^2 + x^2 y^3}{x^2 - y^2}$ **27.** −12   **29.** no solution

31. $\frac{7}{3}$

▮▮ Critical Thinking Exercises

1. If $x = 4$ in this fraction, the numerator equals zero but the denominator does not.

3. The negative sign must be distributed to both the x and the 3 in the numerator.

5. A common denominator must be found; the denominators are not simply added.

7. The negative sign was not distributed to the 7 in the numerator of the second fraction. The correct answer is $\dfrac{2x - 9}{x + 5}$

▮▮ Chapter 4 Achievement Test

1. -1

2. $-5, -1$

3. -3

4. -2

5. $\dfrac{x + 1}{2}$

6. $\dfrac{x^2 - 3x - 4}{x^2 + 3x - 10}$

7. $\dfrac{1}{x + 3}$

8. $\dfrac{x + y}{5xy + 1}$

9. 1

10. 2

11. $\dfrac{3x^2 + 4}{x}$

12. $\dfrac{3 - a}{a - 2}$

13. $\dfrac{2x - 10}{(x - 1)(x + 1)^2}$

14. $\dfrac{-6}{x + 1}$

15. $\dfrac{-x + 3}{2x}$

16. $\dfrac{2a^2 + a}{2a - 4}$

17. $x = 1$

18. $x = 2$

19. $a = -5$

20. no solution

21. $-36, -30$

22. 24 min

Chapter 5

▮▮ Exercise 5.1

1. ± 10

3. ± 8

5. ± 11

7. ± 12

9. 2

11. 1

13. 7

15. not a real number

17. -9

19. -13

21. $|a|$

23. $|3x|$ or $3|x|$

25. $|x|$

27. $|x - 2|$

29. $-|x - 2|$

31. 4

33. 1

35. -2

37. 2

39. ± 2

41. not a real number

▮▮ Exercise 5.2

1. $\sqrt{6}$

3. 6

5. $2\sqrt[3]{5}$

7. $6\sqrt{2}$

9. $2\sqrt[4]{2}$

11. y^2

13. $4x^3$

15. $3z\sqrt[3]{2z^2}$

17. $t^{12}\sqrt{t}$

19. $2ab\sqrt{10\,ac}$

21. $2ab\sqrt[3]{3a}$

23. $4\sqrt{3}$

25. $2\sqrt{7}$

27. x^4

29. $2\sqrt[3]{2}$

31. a^2

33. $4x\sqrt{3y}$

35. $5x^7\sqrt{2x}$

37. $4x^4\sqrt{3x}$

39. $6x^2y^3$

▮▮ Exercise 5.3

1. $5\sqrt{5}$

3. $10\sqrt[3]{x}$

5. cannot be combined

7. $\sqrt{2}$

9. $-\sqrt[3]{2}$

11. $8\sqrt{2} - 9\sqrt{3}$

13. $\sqrt{11}$

15. cannot be combined

17. $-12\sqrt{2}$

19. $16\sqrt{10} - 7\sqrt{5}$

21. $11\sqrt{5ab}$

23. cannot be combined

▮▮ Exercise 5.4

1. $\sqrt{3}$

3. 2

5. $\sqrt[3]{6}$

7. 2

9. $\dfrac{7}{2}$

11. $\dfrac{2}{3}$

13. 2

15. $x\sqrt{5}$

17. $x^2\sqrt{6}$

19. $\dfrac{\sqrt{6}}{2}$

21. $\dfrac{\sqrt{10}}{2}$

23. $2\sqrt{5}$

25. $\sqrt[3]{2}$

27. $\dfrac{\sqrt{35}}{7}$

29. $\dfrac{\sqrt{7}}{7}$

31. $\dfrac{7\sqrt{y}}{y}$

33. $\dfrac{\sqrt{5x}}{x}$

35. $\dfrac{\sqrt{30}}{3}$

37. $\dfrac{5\sqrt{10}}{2}$

39. $\dfrac{3\sqrt{2x}}{x}$

41. $\dfrac{\sqrt{15}}{6}$

43. $\dfrac{\sqrt{6x}}{9x}$

45. $\dfrac{\sqrt[3]{18}}{3}$

47. $\dfrac{\sqrt[3]{10}}{2}$

49. $\dfrac{\sqrt[3]{12x^2}}{2x}$

51. $\dfrac{\sqrt[3]{15x^2}}{3x}$

53. $\dfrac{\sqrt[4]{6}}{2}$

55. $3 + \sqrt{2}$

▮▮ Exercise 5.5

1. $12 + 2\sqrt{3}$

3. $2\sqrt{21} - 14$

5. $6 - 6\sqrt{3}$

7. $3\sqrt{y} + 2y$

9. $16 + 8\sqrt{3}$

11. $2\sqrt{2} - 1$

13. $7 - 3\sqrt{6}$

15. $16 + 7\sqrt{6}$

17. -23

19. $3 - 3\sqrt{2} + \sqrt{6} - 2\sqrt{3}$

21. $28 + 10\sqrt{3}$

23. $7 + 4\sqrt{3}$

25. $3 + \sqrt{7}$

27. $\sqrt{5} - 3\sqrt{6}$

29. $x + \sqrt{5}$

31. $2\sqrt{2} - 2$

33. $3\sqrt{5} - 3\sqrt{3}$

35. $1 + \sqrt{3}$

37. $6 + 3\sqrt{3}$

39. $2 - \sqrt{3}$

41. $\dfrac{13 + 9\sqrt{2}}{7}$

43. $\dfrac{23 - 14\sqrt{5}}{11}$

45. $\sqrt{10} + 2$

▮▮ Exercise 5.6

1. $5^{1/2}$

3. $y^{5/4}$

5. $a^{4/3}$

7. $\sqrt{6}$

9. $\sqrt[3]{25}$

11. $\dfrac{\sqrt[3]{x}}{x}$

13. 2

15. 27

17. -4

19. -3

21. $\dfrac{4}{3}$

23. $\dfrac{1}{27}$

25. $\dfrac{7}{4}$

27. 4

29. $\dfrac{10}{21}$

31. x

33. x

35. $x^{17/12}$

37. $x^{1/4}$

39. $x^{1/2}y^{2/3}$

41. $x^{1/2}y^{1/3}$

43. $x^{1/6}$

45. 1

▮▮ Exercise 5.7

1. $3i$

3. $-5i$

5. $2i\sqrt{5}$

7. $5i\sqrt{3}$

9. $-3i\sqrt{3}$

11. $-4i\sqrt{3}$

13. -1

15. i

17. $-i$

19. -1

21. i

23. $10 - 2i$

25. $3 - 2i$

27. $13 - 5i$

29. $-12 + 4i$

31. -1

33. 0

35. $17 - 5i$

37. $7 + 9i$

39. $-4 + 4i$

41. $4 - 3i$

▮▮ Exercise 5.8

1. $12 + 30i$

3. $-4 + 6i$

5. $4 + 19i$

7. $9 + 2i$

9. $26 - 22i$

11. $-23 + 11i$

13. $7 + 24i$

15. $24 - 10i$

17. 34

19. $3 + 4i$

21. $-4 - i$

23. $-5i$

25. $-3 - 4i$

27. $\dfrac{1}{4} + \dfrac{3}{2}i$

29. $\dfrac{9}{5} - \dfrac{3}{5}i$

31. $-\dfrac{3}{5} - \dfrac{6}{5}i$

33. $2 + 0i$ or 2

35. $\dfrac{25}{17} + \dfrac{2}{17}i$

37. $-\dfrac{1}{5} + \dfrac{8}{5}i$

39. $-\dfrac{3}{10} + \dfrac{1}{10}i$

▪ ▪ Chapter 5 Review Exercises

1. ± 6 **3.** 9
5. $9i$ (not a real number)   **7.** 4
9. $2\sqrt{6}$ **11.** $2x^2\sqrt{3x}$
13. x^2 **15.** $6 - 18\sqrt{3}$
17. cannot be combined   **19.** cannot be combined
21. 2   **23.** $\dfrac{3\sqrt[3]{2}}{2}$
25. $\sqrt{5} + \sqrt{2}$ **27.** $a^{5/2}$
29. $5i$ (not a real number)   **31.** $\frac{5}{2}$
33. $\frac{7}{12}$ **35.** $x^{7/12}$
37. $x^{1/6}$ **39.** $5i\sqrt{2}$
41. $-i$ **43.** -1
45. $4 + 5i$ **47.** $-5 + 4i$
49. $43 - 18i$ **51.** 34
53. $-1 + \frac{2}{3}i$ **55.** $-\frac{17}{10} - \frac{9}{10}i$

▪ ▪ Critical Thinking Exercises

1. The root of a sum is not equal to the sum of the roots.
3. $4^{2/3} = \sqrt[3]{4^2} = 2\sqrt[3]{2}$; not $\sqrt{4^3}$.
5. The outer and inner products were left out of the result.
7. The radical denotes only the principle square root.
9. $x^a \cdot x^b = x^{a+b}$; not x^{ab}.
11. The entire numerator must be divided by the denominator.
13. False: $(1 + i)(1 + i) = 2i$.

▪ ▪ Chapter 5 Achievement Test

1. $9, -9$ **2.** 8
3. $8i$ (not a real number)   **4.** -8
5. 4   **6.** 4
7. $3\sqrt{5}$ **8.** $3x\sqrt{2x}$
9. $3x^2\sqrt{2}$ **10.** $20 - 40\sqrt{3}$
11. $7\sqrt{2}$ **12.** cannot be combined
13. 3   **14.** $3\sqrt{2} - 6$
15. $\sqrt{3}$ **16.** $\dfrac{\sqrt{15x}}{6x}$
17. $2\sqrt[3]{3}$ **18.** $\sqrt{11} + \sqrt{7}$
19. $b^{7/4}$ **20.** -7
21. $6i$ (not a real number)   **22.** $\frac{3}{4}$
23. $-\frac{2}{27}$ **24.** $x^{1/4}$
25. $x^{17/12}$ **26.** $x^{2/15}$
27. $x^{1/4}y^{1/8}$ **28.** $2i\sqrt{11}$
29. $-4 + 7i$ **30.** $7 + 11i$
31. $-\frac{13}{5} - \frac{1}{5}i$

Chapter 6

▪ ▪ Exercise 6.1

1.

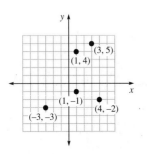

3.

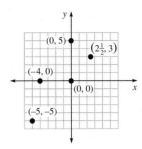

5. $A(3, 5)$; $B(2, -2)$; $C(-4, 3)$; $D(-4, -2)$; $E(4, 0)$; $F(-2, 1)$; $G(0, 0)$; $H(0, -3)$

▪ ▪ Exercise 6.2

1.

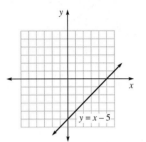

3.

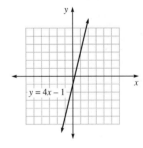

5.

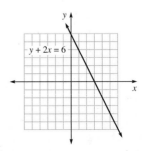

7.

9.

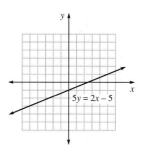

$5y = 2x - 5$

11.

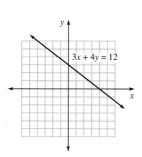

$3x + 4y = 12$

13.

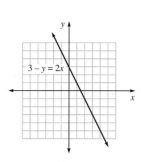

$3 - y = 2x$

15.

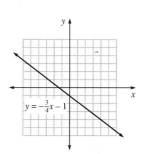

$y = -\frac{3}{4}x - 1$

17.

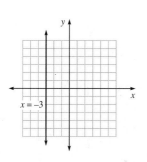

$x = -3$

19.

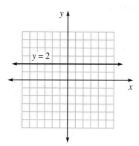

$y = 2$

21.

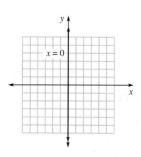

$x = 0$

23.

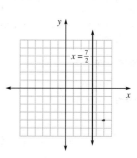

$x = \frac{7}{2}$

■ ■ **Exercise Set 6.3**

1.

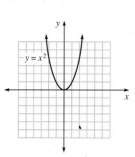

$y = x^2$

3.

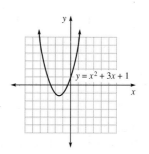

$y = x^2 + 3x + 1$

5.

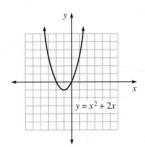

$y = x^2 + 2x$

7.

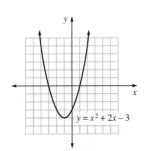

$y = x^2 + 2x - 3$

9.

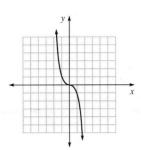

53.

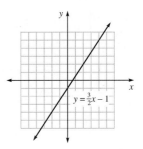

$y = \frac{3}{2}x - 1$

55.

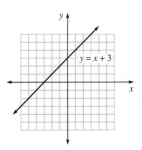

$y = x + 3$

57.

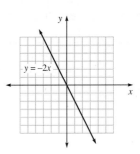

$y = -2x$

59.

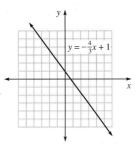

$y = -\frac{4}{3}x + 1$

61.

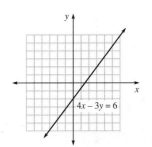

$4x - 3y = 6$

▪ ▪ Exercise 6.4

1. $\frac{1}{2}$ **3.** $-\frac{1}{2}$ **5.** 1

7. 2   **9.** undefined   **11.** 0

13. 0   **15.** $\frac{5}{3}$ **17.** 0

19. -2 **21.** -1 **23.** $-\frac{2}{3}$

25. 4   **27.** $\frac{1}{4}$ **29.** $-\frac{1}{2}$

31. yes   **33.** no   **35.** yes

▪ ▪ Exercise 6.5

1. $y = 2x + 1$ **3.** $y = -3x - 11$

5. $y = x - 5$ **7.** $y = \frac{3}{4}x - \frac{13}{4}$

9. $y = 4$ **11.** $y = \frac{4}{3}x$

13. $y = -x + 5$ **15.** $y = -\frac{1}{2}x + 3$

17. $y = \frac{6}{5}x - \frac{34}{5}$ **19.** $x = -5$

21. $m = 4;\ b = -3$ **23.** $m = -\frac{3}{4};\ b = 2$

25. $m = \frac{4}{3};\ b = 2$ **27.** $m = \frac{1}{3};\ b = 2$

29. $m = \frac{6}{5};\ b = 1$ **31.** $m = 0;\ b = -4$

33. m is undefined; no y-intercept

35. $y = \frac{2}{3}x + 4$ **37.** $y = -4x - 2$

39. $x = 2$ **41.** $y = 3$

43. $x = -2$ **45.** $y = \frac{2}{3}x + b;\ \ b \neq -4$

47. $y = -\frac{3}{2}x + b$ **49.** $y = \frac{2}{3}x + b;\ \ b \neq \frac{5}{3}$

51. **(a)** perpendicular **(b)** neither **(c)** parallel **(d)** perpendicular

63.

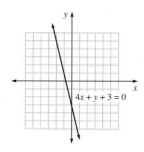

$4x + y + 3 = 0$

23. $y = 3x - 4$: function

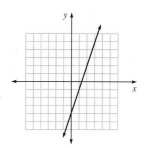

▮▮ Exercise 6.6

1. 5 **3.** 13 **5.** $\sqrt{65} \approx 8.062$

7. $2\sqrt{10} \approx 6.325$ **9.** $(4, 3)$ **11.** $(-2, 2)$

13. $\left(-\frac{7}{2}, 3\right)$ **15.** $(0, 0)$ **17.** 17, 10, 21

19. $(5, 0)$ to $(8, -3)$ is $3\sqrt{2}$; $(2, -3)$ to $(5, 0)$ is $3\sqrt{2}$

21. The midpoint is $(3, 2)$; the distance to the vertices is $\sqrt{13}$.

23. The midpoint of $\overline{AB}$ is $\left(\frac{7}{2}, 3\right)$; the midpoint of $\overline{CD}$ is $\left(\frac{7}{2}, 3\right)$

▮▮ Exercise 6.7

1. function: domain $= \{1, 3, 5\}$; range $= \{2, 4, 6\}$

3. function: domain $= \{-2, 4\}$; range $= \{0, 2\}$

5. not a function

7. function: domain $= \{-2, 0, 4\}$; range $= \{-1, 3\}$

9. not a function **11.** function

13. function **15.** not a function

17. not a function

19. $y = x^2$: function

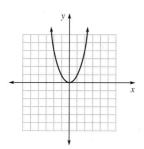

21. $x^2 + y^2 = 4$: not a function

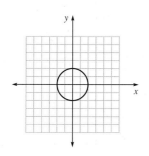

▮▮ Exercise 6.8

1. (a) 9 (b) -1 (c) 5 (d) 6

3. (a) 7 (b) 19 (c) $\frac{13}{4}$ (d) $a^2 + 2ab + b^2 - 2a - 2b + 4$

5. all real numbers

7. all real numbers

9. all real numbers except $x = 1$

11. all real numbers except $x = -\frac{1}{2}$

13. all real numbers except $x = 5$

15. all real numbers except $x = -1, 3$

17. all real numbers except $x = -4, 1$

19. all real numbers except $x = -1, \frac{3}{2}$

21. all real numbers ≥ 4 **23.** all real numbers ≥ 8

25. all real numbers $\geq \frac{1}{2}$ **27.** all real numbers $\geq -\frac{1}{5}$

29. all real numbers $\leq \frac{1}{4}$ **31.** all real numbers $\leq \frac{1}{3}$

33. all real numbers > 5

35. (a) 95 (b) $95 (c) 4 hours

▮▮ Exercise 6.9

1. $f^{-1} = \{(3, 1), (7, 2), (-2, 4)\}$; f^{-1} is a function.

3. $f^{-1} = \{(3, 1), (3, 2), (3, 3)\}$; f^{-1} is not a function.

5. $f^{-1} = \{(0, 0), (1, 1)\}$; f^{-1} is a function.

7. $f^{-1}(x) = \frac{1}{3}x + \frac{2}{3}$

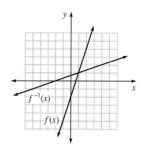

$f^{-1}(x)$

$f(x)$

9. $f^{-1}(x) = \frac{3}{2}x + 3$

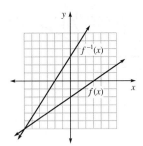

11. $f^{-1}(x) = -\frac{3}{2}x + \frac{3}{2}$

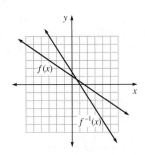

13. $f^{-1}(x) = -\frac{4}{3}x - \frac{8}{3}$; $f^{-1}(x)$ is a one-to-one function.

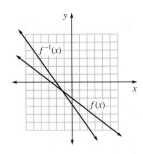

15. $f^{-1}(x) = \pm\sqrt{x + 4}$; $f^{-1}(x)$ is not a function.

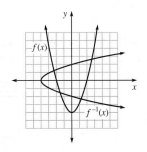

17. $f^{-1}(x) = \sqrt[3]{x}$; $f^{-1}(x)$ is a one-to-one function.

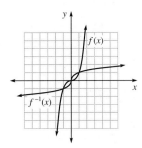

▮▮ Chapter 6 Review Exercises

1.

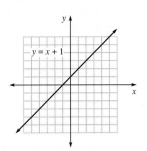

3.

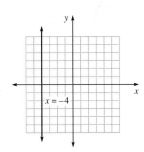

5.

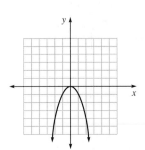

7. $-\frac{5}{4}$ **9.** 1   **11.** 0

13. $y = -\frac{2}{3}x$ **15.** $y = -\frac{5}{2}x + \frac{27}{2}$ **17.** $m = -8$; $b = 6$

19. $\sqrt{34} \approx 5.831$

21. function: domain = $\{0, 2, 6\}$; range = $\{-8, 4\}$

23. function, not one-to-one

25. 0   **27.** -1 **29.** $2b^2 + 3b$

31. all real numbers except ± 2

33. $f^{-1} = \{(3, 2), (-1, 4), (3, 6)\}$; f^{-1} is not a function

35. $f^{-1}(x) = \frac{1}{3}x - \frac{1}{3}$; f^{-1} is a function; f is one-to-one

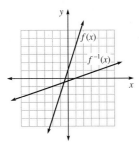

Critical Thinking Exercises

1. The graph of $y = 2$ is parallel to the x-axis, not the $y =$ axis.
3. The graph of $y + 2x = -6$ has slope -2.
5. $\{(1, 3), (2, 3), (3, 3)\}$ is a function because no first elements are the same.
7. $x = -3$ also makes the denominator zero
9. The x-intercept occurs where $y = 0$.

Chapter 6 Achievement Test

1.

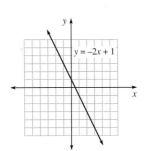

2.

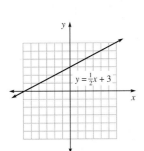

3. $-\frac{2}{3}$ **4.** 0 **5.** undefined
6. $y = 4x - 11$ **7.** $y = \frac{1}{2}x - \frac{11}{2}$ **8.** $x = 2$
9. $y = -\frac{3}{7}x + \frac{5}{7}$ **10.** $x = -2$ **11.** $y = -1$
12. $m = -5, b = -3$ **13.** $m = \frac{3}{4}, b = 3$ **14.** $m = 0, b = 3$
15. m is undefined and there is no y-intercept
16. $\sqrt{34} \approx 5.831$ **17.** $(-5, 3)$ **18.** not a function
19. function   **20.** -3
21. $a^2 + 2ab + b^2 - 4a - 4b + 1$
22. all real numbers except 3   **23.** all real numbers ≥ 1

24. (a) $\{(-1, 2), (2, 3), (-1, 4)\}$; (b) not a function
25. (a) $f^{-1}(x) = \frac{1}{2}x + \frac{1}{2}$
(b)

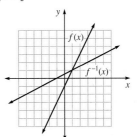

(c) yes (d) yes

Cumulative Review Exercises—Chapters 4, 5, and 6

1. $x = \frac{1}{2}$ **3.** $\frac{1}{2}$ **5.** $\dfrac{4a^2 + ab}{2a^2 - 2b^2}$
7. 4 hr   **9.** 5   **11.** $4\sqrt{6}$
13. $2x^4\sqrt{6xy}$ **15.** $\sqrt{3} - 4$ **17.** $-\sqrt{2}$
19. $\sqrt{2}$ **21.** 16   **23.** $x^{5/12}$
25. $6 - 2i$ **27.** $-8 - i$
29.

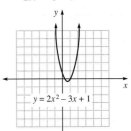

31. $y = -3x + 3$ **33.** $3\sqrt{10}$ **35.** $g^{-1}(x) = 3x + 6$

Chapter 7

Exercise 7.1

1. $4x^2 - 3x + 2 = 0$; $a = 4, b = -3, c = 2$
3. $3x^2 - x + 1 = 0$; $a = 3, b = -1, c = 1$
5. $x^2 - 2x + 6 = 0$; $a = 1, b = -2, c = 6$
7. $5x^2 + x + 2 = 0$; $a = 5, b = 1, c = 2$
9. $x = -3, 4$ **11.** $x = -2, -8$ **13.** $y = -4, 3$
15. $x = 0, 14$ **17.** $x = \frac{4}{5}, \frac{1}{2}$ **19.** $x = -\frac{3}{2}, \frac{1}{2}$
21. $a = \frac{1}{3}, -2$ **23.** $x = \pm\frac{5}{2}$ **25.** $y = \pm 3\sqrt{2}$
27. $x = \frac{1}{6}, -\frac{2}{3}$ **29.** $t = \pm\dfrac{2\sqrt{6}}{7}$

Exercise 7.2

1. $x^2 + 4x + 4 = (x + 2)^2$ **3.** $y^2 - 8y + 16 = (y - 4)^2$
5. $x^2 - 7x + \frac{49}{4} = \left(x - \frac{7}{2}\right)^2$ **7.** $x^2 + \frac{1}{4}x + \frac{1}{4} = \left(x + \frac{1}{2}\right)^2$
9. $x = -2, -4$ **11.** $t = 3, 5$
13. $x = -3, -7$ **15.** $y = \dfrac{3 \pm \sqrt{29}}{2}$
17. $y = \dfrac{-1 \pm \sqrt{29}}{2}$ **19.** $m = -1 \pm \sqrt{6}$

21. $x = \dfrac{-3 \pm \sqrt{17}}{2}$ **23.** $x = \dfrac{5 \pm \sqrt{65}}{4}$

25. $x = -\frac{4}{3}, \frac{1}{2}$

▌▌ Exercise 7.3

1. $x = -4, -1$ **3.** $y = \dfrac{4 \pm \sqrt{10}}{2}$

5. $x = -3 \pm \sqrt{6}$ **7.** $x = \dfrac{-1 \pm \sqrt{5}}{2}$

9. $x = \dfrac{1 \pm i\sqrt{2}}{3}$ **11.** $x = 1 \pm 2i$

13. $x = 1 \pm \sqrt{7}$ **15.** $t = -2 \pm \sqrt{5}$

17. $x = \dfrac{1 \pm i}{2}$ **19.** $x = -5 \pm i\sqrt{15}$

21. $x = \dfrac{1 \pm 2i\sqrt{5}}{3}$ **23.** $x = \dfrac{-3 \pm i\sqrt{11}}{2}$

25. $x = -3 \pm \sqrt{13}$ **27.** $x = \dfrac{-5 \pm \sqrt{33}}{2}$

29. two real   **31.** one real

33. two complex   **35.** two real

37. two complex   **39.** two complex

41. $k = \pm 12$ **43.** $k = 4$

▌▌ Exercise 7.4

1. $x = 6, -2$ **3.** $x = \pm\dfrac{\sqrt{2}}{2}$

5. $x = 1$ **7.** $x = -\frac{5}{2}, 1$

9. $x = 6$ **11.** $x = -1, 3$

13. $x = \pm 1, \pm 2$ **15.** $x = \pm 1, \pm i\sqrt{5}$

17. $x = \dfrac{\pm\sqrt{2}}{2}, \pm\sqrt{2}$ **19.** $x = 0, 1$

21. $x = 2, 3$ **23.** $x = \pm 3$

25. $x = 12$ ($x = 7$ does not check)

27. $x = -2, -1$ **29.** $x = -3$

31. $x = 5, 8$ **33.** $x = 0$

35. $x = 4$

▌▌ Exercise 7.5

1. 7, 8   **3.** 9 cm, 12 cm

5. base is 6 in.; altitude is 5 in.   **7.** 12 by 16

9. 4 units on a side   **11.** 2 cm

13. 10 mph   **15.** 6 sides

17. $\frac{4}{3}$ or $\frac{3}{4}$ **19.** $2 \pm \sqrt{3}$

21. length: 30 in.; width: 10 in.   **23.** length: 30 ft; width: 10 ft

25. length: 6 ft; width: 4 ft   **27.** 6 in. × 6 in.

29. 9 mph

▌▌ Exercise 7.6

1. vertex: $(0, 0)$; x-intercept: $(0, 0)$

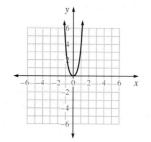

3. vertex: $(0, 0)$; x-intercept: $(0, 0)$

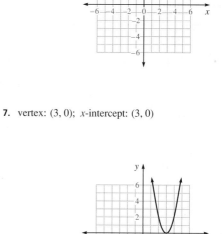

5. vertex: $(3, 1)$; no x-intercepts

7. vertex: $(3, 0)$; x-intercept: $(3, 0)$

9. vertex: $(1, 0)$; x-intercept: $(1, 0)$

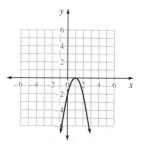

11. vertex: $(-2, 1)$; no x-intercepts

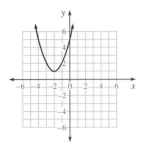

13. vertex: $(-2, -9)$; x-intercepts: $(-5, 0)$, $(1, 0)$

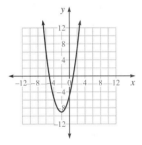

15. vertex: $(-1, 4)$; x-intercepts: $(-3, 0)$, $(1, 0)$

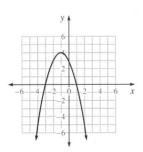

17. vertex: $(1, 4)$; x-intercepts: $(-1, 0)$, $(3, 0)$

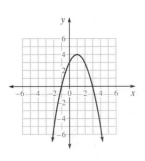

19. vertex: $(2, 5)$; x-intercepts: $(2 \pm \sqrt{5}, 0)$

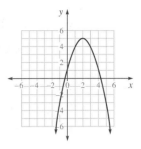

21. vertex: $(0, 4)$; x-intercepts: $(\pm 2, 0)$

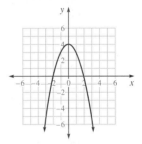

23. vertex: $\left(-\frac{2}{3}, \frac{2}{3}\right)$; no x-intercepts

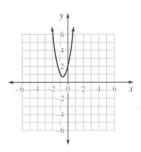

27. **(a)** upward; same as $y = x^2$
(b) downward; narrower than $y = x^2$
(c) downward; narrower than $y = x^2$
(d) upward; wider than $y = x^2$
(e) downward; wider than $y = x^2$

∎ ∎ **Section 7.7**

1. 625 ft²; 25 ft × 25 ft   **3.** 256; 16 and 16
5. -9; 3 and -3 **7.** 80 hot dogs; $40
9. 18 cameras; $2,548   **11.** 64 ft after 2 sec
13. 14 in.; 196 in²

▮▮ Exercise 7.8

1.

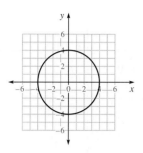

3.

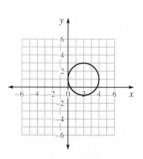

5.

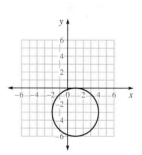

7.

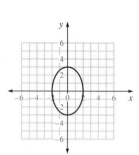

9.

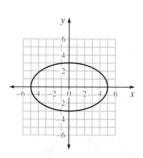

11.

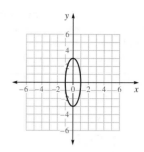

13.

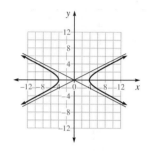

15.

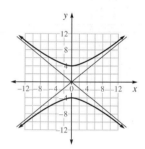

17.

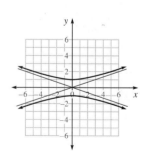

▮▮ Chapter 7 Review Exercises

1. $3x^2 + 2x - 9 = 0$ **3.** $2x^2 - 9x - 24 = 0$

5. $x = -5, 3$ **7.** $x = -\frac{5}{2}, \frac{1}{2}$

9. $a = \pm 9$ **11.** $x = \pm 2\sqrt{5}$

13. $t = \pm 2\sqrt{7}$ **15.** $y^2 - 5y + \frac{25}{4} = \left(y - \frac{5}{2}\right)^2$

17. $x = -1, 5$ **19.** $y = \dfrac{1 \pm \sqrt{5}}{2}$

21. $m = -\frac{1}{2}, 2$ **23.** $x = -3, 4$

25. $x = -2 \pm i\sqrt{5}$ **27.** $x = \pm \sqrt{10}$

29. $x = -2, 1$ **31.** two real solutions

33. two real solutions   **35.** $x = -3, 1$

37. $x = \pm i$

39. $x = \pm 3, \pm i\sqrt{2}$

41. $x = 0, 4$

43. no solution

45. length: 12 in.; width: 9 in.

47. 10

49. vertex: $\left(-\frac{3}{2}, -\frac{25}{4}\right)$; x-intercepts: $(-4, 0), (1, 0)$

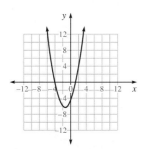

20. vertex: $(-2, -2)$; x-intercepts: $(-2 \pm \sqrt{3}, 0)$

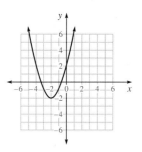

21. vertex: $\left(\frac{1}{2}, \frac{25}{4}\right)$; x-intercepts: $(-3, 0), (2, 0)$

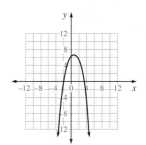

51.

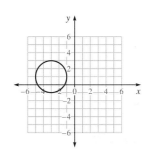

22.

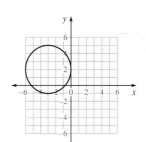

53.

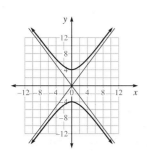

23.

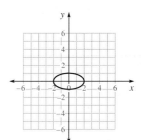

▪ ▪ Chapter 7 Acheivement Test

1. $3x^2 + 9x + 2 = 0$

2. $4x^2 - 3x - 30 = 0$

3. $x = -4, 3$

4. $x = \frac{1}{3}, -\frac{5}{2}$

5. $x = \pm 2\sqrt{5}$

6. $x^2 - 4x + 4 = (x - 2)^2$

7. $x = 4 \pm \sqrt{15}$

8. $x = \dfrac{3 \pm \sqrt{17}}{4}$

9. $x = 2 \pm i\sqrt{2}$

10. one real solution

11. two complex solutions

12. $x = -4, 2$

13. $x = \pm 2, \pm i\sqrt{2}$

14. $x = 5$

15. no solution

16. 12 in. $\times$ 5 in.

17. $\frac{2}{9}$ or $\frac{9}{2}$

18. 7 mph

19. 100 ft

24.

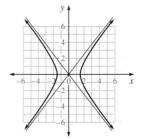

Chapter 8

▮▮ Exercise 8.1

1. $(5, -2)$ **3.** $(-4, 1)$
5. parallel lines; no solution   **7.** parallel lines; no solution
9. $(3, -8)$ **11.** parallel lines; no solution
13. $(4, 6)$ **15.** $(2, 1)$
17. $(-1, 4)$ **19.** parallel lines; no solution
21. coincident lines, infinitely many solutions
23. $(3, 4)$
25. coincident lines, infinitely many solutions
27. Place both equations in the $y = mx + b$ form. If the slopes are different, the system is consistent. If the slopes are the same and the y-intercepts are different, the system is inconsistent. If both the slopes and the y-intercepts agree, the system is dependent.

▮▮ Exercise 8.2

1. $(4, 7, 2)$ **3.** $(2, 1, 3)$ **5.** $(-1, -1, 3)$
7. $(-3, 2, 1)$ **9.** no solution   **11.** $(-3, -4, 2)$
13. $(5, -4, 6)$ **15.** $(6, 0, -6)$

▮▮ Exercise 8.3

1. 26, 32   **3.** 13, 39
5. 32, 20   **7.** 2, 3, 8
9. 40°, 60°, 80°   **11.** 20 nickels, 25 dimes
13. 6 quarters, 9 nickels   **15.** 65 nonstudents, 182 students
17. 422 students, 188 nonstudents
19. 159
21. 2 gal of the 70% solution, 10 gal of the 82% solution
23. box A: 4 oz; box B: 3 oz; box C: 6 oz
25. 160 hot dogs, 95 sodas
27. 700 tickets at $14.50, 800 tickets at $11.50
29. 10°, 30°, 140°
31. 6 birds on the first wire, 3 birds on the second, 18 birds on the third

▮▮ Exercise 8.4

1. -6 **3.** -2 **5.** 3
7. 0   **9.** 0   **11.** -4
13. 1, 3   **15.** 11   **17.** -7
19. 7   **21.** 0   **23.** 0
25. -21

▮▮ Exercise 8.5

1. $(2, 1)$ **3.** $\left(\frac{9}{7}, -\frac{8}{7}\right)$ **5.** $\left(\frac{2}{41}, -\frac{17}{41}\right)$
7. $(0, -2)$ **9.** $(4, -3, 3)$ **11.** $(1, 2, 3)$
13. $(-2, 2, 0)$ **15.** $\left(\frac{1}{2}, 2, -1\right)$

▮▮ Exercise 8.6

1.

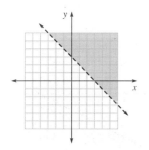

3.

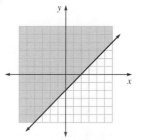

5.

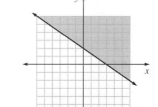

7.

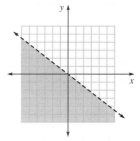

9.

11.

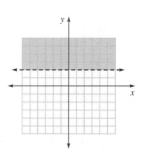

13. $y > 2x + 3$

15. $y \le x + 4$

▮▮ Exercise 8.7

1.

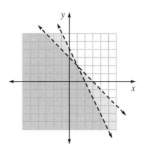

3.

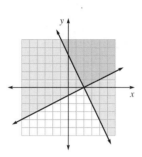

5.

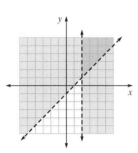

7.

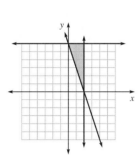

9.

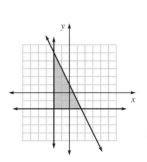

▮▮ Chapter 8 Review Exercises

1. $(1, 1)$

3. coincident lines; infinitely many solutions

5. $(3, 2)$

7. parallel lines; no solution

9. $(1, -1, 2)$

11. 31, 13

12. 140 lbs of the \$4.80 seed 60 lbs of the \$3.60 seed

15. $x = 2$

17. $\left(-\frac{14}{19}, -\frac{17}{19}\right)$

19.

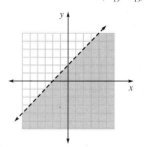

21.

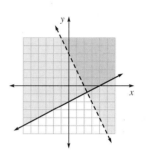

▮▮ Chapter 8 Achievement Test

1. $(2, 1)$

2. coincident lines; infinitely many solutions

3. parallel lines; no solution

4. $(-2, -1)$

5. $(1, 0, 2)$

6. $64°, 26°$

7. 84 students, 36 nonstudents

8. eight 19¢ stamps, ten 30¢ stamps, twenty 29¢ stamps

9. 14

10. -15

11. $\left(\frac{1}{2}, -\frac{1}{2}\right)$

12. $(-1, 3, 2)$

13.

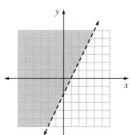

14.

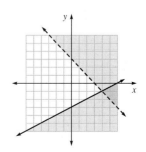

15.

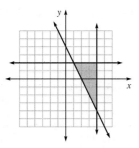

Chapter 9

▮▮ **Exercise 9.1**

1.

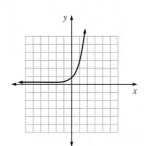

3.

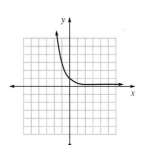

5.

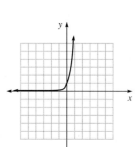

▮▮ **Exercise 9.2**

1. $\log_2 8 = 3$ **3.** $\log_5 125 = 3$ **5.** $\log_2 \left(\frac{1}{32}\right) = -5$

7. $\log_6 1 = 0$ **9.** $\log_{\frac{1}{2}} 32 = -5$ **11.** $\log_a y = r$

13. $2^3 = 8$ **15.** $10^3 = 1000$ **17.** $b^0 = 1$

19. $8^{\frac{2}{3}} = 4$ **21.** $a^{-1} = \frac{1}{a}$ **23.** $3^{-3} = \frac{1}{27}$

25. 16   **27.** $\frac{1}{16}$ **29.** 3

31. 12   **33.** 2   **35.** 3

▮▮ **Exercise 9.3**

1. $\log_3 5 + \log_3 x$ **3.** $\log_3 5 - \log_3 x$

5. $5 \log_3 x$ **7.** $3 \log_b x + 2 \log_b y$

9. $\frac{1}{5} \log_{10} y$ **11.** $\log_2 2 + \log_2 z - \log_2 y$ or $1 + \log_2 z - \log_2 y$

13. $\frac{1}{3} \log_3 x - \frac{1}{3} \log_3 y$ **15.** $\log_a x - \frac{1}{6} \log_a y$

17. $\log_3 \frac{5}{x}$ **19.** $\log_b c$

21. $\log_a \frac{\sqrt{x}}{y^2}$ **23.** $\log_a \frac{z\sqrt{x}}{\sqrt{y}}$

25. 9   **27.** 4

29. false   **31.** false

33. true

▮▮ **Exercise 9.4**

1. 3.1931   **3.** 1.3345   **5.** 4.8531

7. -2.2000 **9.** 0.2253   **11.** 3.97

13. 959   **15.** 0.0413   **17.** 0.0457

19. $0.0969 - 3$ **21.** $0.8248 - 4$ **23.** $0.5888 - 3$

25. 3,213,000   **27.** 8.623   **29.** 969.3

31. 74.05   **33.** 13.1

▮▮ **Exercise 9.5**

1. 2.322   **3.** 3.358   **5.** 1

7. 2.4406   **9.** $\frac{5}{2}$ **11.** -1.6396

13. 1.5439   **15.** 3.3223   **17.** $\frac{2}{3}$

19. 1.0566   **21.** 1.3687   **23.** 3.0226

▮▮ **Exercise 9.6**

1. 7 years   **3.** 35 years

5. 6.40   **7.** (a) 74 (b) 82 (c) 9

9. $19,723.43   **11.** 17.67 years

▮▮ **Chapter 9 Review Exercises**

1. $\log_{10} 100 = 2$ **3.** $\log_{\frac{1}{2}} \left(\frac{1}{4}\right) = 2$

5. $10^4 = 10,000$ **7.** $10^{-2} = 0.01$

9. 27   **11.** 3

13. $\log_5 4 + \log_5 x$ **15.** $\log_5 3 + \log_5 x - \log_5 y$

17. $\log_3 24$ **19.** $\log_{10} 0.3$

21. 1.2718   **23.** -1.0241

25. 75.0   **27.** 0.0397

29. 6410   **31.** 2.7268

33. 1.6131   **35.** 18.8 years

▮▮ **Achievement Test—Chapter 9**

1. $\log_5 125 = 3$

2. $3^4 = 81$

3. $x = 3$

4. $2 \log_5 x - \log_5 y$

5. $\log_4 54$

6. 2.2695

7. $0.0531 - 1$

8. 25

9. 7.9158

10. 2.6572

11. 2.2619
12. 22.1 years

Cumulative Review Exercises—Chapters 7, 8, and 9

1. $-5, \frac{1}{2}$ **3.** $3 \pm i$ **5.** $-3, 2$

7. 50 mph going; 60 mph returning

9.

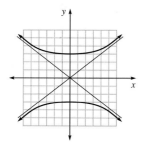

11. $(4, -2)$ **13.** 140 adults, 35 children
15. $4^2 = 16$ **17.** 1.8614

Appendix

Appendix B

1. 17.1   **3.** 11.5   **5.** 5.8
7. $\angle A = 26°$; $\angle B = 74°$; $\angle C = 90°$; $a = 38.8$; $b = 125$; $c = 130$
9. $\angle A = 23.6°$; $\angle B = 66.4°$; $\angle C = 90°$; $a = 20$; $b = 45.8$; $c = 50$

11. $\angle A = 77.9°$; $\angle B = 12.1°$; $\angle C = 90°$ $a = 32.65$; $b = 7$; $c = 33.39$
13. 564.3 ft   **15.** 70.7°
17. 1.13 miles

Exercise D.1

1. 9250   **3.** 35.146   **5.** 7071
7. $-20,211$ **9.** -11.622 **11.** 2924
13. 208.008   **15.** 53,458.35   **17.** 554.92308
19. 0.32597   **21.** 14.555556   **23.** 0.5287671
25. -2157.6 **27.** 14,304   **29.** 32.649655
31. 29.274562   **33.** 94.074439   **35.** 4.64758
37. 441   **39.** 71,824   **41.** 987.2164
43. 0.16

Exercise D.2

1. 440   **3.** 504   **5.** 10.954451
7. 4.8989795   **9.** 216   **11.** 625
13. 59,049   **15.** 32.768   **17.** 8008.7465
19. 0.1785714   **21.** 560   **23.** 1.1138462
25. 1.9381443   **27.** 0.32   **29.** 7.4431707
31. 960   **33.** -14.994 **35.** -39.304
37. 44.718947   **39.** 4.712389

Additional Calculator Exercises

1. Estimate: 21,000; Calculator: 20,908
3. Estimate: 60,000; Calculator: 59,074.84
5. Estimate: $-24,000$; Calculator: $-24,336$
7. Estimate: 4; Calculator: 3.9051248
9. 1495   **11.** 16.248077   **13.** 5098.317
15. -185.193 **17.** 1.3541053   **19.** $19.96

Index